Junipers of the World: The genus *Juniperus*
4th Edition

Dr. Robert P. Adams, Prof.
Biology Department
Baylor University
Waco, TX USA

Trafford Publishing Co., Bloomington, IN

Order this book online at www.trafford.com
or email orders@trafford.com

Most Trafford titles are also available at major online book retailers.

Printed in the United States of America.

ISBN: 978-1-4907-2325-9 (sc)
ISBN: 978-1-4907-2326-6 (e)

Library of Congress Control Number: 2014900169

Trafford rev. 01/03/2014

Trafford PUBLISHING® www.trafford.com
North America & international
toll-free: 1 888 232 4444 (USA & Canada)
fax: 812 355 4082

CONTENTS

Preface and Acknowledgements

My introduction to *Juniperus* (in a formal sense) occurred in 1966 at the University of Texas in Austin, TX. As an undergraduate, I majored in mathematics and minored in physics, then entered graduate school in botany, so I had much to learn. Prof. Billie L. Turner and I decided that it would be interesting to examine geographic variation in a species that has a widespread distribution in order to take advantage of my knowledge of computer graphic methods to analyze spatial patterns. One day, Prof. Turner (just "Billie" to his colleagues) said "You need to study a large plant so you can easily obtain samples at regular intervals." So he took me to the herbarium to examine some *Cercis canadensis* (red bud). But he concluded that "so many have been planted along the roadsides in Oklahoma, you would never be sure if you sampled a natural or a cultivated tree, but there is an interesting problem in *Juniperus*." Not long after that, Dr. Ernst von Rudloff came to the University of Texas and spent several months teaching graduate students to extract leaf essential oils and gas chromatography. And, of course, *Juniperus* has lots of diversity in its essential oils that could be exploited in geographic studies. A sabbatical in 1976 at the National Research Council of Canada with Dr. von Rudloff and Lawrence Hogge was extremely valuable in introducing me to mass spectroscopy of terpenoids. I was fortunate to obtain a position at Colorado State University and secure a NSF grant to study the junipers of Mexico/ Guatemala. Tom Zanoni (now at New York Botanic Garden) came and worked on those junipers (and, from whom, I learned some much needed classical nomenclature). Other students, Walt Kelly, James Kistler, etc. continued work on *Juniperus*.

But the really big opportunity to study *Juniperus*, world-wide, came in the late 1980's. Sometimes a "window of opportunity" becomes available in science. Such a "window" opened in the late 1980's with the breakup of the Soviet Union. Heretofore, collaborations and field collections were limited (such as short trips as provided with the International Botanical Congress, 1975, Leningrad). However, the dissolution of the Soviet Union and the concurrent opening of collaboration in China provided just such a "window" for me. Never before have we had the technology for transportation, coupled with the accessibility (politically) to the northern hemisphere, plus the ease of internet and FAX communications. In 1990, under the sponsorship of NSF, I spent 6 weeks with Prof. Ge-lin Chu in Gansu, China collecting *Juniperus* (all preserved in silica gel for subsequent DNA analysis). Following this study, collaborations and field collections were conducted in Armenia, China (Qinghai, Sichuan, Xinjiang, Yunnan), Georgia (CIS), Kazakhstan, Kyrgyzstan, Mongolia, Russia, Taiwan, and Turkmenistan. These collaborations, many taking several years to develop, resulted in field collections and observations for all the *Juniperus* taxa in central Asia. Supplemental trips to the Mediterranean, Ethiopia, Kenya, Morocco, and Sweden plus recollections of all the *Juniperus* taxa from the western hemisphere has resulted in the assemblage of DNA and essential oils for the entire set of *Juniperus* taxa of the world. During the past 47 years, I have collected over 10,000 *Juniperus* specimens. I have been fortunate to have the "window of opportunity" to visit and collect all of *Juniperus* in their native habitats. Hopefully, this treatment of *Juniperus* will be useful to the many colleagues who have contributed to these trips and a lifetime of the study of *Juniperus*. The reader is also encouraged to visit **www.juniperus.org** to obtain additional information about *Juniperus*, reprints of articles and updated keys.

This research has been supported by funds from NSF (USA), UNESCO, USDA, Wallace Genetic Foundation, Stanley Smith Horticultural Trust, and Baylor University.

Finally, I would like to thank by wife, Janice, for assistance in proofing the book, and my daughter, Jessica, for assisting in field collections in Armenia, Morocco and Turkmenistan and proofing.

Below is a list of collaborators who deserve great thanks:
Altarejos, Joaquin, Chemistry Dept., University of Jaen, Jaen, Spain.
Arista, Montserrat, Univ. de Sevilla, Sevilla, Spain/
Baker (Flournoy), Lori, E., Forensic Science, Baylor University, Waco, TX.
Bartel, James, US Dept. of Interior, Carlsbad, CA
Baser, K. Husnu C., Anadolu Univ, Eskisehir, Turkey.
Boratynski, Adam, Polish Acad. Sci., Kornik, Poland
Casanova, Joseph, Equipe Chimie et Biomass, University of Corse, Ajaccio, France.
Chaudhary, Ram P., Botany Dept., Tribhuvan University, Kathmandu, Nepal.
Chibbar, Ravi (sabbatical leave 1989), National Research Council, Plant Biotechnology Inst., Canada.
Chu Ge-Lin, Northwest Normal University, Lanzhou, China.
Dignard, Norman, Ministere des Resources Naturelles, Sainte-Foy, Québec, Canada.
Farjon, Aljos, Kew Gardens, London, UK
Fei, Yang, Kunming Botanical Institute, Kunming, China (now deceased).
Fontinha, Susana S., Parque Natural de Madeira, Madeira, Portugal
Gonzales, Socorro E., Herbario CIIDIR IPN Unidad, Durango, Mexico
Hoegh, Ken, Greenland Agricultural Advisory Service, Qaqotoq, Greenland.
Hsieh, Chang-Fu, Dept. of Botany, National Taiwan University, Taipei, Taiwan.
Johnson, Hiram J., USDA-ARS, Grasslands Lab, Temple, TX.
Lara, Armando, Chemistry Dept., Univ. of Granada, Granada, Spain.
Le Duc, Alice, 10180 Santa Ana Rd., Atascadero, CA.
Leschner, Hagar, HUJ Herbarium, The Hebrew Univ., Jerusalem, Israel.
Leverenz, Jerry W., Arboretum Royal Veterinary & Agric. University, Horsholm Denmark.
Liber, Zlatko, Univ. of Zagreb, Zagreb, Croatia.
Little, Damon, L.H. Bailey Hortorium, Cornell University, Ithaca, NY.
Liu, Jianquan, Northwest Plateau Inst. of Biology, Chinese Acad. Sci., Xining, China
Mao, Kangshan, Northwest Plateau Inst. of Biology, Chinese Acad. Sci., Xining, China
Mataraci, Tugrul, Istanbul, Turkey
Minissale, Pietro, Univ. de Catania, Catania, Italy.
Morris, Julie, Dept. of Biological Sciences, Univ. of Texas at Brownsville, TX.
Murata, Jin, Graduate School of Science, University of Tokyo, Tokyo, Japan.
Nogales, Manuel, Island Ecology and Evolution Res. Group, Tenerife, Canary Islands
Pandey, Naresh Ram, RECAST, Tribhuvan University, Kathmandu, Nepal.
Sanko Nguyen, University of Oslo, Oslo, Norway
Rieseberg, Loren H., Biology Dept., Indiana University, Bloomington, IN.
Rumeu, Beatriz (Bea), Island Ecology and Evolution Res. Group, Tenerife, Canary Islands
Rushforth, Keith, Arborist, The Shippen, Cullompton, Devon, UK
Schwarzbach, Andrea E., Dept. of Biological Sciences, Univ. of Texas at Brownsville, TX.
Shao-Zhen, Zhang, Northwest Normal University, Lanzhou, China.
Shatar, Sanduin, Chemistry Institute, Ulan Batar, Mongolia.
Singh, Ram, Botany Dept., Tribhuvan University, Kathmandu, Nepal.
Tashev, Alexander, University of Forestry, Sofia, Bulgaria
Thorfinnsson, Thor, Iceland Forestry Service, Egilsstadir, Iceland.
Turuspekov, Yerlan, Institute of Genetics, Alma Ata, Kazakhstan.
von Rudloff, Ernst, emeritus, Prairie Regional Lab, Natl. Res. Council, Saskatoon, Sask., Canada.
Zanoni, Tom A., New York Botanical Garden, Bronx, NY.
Zhong, Ming, Agronomy Dept, August 2nd University, Urumqi, Xinjiang, China

Junipers of the World: The Genus Juniperus, 4th Edition

Dr. Robert P. Adams, Professor
Biology Dept., Baylor University, Waco, TX USA

Chapter 1: Introduction

Juniperus is one of most diverse genera of the conifers. *Juniperus* species are found from sea level (several, *J. virginiana* var. *silicicola*, *J. taxifolia* var. *lutchuensis*, *J. procumbens*) to above timberline (*J. zanonii*). Although limestone is the preferred substrate for many *Juniperus* species, other taxa grow on sand dunes, granite, and sandstone and range from deserts (*J. californica*, *J. osteosperma*) to bogs (*J. communis* var. *charlottensis* on Queen Charlotte Island, B.C.). In North America, *Juniperus* species have become weedy and invaded millions of acres of rangeland and abandoned farms (Adams et al., 1998).

Juniperus species have evolved a fleshy female cone in which the cone scales are fused. These are often called "fruits" or "berries". This reproductive structure is especially consumed by birds and small mammals (Phillips, 1910; McAtee, 1947; Holthuijzen and Sharik, 1985; Santos et al., 1999). In fact, the long distance dispersal by birds has resulted in *Juniperus* being established on Atlantic islands such as the Azores, Bermuda, the Caribbean and Canary Islands.

The genus *Juniperus* - Evergreen shrubs or trees. Branchlets terete, 4-6 angled, variously oriented, but not in flattened sprays (except in *J. thurifera* and occasionally in *J. flaccida*). Leaves in decussate (alternating) opposite pairs in 4 ranks or in alternating whorls of 3. Adult leaves either acicular (subulate) or closely appressed (scale like) to divergent tips (whip leaves); abaxial gland visible or not, elongate to hemispheric (*J. ashei*), sometimes exuding white crystalline deposit. Pollen cones with 3-7 pairs or trios of sporophylls, each sporophyll with 2-8 pollen sacs. Seed cones maturing in 1 or 2 years (to 3 yrs. in *J. communis*), globose to ovoid and berrylike, 3-20 mm (to 25 mm in *J. drupacea*), remaining closed, often glaucous; scales persistent, 1-5 pairs or whorls of three, peltate or valvate, tightly coalesced and fused together, thick and fleshy or fibrous to obscurely woody; some sweet (e.g. *J. pinchotii*, *J. coahuilensis*), many bitter and/or resinous. Seeds 1-3 per scale, round to faceted, wingless; cotyledons 2-6. Seed dispersal by frugivorous birds, that swallow the cones whole, digest the fleshy scales and pass the hard-shelled seeds undamaged through the gut; the bitter taste of many species may be related to discouraging mammalian predators of the seeds. 2n = 22 but reports of tetraploids (4n = 44) in *J. chinensis* and *J. thurifera* (Romo et al. (2012).

The phylogenetic position of *Juniperus* (and Cupressaceae) in the plant kingdom (Fig. 1.1) indicates *Juniperus* as a terminal clade and as one of the most advanced conifer genera (Rai et al., 2008).

A simplified Bayesian tree with genera collapsed shows (Fig. 1.2) the relationship for *Juniperus* to *Cupressus* and *Hesperocyparis* (adapted from Mao et al., 2010).

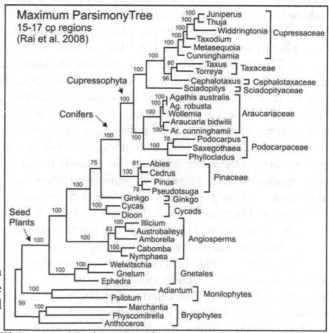

Figure 1.1. Maximum parsimony tree showing the position of *Juniperus*. Adapted from Rai et al. (2008).

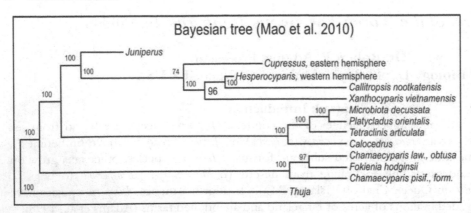

Figure 1.2. Simplified Bayesian tree with genera collapsed showing *Juniperus* relationship to *Cupressus* and *Hesperocyparis*. Adapted from Mao et al. (2010).

The phylogeny of Junipers is depicted in Figure 1.3. The three sections are in well supported clades. *Juniperus phoenicea* and *J. turbinata* stand loosely affiliated with sect. Sabina (Fig. 1.3). These taxa have small serrations on the leaf margins, but are denoted as 'pseudoserrate' (Fig. 1.3). It seems unlikely that serrate leaf margins in the eastern and western hemispheres is a homologous character, but has arisen independently as *J. phoenicea - J. turbinata* is not in the clade with the serrate, semi-arid junipers of the western hemisphere. It is interesting that *J. erectopatens* and *J. microsperma* form an unusual clade that does not nest into the *J. chinensis* clade (Fig. 1.3).

The genus *Juniperus* is comprised of approx. 75 species in 3 sections: **sect *Caryocedrus*** Endlicher, 1847, the type and only species, *J. drupacea* Labill., syn: genus *Arceuthos* Antoine, 1854 with large, blue, woody, 3-seeded cones, showing the fusing of 3 cone scales, with an Old World Mediterranean distribution (Fig. 1.3.1), see Adams, 2011, Adams and Schwarzbach, 2012a); **sect *Juniperus*** (syn: sect. *Oxycedrus* Spach, Ann. Sci. Nat. Ser. 2, 16: 288 (1841), 14 species, 12 only in the eastern hemisphere, one endemic to North America (Fig. 1.3.2) and one species, *J. communis*, being circumboreal, seed cones blue or red, often with 3 seeds (Adams and Schwarzbach, 2012a) and **sect *Sabina*** (Miller) Spach, Ann. Sci. Nat. Ser. 2, 16: 291 (1841), **sect. *Juniperus,. Sabina*** syn: genus *Sabina* Miller, 1754 (approx. 60 species), with species about equally divided between the eastern and western hemispheres, seed cones with 1 to 13 seeds, blue, red-copper, rose, or brown (Adams, 2011).

The leaf types in the three sections are shown in Figures 1.3.3.1, 1.3.3.2, and 1.3.3.3: Left: acicular leaf, the only kind of leaves in sections *Caryocedrus* and *Juniperus*. Notice the abscission layer (leaf drop) is on the stem (as in the Pinaceae). Center: decurrent leaf, found only in section *Sabina*. Right: scale-like leaves (adult foliage), present only in sect. *Sabina* on mature trees (after 3-5 years). The various types of leaves are shown diagrammatically in Figure 1.3.3.4. It should be noted that in section *Sabina*, some decurrent leaves are always present at the terminal branchlet tips when the foliage is growing fast. Decurrent leaves are the juvenile leaves in section *Sabina*. After a few years (3-5 years) most species in sect. *Sabina*, will begin to produce scale-like (adult) leaves (Fig. 1.3.3.3). However, some taxa, such as *J. gracilior* var. *silicicola* (Britton and P. Wilson) R. P. Adams, *J. coxii* A. B. Jacks, *J. recurva* Buch.-Ham. ex D.Don, *J. fargesii* (Rehder & Wils.) Kom., and *J. morrisonicola* Hayata (see Fig. 3.3.5) are frozen in the juvenile leaf state for life (neoteny).

Section *Caryocedrus*, consisting of one species, *J. drupacea*, is confined to the region from Greece to Turkey (Fig. 1.3.1) and is considered the most ancestral section of the genus (Adams and Schwarzbach, 2013b). Dioecious. Trees to 30 m or more. Leaves acicular (subulate, jointed at the base, Fig. 1.3), broad (2-3.5 mm wide); 10-25 mm long. Scale-leaves absent. Cones axillary on shoot, female cones 18 - 25 mm at maturity with 3 cone scales visible (see Fig. 1.3.3.1), three seeds fused together to make a drupe, female cones mature in 18 mos. to 2 years.

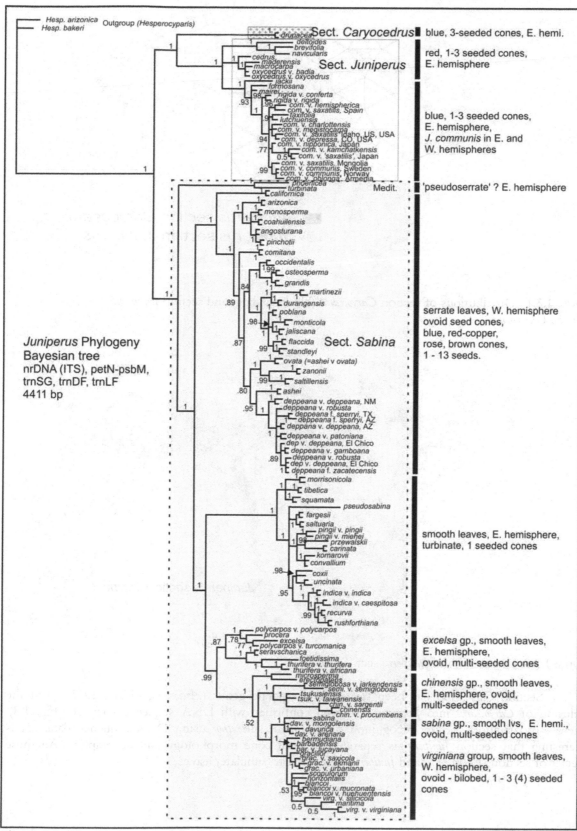

Figure 1.3. The Phylogeny of *Juniperus*. Numbers branch points are posterior probabilities. Adapted from Adams and Schwarzbach (2013e).

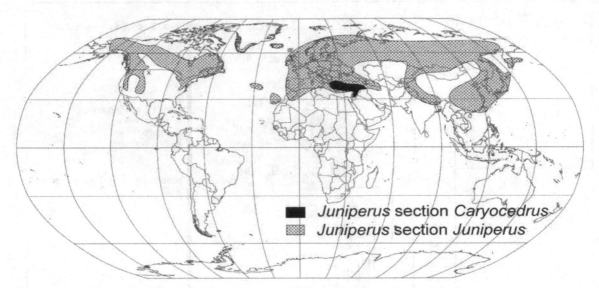

Figure 1.3.1. Distributions of section *Caryocedrus* (black shading) and section *Juniperus* (= sect. *Oxycedrus*).

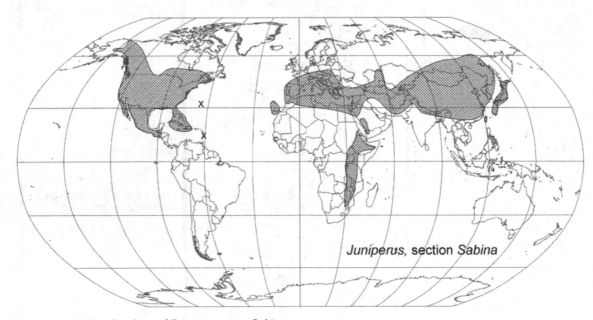

Figure 1.3.2. Distribution of *Juniperus* sect. *Sabina*.

Section *Caryocedrus* has been treated as a separate genus (*Arceuthos*, Florin, 1963). But the inclusion of *Caryocedrus* into *Juniperus* has been confirmed with DNA sequence data (see Fig. 1.3). There is no support for the recognition of the genus *Arceuthos* using DNA sequence data. It is interesting that section *Juniperus* is separated by seed cone morphology and geography. All these species of sections *Caryocedrus* and *Juniperus* have acicular (subulate) leaves.

J. drupacea   *J. communis* var.   *J. bermudiana*
                        communis

Figure 1.3.3.1. The leaf scales visible on the seed cone. Also note the attachment of the acicular leaves to the stem is shown.

Figure 1.3.3.2. Note the acicular (subulate) leaves that are jointed at the base. The abscission layer is at the stem (as in spruce).

Figure 1.3.3.3. Scale-like leaves *J. bermudiana*, sect. *Sabina*. Leaves are decussate (opposite) giving a 'string of beads' look.

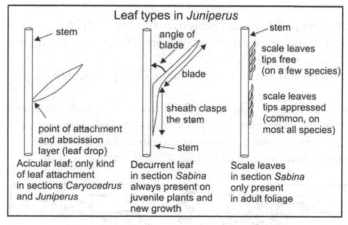

Figure 1.3.3.4. Illustration of leaf types in *Juniperus*. See text for discussion.

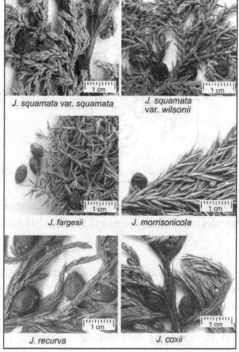

Figure 1.3.3.5. Variation among species with only decurrent leaves. The leaves of *J. squamata* v. *squamata*, *J. coxii*, and *J. recurva* may represent a fourth kind of leaf in *Juniperus*.

The morphological characteristics of the three sections are shown in Table 1.1.

Table 1.1. Characteristics of the three sections of *Juniperus*.

Caryocedrus	*Juniperus* (= *Oxycedrus*)	*Sabina*
1 species	14 species	60 species
Mediterranean	1 circumboreal sp.+ 1 N Am + 12 Mediterranean, Asia	Northern Hemisphere
leaves acicular (as in spruce)	leaves acicular (as in spruce)	leaves scale like &/or decurrent
dioecious	dioecious	dioecious or monoecious
cones 18-25 mm	cones 8-18 mm	cones 6-15 mm
3 seeds/ cone	usually 3	1 to 10 or 12.
woody cones	woody or resinous cones	mostly juicy or resinous cones
purple cones	dark blue, bluish-black, brown, reddish-brown, red, copper	blue, black, red, copper, brown, pink, green
leaf margins entire	leaf margins entire	leaf margins entire and serrate

Section *Sabina* is divided into three major clades (Mao et al., 2010, Adams and Schwarzbach, 2013a):
1. Serrate-leaf junipers of North America (21 species, Adams and Schwarzbach, 2011, 2013d),
2. Turbinate-seed cones, single-seeded, entire-leaf junipers, eastern hemisphere (16 species, Adams and Schwarzbach, 2012b, 2013a, Zanoni and Adams, 1976, 1979) and
3. Multi-seeded, entire-leaf junipers, both eastern and western hemispheres (23 species, Adams and Schwarzbach, 2012c, 2013b).

Mao et al. (2010) used three *Juniperus* fossil dates: *J. pauli* (ca. ≥ 33.0 mya, cf. extant *J. sabina* and allies), *J. creedensis* (ca. ≥ 23.0 mya, cf. *J. californica / J. osteosperma*), and *J. desatoyana* (ca. ≥ 16.0 mya, cf. *J. occidentalis / J. osteosperma*). They postulated the serrate, semi-arid junipers migrated from the eastern to the western hemisphere via the North American Land Bridge (NALB) ca. 47 - 30.3 mya (Fig. 1.4).

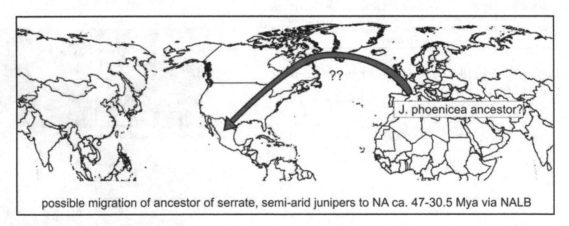

Figure 1.4. Possible migration of ancestor of the serrate, semi-arid junipers from the Mediterranean to the western hemisphere via the NALB (adapted from Mao et al., 2010).

The fossil *J. creedensis* of the Creede geoflora (ca. ≥ 23.0 mya) bears a striking resemblance to present-day *J. californica* (Fig. 1.5). Because the present-day *J. californica* appears little changed from the fossil, *J. creedensis*, it may be that the serrate junipers in North America are much older than thought. It might be noted that Axelrod (1987) described a second juniper from the Creede geoflora as *J. gracillensis* that he thought was similar to extant *J. flaccida*, but Wolfe and Schorn (1990) have identified the specimen as *Eleopoldia lipmanii* (Rosaceae).

Fig. 1.5. *Juniperus creedensis* Axelrod paratype and present-day *J. californica.*

The Madrean-Tethyan vegetation belts in Eurasia and North America may have been continuous during the Eocene and Oligocene (Axelrod, 1975; Wen and Ickert-Bond, 2009), such that *Juniperus* section *Sabina* might have had a wider distribution (Fig. 1.6). So it is possible that the serrate-leaf junipers may have existed in the Madrean-Tethyan vegetation belts in both Eurasia and North America during the same period (Fig. 1.6), and there may have been exchanges via the North Atlantic Land Bridge (NALB). Although the Madrean-Tethyan vegetation, depicted in Figure 1.6, predates the ages of any known juniper fossils, because only a few *Juniperus* fossils are known, older fossils likely exist.

It is unfortunate that the serrate-leaf species (*J. phoenicea, J. turbinata*) that are extant in the eastern hemisphere, have DNA so different that they are poorly grouped with any clade (Fig. 1.3). It is quite removed from the serrate junipers clade (Fig. 1.3). At present, it seems appropriate to consider *J. phoenicea* and *J. turbinata* as 'pseudoserrate' and of a different lineage than the serrate junipers of North America. If the serrate leaves of *J. phoenicea/ J. turbinata* are not homologous to the serrate leaves of junipers in North America, then we are left with no extant (or known fossils) of truly serrate junipers in the eastern hemisphere.

The migration of the smooth leaf members of sect. *Sabina* to the western hemisphere is thought to be more recent (17.6-5.5 mya, Mao et al., 2010) and those dates are younger than the fossil *J. creedensis* of the Creede geoflora (Axelrod, 1987). Since *J. phoenicea/ J. turbinata* do not appear to be true members of the serrate junipers and no serrate juniper fossils have been found in the eastern hemisphere, the serrate junipers may be endemic to the western hemisphere. Undoubtedly, additional fossils will be found some day to help resolve the question.

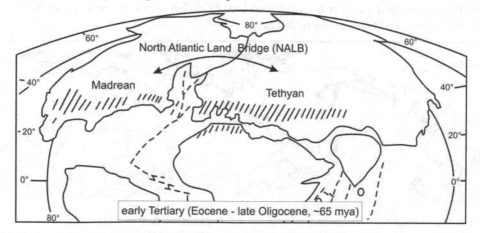

Figure 1.6. Madrean - Tethyan vegetation zones of Axelrod (adapted from Axelrod (1975) and Wen and Ickert-Bond (2009).

Mao et al. (2010) argues that the movement of sect. *Sabina* to the western hemisphere (17.6-5.5 mya) is too young for migration across the North Atlantic Land Bridge (NALB), but possible via the Bering Land Bridge (BLB). Because sect. *Sabina* species such as *J. sabina* and *J. davurica* are quite cold adapted; they (or an ancestor) could have migrated to produce the ancestors that gave rise to the current, cold climate, western hemisphere species such as *J. horizontalis* and *J. scopulorum*. *Juniperus davurica* is the northeastern-most species in northeast Asia (in sect. *Sabina*) and could have provided ancestral stock to migrate across the Bering Land Bridge (Fig. 1.7). Notice that *J. davurica* / *J. sabina* are in a sister clade to the smooth-leaf juniper of North America (Fig. 1.3), supporting the concept of migration from northeastern Asia via the BLB (Fig. 1.7.)

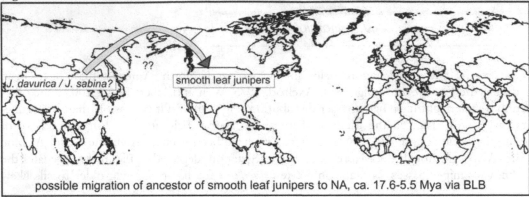

Figure 1.7. Possible migration of ancestor of smooth leaf junipers to North America. (cf. Mao et al., 2010).

The North America *communis* group is equally linked (Fig. 1.11.2) between the Japan and Europe-Central Asia groups. Thus, the linkage map gives equal support to the Bering Land Bridge and North Atlantic island hopping model for the origin of *J. communis* in North America. The situation was previously more unclear when *J. jackii* was included in *J. communis* (*J. c.* var. *jackii*). However, *J. jackii* is clearly quite differentiated (20 MEs from *J. mairei*, Gansu, China; 21 MEs from *J. c.* var. *megistocarpa*, NA and Fig. 1.11.2), but the data is equivocal as to whether its origin is from the BLB or North Atlantic island-hopping model. *Juniperus jackii* grows on serpentine and volcanic basalt of quite recent origin in the Cascade Range of western Oregon/ northern California (but occasionally on granite in the Trinity Alps, CA).

A diagrammatic representation of the possible migrations of *J. communis* (and *J. jackii*) is shown in Figure 1.8. The migration dates proposed by Mao et al. (2010) seem consistent with the recent habitat availability for *J. jackii* and support the observed lack of differentiation among morphological varieties of *J. communis* (Adams and Schwarzbach, 2012a).

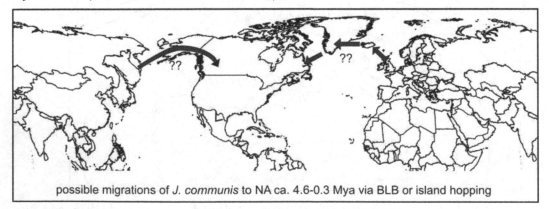

Figure 1.8. Two possible migrations of *J. communis* to North America. Based on data in Mao et al. (2010).

Wen and Ickert-Bond (2009) summarized data from 17 studies concerning Madrean-Tethyan disjunctions. Their summaries are useful in the present discussion. They concluded (Fig. 1.9) that: 53% of the inter-continental migrations was by the North Atlantic Land Bridge; 40% was by long distance dispersal and 7% by the Bering Land Bridge (BLB). Their summary of the directional data indicated the origins as: 86% from eastern to western hemisphere; 7% from western to eastern hemisphere and for 7% the direction was uncertain (Fig. 1.9).

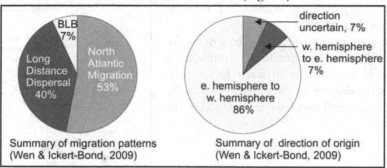

Figure 1.9. Summary of trends based on 17 studies of Madrean-Tethyan disjunctions. Adapted from Wen and Ickert-Bond (2009).

Section *Juniperus* (= section *Oxycedrus* of other treatments, type species *J. communis* L.). Trees or shrubs, Dioecious (male and female cones on different plants); Leaves acicular (subulate, Fig. 1.3.3.2) as in section *Caryocedrus*, with a basal abscission area, in whorls of three. Both seed and pollen cones axillary on shoot, on a very short 0.3-1 mm peduncle (appearing sessile); Mature seed cones, 6-15 mm, scales valvate in 1(-2) whorls of three, usually 3 - seeded, but seeds not fused together; mature in 2 (-3) years. Eleven species are recognized in this treatment.

Section *Juniperus* can be divided into 2 groups (Fig. 1.10): a northern (and far eastern) group associated with *J. communis* with blue or blue-black mature seed cones and one stomatal band on the adaxial leaf surface; and the *J. oxycedrus* allies of the Mediterranean region that have 2 stomatal bands and red, red-copper, reddish-brown to reddish-purple mature seed cones.

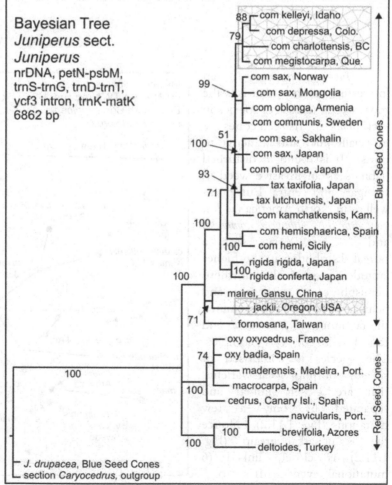

Figure 1.10. Bayesian tree of *Juniperus* sect. *Juniperus*. Adapted from Adams and Schwarzbach (2012a).

The phylogeny of *Juniperus* sect. *Juniperus* is shown figure 1.10. The varieties of *J. communis* are not well resolved, which indicates the closeness of this group.

The red colored seed cone species occur around the Mediterranean. The DNA sequence data (Adams and Schwarzbach, 2012a) support the recognition of several species: *J. navicularis, J. brevifolia, J. deltoides, J. cedrus, J. maderensis, J. oxycedrus* and *J. macrocarpa* in this group (Fig. 11.1).

Table 1.2. shows the support from DNA sequencing for various species and varieties in section *Juniperus*. It might be noted that not every case is concordant between DNA data, morphology, ecology and secondary plant chemistry. No doubt, the reality of species delimitation will be better understood in the future.

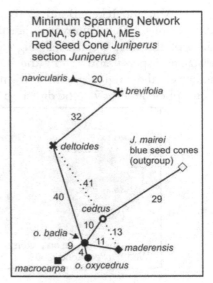

Fig. 1.11.1. Minimum spanning network of red seed cone taxa. Numbers on the links are MEs (mutational events). Adapted from Adams and Schwarzbach (2012a).

Juniperus communis is an interesting taxon in that it is the most weedy (invasive) species in sect. *Juniperus*. Its seed cones are especially juicy and attractive to birds. It is found in disturbed habitats as an invasive weed in Hungary and central Europe as well as in North America (as *J. c.* var. *depressa*). *Juniperus communis* and *J. c.* var. *depressa* form a boreal distribution in the higher latitudes around the northern hemispheres. Adams and Schwarzbach (2012a) examined the taxonomy of *J. communis* and found it to be very complex. The species is comprised of several morphological varieties that are closely linked and separated by only a few mutations (Fig. 1.11.2). Notice the Kamchatka group (Fig. 1.11.2) is closely linked (6 mutational events, MEs) to *J. communis* var. *nipponica*, Japan, thence to *J. c.* var. *megistocarpa* (NA, 5 MEs).

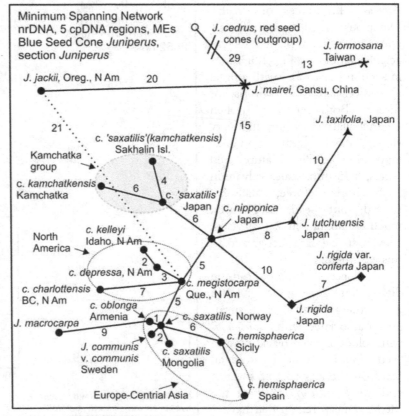

Figure 1.11.2. Minimum spanning network of blue seed cone junipers. Numbers on the links are MEs (mutational events). Adapted from Adams and Schwarzbach (2012a).

Table 1.2.   Comparison of support using DNA sequencing data versus the taxonomies of Adams (2011) and Farjon (2010).   Support levels: ++ = very strong, + = strong, +/- = not strong, - = not supported.

Adams(2011)	Farjon (2005, 2010)	Supported (Adams and Schwarzbach, 2012a
J. brevifolia	*J. brevifolia*	++ *J. brevifolia*
J. cedrus	*J. cedrus*	++ *J. cedrus*
J. communis	*J. communis*	++ *J. communis*
v. *charlottensis*	v. *saxatilis*	++ v. *charlottensis*
v. *depressa*	v. *depressa*	++ v. *depressa*
v. *hemisphaerica*	v. *communis*	+ v. *hemisphaerica*
v. *megistocarpa*	v. *megistocarpa*	++ v. *megistocarpa*
v. *nipponica*	v. *nipponica*	++ v. *nipponica*
v. *oblonga*	v. *communis*	++ v. *communis*
v. *saxatilis* (Europe, central Asia)	v. *saxatilis*	++ v. *saxatilis* (Europe, cent. Asia)
v. *saxatilis* (North America)	v. *saxatilis*	++ v. *kelleyi*
v. *saxatilis* (Japan, Sakhalin Isl.)	v. *saxatilis*	+ v. *nipponica*? or *kamchatkensis*
v. *saxatilis*(Kamchatka)	v. *saxatilis*	++ v. *kamchatkensis*
v.*jackii*	v. *saxatilis*	++ *J. jackii*
J. deltoides	*J. oxycedrus*	++ *J. deltoides*
J. formosana	*J. formosana*	++ *J. formosana*
v. *mairei*	*J. formosana*	++ *J. mairei*
J. macrocarpa	*J. oxycedrus.* subsp. *macrocarpa*	++ *J. macrocarpa*
J. navicularis	*J. oxycedrus.* subsp. *transtagana*	++ *J. navicularis*
J. oxycedrus	*J. oxycedrus*	++ *J. oxycedrus*
v. *badia*	subsp. *badia*	+ ⊦ *J. oxycedrus*
J. rigida	*J. rigida*	++ *J. rigida*
v. *conferta*	subsp. *conferta*	+ *J. rigida* v. *conferta*
J. taxifolia	*J. taxifolia*	++ *J. taxifolia*
v. *lutchuensis*	*J. taxifolia*	+ *J. t.* v. *lutchuensis*

   Section *Sabina* (type species *J. sabina* L.), Trees or shrubs, Dioecious or rarely monoecious species; Leaves not acicular (subulate) but decurrent (with a blade that may be appressed to the stem or free), and a sheath that clasps the stem.   Leaves on juvenile plants (or the tips of rapidly growing shoots) are composed of only whip leaves (decurrent leaves with the tips free); occasionally, an adult tree may have only juvenile leaves.   Some species have only juvenile leaves (*J. carinata, J. coxii, J. morrisonicola, J. pingii, J. procumbens, J. recurva, J. gracilior* var. *saxicola, J. squamata*).   Leaves are without an abscission zone, decurrent (in pairs) or whorls of 3.   Most species have only scale-like leaves on adult foliage (except at growing tips) as seen in figure 1.3.3.3.

   Unfortunately, scale and whip leaves can be very variable in length and this is a source of confusion in many keys.   Both scale and whip leaves are characteristically found on adult foliage in *J. chinensis* (Fig. 1.12.1).   Most junipers of sect. *Sabina* have only scale-type leaves on adult foliage as shown on *J. ashei* (Fig. 1.12.2).   The odd scale leaves in *J. squamata* (Fig. 1.12.3) and *J. pingii* may represent a different type leaf.   The branchlets may appear square when the scale-leaves are opposite (Fig. 1.12.4).

Figure 1.12.1. Both scale and whip (decurrent) leaves on adult foliage of *J. chinensis* var. *sargentii*.

Figure 1.12.2. Scale leaves of *J. ashei*. Old, persistent whip leaves are on the woody stem. These are dead, juvenile leaves.

Figure 1.12.3. Unusual scale leaves of *J. squamata*.

Figure 1.12.4. Opposite ranked scale leaves of *J. bermudiana* result in square branchlets.

Both pollen and seed cones are borne on terminal branchlets. Seed cones are from 4 to 20 mm, 1 -12 seeded, not fused, and mature in 1 to 2 yrs.

The truly serrate leaf margined junipers are confined to the western hemisphere, although Gaussen (1968) scored several eastern hemisphere species as serrate leafed. Gaussen (1968) follows the classical treatment in dividing the genus into 3 sub-genera and he further divided *Sabina* into two series: denticulees (denticulate) and entires (entire) leafed species. Gaussen even subdivides these series into sections. The section *phoenicioides* is put in the series with serrate leaf margins, but the leaves of all the taxa listed in Gaussen's treatment (p. 81) have very minor serrations on the leaf margins at 40 X. For example, the minor serrations near the leaf base of *J. phoenicea* do not appear to be the same kind as found on *Juniperus* in the western hemisphere (Figs. 1.13.1-1.13.4). From DNA sequencing (Fig. 1.3), one can see that *J. phoenicea/ J. turbinata* are clearly not allied to either the serrate or entire leaf junipers of section *Sabina*..

Figure 1.13.1. *J. flaccida* with faint, wavy serrations.

Figure 1.13.2 *J. phoenicea* with a few serrations.

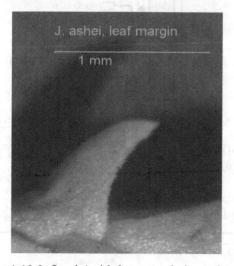

Figure 1.13.3. *J. ashei* with large teeth (serrations) on leaf margin.

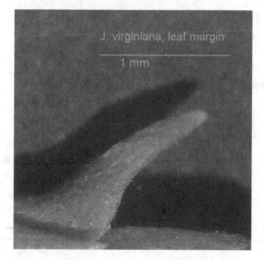

Figure 1.13.4. *J. virginiana* with smooth leaf margins.

Furthermore, DNA evidence shows that Gaussen's *phoenicioides* is not related to the serrate junipers of the western hemisphere (Fig. 1.3). The recent work using DNA markers has not supported Gaussen's treatment, except in very general terms.

Section *Sabina* is divided into three major clades (Fig. 1.3 above and Adams and Schwarzbach, 2013a):

1. Serrate-leaf junipers of North America (22 species, Adams and Schwarzbach, 2011, 2013d),
2. Turbinate-seed cones, single-seeded, entire-leaf junipers, eastern hemisphere (16 species, 3 varieties, Adams and Schwarzbach, 2012b, 2013a, Zanoni and Adams, 1976, 1979) and
3. Multi-seeded, entire-leaf junipers, both eastern and western hemispheres (23 species, 16 varieties, Adams and Schwarzbach, 2012c, 2013b).

1. Serrate-leaf junipers of North America (22 species, 4 varieties., 7 forms, Adams and Schwarzbach, 2013a, b).

The serrate-leaf junipers of North America form a well supported clade (Figs. 1.3, 1.14). A minimum spanning network shows the magnitude of differences between taxa (Fig. 1.14.1).

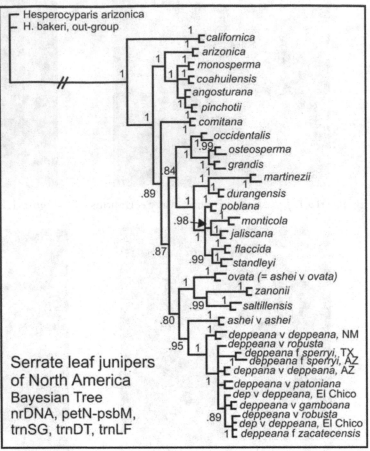

Fig. 1.14. Bayesian tree for the serrate-leaf junipers of North America. Numbers branch points are posterior probabilities. From Adams and Schwarzbach (2013a, b).

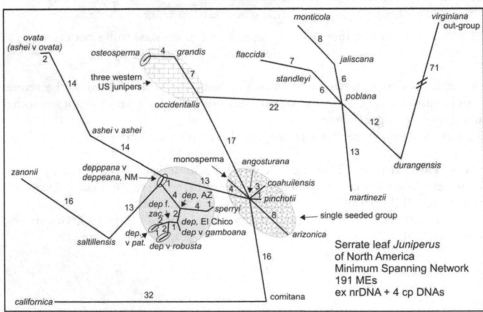

Figure 1.14.1. Minimum spanning network of serrate leaf *Juniperus* of North America. Numbers on the lines are the number of MEs (Mutational Events).

Adams and Schwarzbach (2013b) found considerable support for most of the recognized taxa in this group (Table 1.3). However, some species such as *J. coahuilensis*, *J. monosperma* and *J. pinchotii* are not well supported (Fig. 1.14.1 above), yet are well characterized both morphologically and chemically.

Table 1.3. Support using DNA sequencing data versus the taxonomies of Adams (2011) and Farjon (2010). Support levels: ++ = very strong, + = strong, +/- = not strong, - = not supported.

Adams (2011)	Farjon (2005, 2010)	Supported by DNA
J. angosturana R. P. Adams	*J. angosturana*	+/- *J. angosturana*
J. arizonica (R. P. Adams) R. P. Adams	*J. arizonica*	++ *J. arizonica*
J. ashei Buchholz	*J. ashei*	++ *J. ashei*
var. *ovata* R. P. Adams	var. *ovata*	++ *J. ovata*
J. californica Carriere	*J. californica*	++ *J. californica*
J. coahuilensis (Martinez) Gaussen ex R. P. Adams	*J. coahuilensis*	+/- *J. coahuilensis*
J. comitana Martinez	*J. comitana*	++ *J. comitana*
J. deppeana Steudel var. *deppeana*	*J. d.* var. *deppeana*	+ *J. d.* var. *deppeana*
J. deppeana Steudel var. *deppeana*	*J. d.* var. *pachyphlaea*	+ *J. d.* var. *deppeana*
forma *elongata* R. P. Adams	--	+ *f. elongata*
forma *sperryi* (Correll) R. P. Adams	var. *sperryi*	+ *f. sperryi*
forma *zacatacensis* (Mart.) R. P. Adams	var. *zacatacensis*	+/-f./var. *zacatacensis*?
var. *gamboana* (Mart.) R. P. Adams	*J. gamboana*	++ var. *gamboana*
var. *patoniana* (Martinez) Zanoni	var. *robusta*	+/-f. or var. *patoniana*?
var. *robusta* Martinez	var. *robusta*	+/-f. or var. *robusta* ?
J. durangensis Martinez	*J. durangensis*	++ *J. durangensis*
var. *topiensis* R. P. Adams & S. Gonzalez	--	++ *v. topiensis*
J. flaccida Schlecht.	*J. flaccida*	++ *J. flaccida*
J. grandis R. P. Adams	*J. occidentalis* var. *australis*	+ *J. grandis*
J. jaliscana Martinez	*J. jaliscana*	++ *J. jaliscana*
J. martinezii Perez de la Rosa	*J. flaccida* var. *martinezii*	++ *J. martinezii*
J. monosperma (Engelm.) Sarg.	*J. monosperma*	+ *J. monosperma*
J. monticola Martinez forma *monticola*	*J. monticola*	++ *J. monticola*
forma *compacta* Martinez	forma *compacta*	++ *J. zanonii* in part
forma *orizabensis* Martinez	forma *orizabensis*	+ *J. m. f. orizabensis*
J. occidentalis Hook.	*J. occidentalis*	++ *J. occidentalis*
J. occidentalis f. *corbetii* R. P. Adams	--	+ *J. o. f. corbetii*
J. osteosperma (Torr.) Little	*J. osteosperma*	+ *J. osteosperma*
J. pinchotii Sudworth	*J. pinchotii*	+/- *J. pinchotii*
J. poblana (Martinez) R. P. Adams	*J. flaccida* var. *poblana*	++ *J. poblana*
J. saltillensis M. T. Hall	*J. saltillensis*	++ *J. saltillensis*
J. standleyi Steyermark	*J. standleyi*	++ *J. standleyi*
J. zanonii R. P. Adams	*J. monticola* f. *compacta*	++ *J. zanonii*

2. Turbinate-seed cones, single-seeded, entire-leaf junipers, eastern hemisphere (16 species, 3 varieties, Adams and Schwarzbach, 2012b, 2013a, Zanoni and Adams, 1976, 1979)

The second major clade is composed of the single-seeded, turbinate-shaped seed cones, entire-leaf junipers of the eastern hemisphere and these taxa comprise a relatively uniform clade (Figs. 1.15, 1.15.1). Variations in leaf types are extreme, hybridization is likely to be common confounding the phylogeny, and it is very difficult to make systematic collections in their natural habitat in central Asia and western China. Undoubtedly, new graduate students will take a closer look at this group to resolve the taxa and their phylogeny.

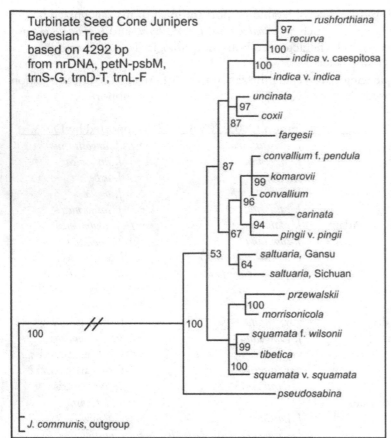

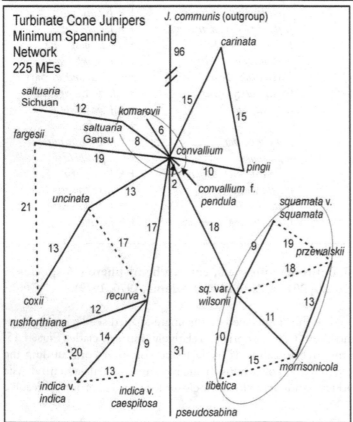

Figure 1.15. Bayesian tree for the turbinate seed cone taxa, sect. *Sabina*. Numbers at the branch points are posterior probabilities (as percent).
Nomenclature follows Adams and Schwarzbach (2013c).

Figure 1.15.1. Minimum spanning network based on 225 MEs. Numbers at branch points are the number of mutational events (MEs).
Nomenclature follows Adams and Schwarzbach (2013c).

Table 1.3.   Comparison of support using DNA sequencing data versus the taxonomies of Adams (2011) and Farjon (2010). Support levels: ++ = very strong, + = strong, +/- = not strong, - = not supported.

Adams(2011)	Farjon (2005, 2010)	Supported by DNA data
J. convallium	*J. convallium*	+ *J. convallium*
J. coxii	*J. recurva* var *coxii*	++ *J. coxii*
J. indica	*J. indica*	++ *J. indica*
J. indica. var. *caespitosa*	*J. indica* var. *caespitosa*	+ *J. indica* var. *caespitosa?*
J. indica var. *rushforthiana*	*J. indica*	++ *J. rushforthiana*
J. komarovii	*J. komarovii*	+ *J. komarovii*
J. morrisonicola	*J. squamata*	++ *J. morrisonicola*
J. pingii	*J. pingii*	++ *J. pingii*
J. p. var. *carinata*	*J. p.* var. *wilsonii*	++ *J. carinata*
J. przewalskii	*J. przewalskii*	++ *J. przewalskii*
J. p. f. *pendula*	*J. przewalskii*	+ *J. convallium* f. *pendula*
J. pseudosabina	*J. pseudosabina*	++ *J. pseudosabina*
J. recurva	*J. recurva*	++ *J. recurva*
J. recurva var. *uncinata*	*J. recurva?*	++ *J. uncinata*
J. saltuaria	*J. saltuaria*	++ *J. saltuaria*
J. squamata	*J. squamata*	++ *J. squamata*
J. squamata var. *fargesii*	*J. squamata*	++ *J. fargesii*
J. squamata f. *wilsonii*	*J. pingii* f. *wilsonii*	+ *J. s.* var. *wilsonii*
J. tibetica	*J. tibetica*	++ *J. tibetica*

3. Multi-seeded, entire-leaf junipers, both eastern and western hemispheres (23 species, Adams and Schwarzbach, 2012c, 2013d).

The third major clade of section *Sabina* is the multi-seeded, entire-leaf junipers, both eastern and western hemispheres. The phylogeny (Fig. 1.16.1) shows a clade of taxa from the western hemisphere (North America) that is sister to *J. sabina - davurica* from central Asia. Sister to these clades is the *J. chinensis* complex (Fig. 1.16.1). Two rather different species, *J. erectopatens* and *J. microsperma*, both from western China, form another distinct clade. The final sister clade is large with two sub-clades: *J. thurifera - foetidissima - seravschanica*, and *J. polycarpos - excelsa - procera* (Fig. 1.16.1).

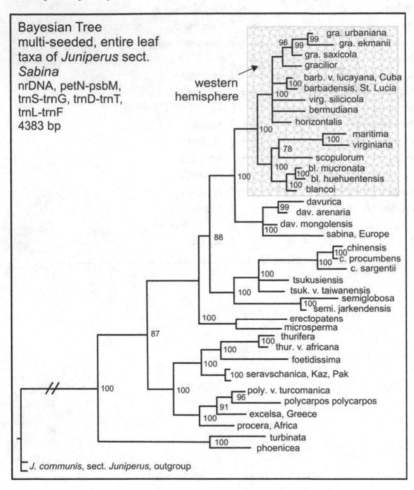

Figure 1.16.1. Bayesian tree of the multi-seeded, entire leaf taxa, sect. *Sabina*. Numbers at the branch points are posterior probabilities (as percent).
Nomenclature follows Adams and Schwarzbach (2013d).

The taxonomy of the Caribbean junipers (upper clade, Fig. 1.16.1) is challenging due to the fact that *J. gracilior* (and varieties), *J. barbadensis*, *J. bermudiana*, *J. horizontalis*, *J. virginiana* var. *silicicola* and *J. horizontalis* all have the same chloroplast DNA, but *J. virginiana* var. *silicicola* and *J. horizontalis* have nuclear DNA (ITS or nrDNA) like continental species (*J. blancoi*, *J. maritima*, *J. scopulorum* and *J. virginiana*, respectively). The Bayesian tree has some support for *J. virginiana* var. *silicicola* being a separate species from *J. virginiana*.

The magnitude of the differences among taxa is seen in the minimum spanning network of an overview of the groups (Fig. 1.16.2). Notice that the 'pseudoserrate' junipers of the eastern hemisphere (*J. phoenicea*, *J. turbinata*) are distantly linked (66 MEs) to the *chinensis* group. These two taxa have been treated as *J. phoenicea* var. *phoenicea* and var. *turbinata* (or subsp. *turbinata*) in Adams (2011) and Farjon (2005, 2010). However, these data (23 MEs between taxa) strongly support their recognition at the specific level.

Juniperus sabina (Europe and central Asia) is separated by 25 MEs from the *davurica* group (*J. d.* var. *davurica*, var. *arenaria*, var. *mongolensis*) (Fig. 1.16.2).

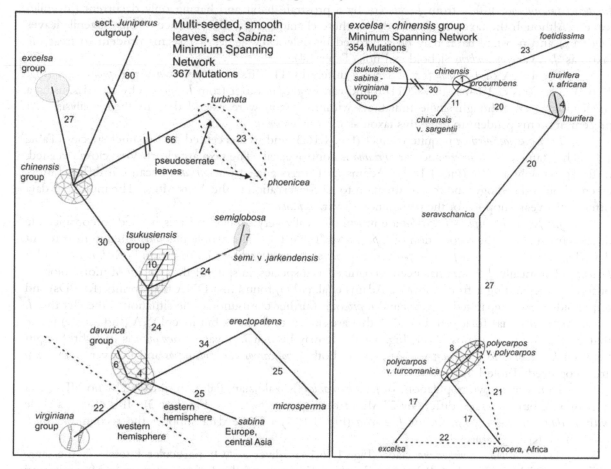

Figure 1.16.2. Minimum spanning network of the major groups of the multi-seeded, smooth leaf junipers of sect. Sabina.

Figure 1.16.3. Minimum spanning network of the *excelsa - chinensis* group.

Note for Figs. 1.16.2-3: Numbers at branch points are the number of mutational events (MEs). Nomenclature follows Adams and Schwarzbach (2013d).

The *Juniperus semiglobosa* / var. *jarkendensis* group is separated by 24 MEs from the *tsukusiensis* group (Fig. 1.16.2), but *J. jarkendensis* and *J. semiglobosa* differ by only 7 out of the total of 354 MEs. Furthermore, the taxa are very similar in morphology. The present data support recognition of *J. jarkendensis* as a variety of *J. semiglobosa*:

The *Juniperus tsukusiensis* group (var. *tsukusiensis*, var. *taiwanensis*) was treated as *J. chinensis* var. *tsukusiensis* by Adams and Farjon (Table 1.4), but later treated as *J. tsukusiensis* (Adams et al., 2011). This group is separated from *J. s.* var. *jarkendensis* by 24 MEs, from the *J. davurica* group by 24 MEs and from *J. chinensis* by 30 MEs (Fig. 1.16.2). The decision by Adams et al. (2011) to recognize *J. tsukusiensis* and *J. t.* var. *taiwanensis* as separate taxa is strongly supported by the molecular data as well as morphological differences (Table 1.4).

The position (Fig. 1.16.3) of *J. chinensis* var. *sargentii*, 11 MEs from *J. chinensis,* supports the recognition of *J. c.* var. *sargentii*, but additional research seems warranted as I have only a few samples from Japan and they may not be representative.

The *Juniperus excelsa - chinensis* group is examined in detail in Figure 1.16.3. These groups differ by 20 MEs (Fig. 1.16.3). The *J. excelsa - chinensis* group is diverse (Fig. 1.16.3) but *J. chinensis* and *J.*

procumbens differ by only 1 ME (of 354). Although both Adams and Farjon (Table 1.4) recognized *J. procumbens*, the current sequence data do not support the recognition of this taxon at the species level. *Juniperus procumbens* differs from *J. chinensis* by a prostrate habit and having only decurrent (juvenile) leaves. Although the taxon is distinct due to these characteristics, the presence of only juvenile leaves (neoteny) and prostate habit may be controlled by only a few genes. It seems prudent to treat the taxon as *J. c.* var. *procumbens* Siebold ex Endl. (Table 1.4).

The DNA of *J. chinensis* var. *sargentii* differs by 11 MEs from *J. chinensis* var. *procumbens* (Fig. 1.16.3). Previously, Adams et al. (2011) found var. *sargentii* to differ from *J. chinensis* by 10 MEs, using a smaller data set, although single mutations within a taxon were excluded from their analysis. At present, it seems prudent to leave this taxon as *J. c.* var. *sargentii*.

Juniperus foetidissima is quite distinct (Fig. 1.16.3) and well accepted as a distinct species (Table 1.4). The status of *J. thurifera* and var. *africana* is controversial. The taxa are clearly very closely related, differing by only 4 MEs (Fig. 1.16.3). Adams (2011) recognized var. *africana* because it differed in its essential oil and ecology, and to call attention to its conservation in the Atlas Mtns. The molecular data offers only weak support for the recognition of var. *africana*.

Juniperus polycarpos - J. excelsa are morphologically very similar and this has led to considerable disagreement about the recognition of *J. polycarpos* (Table 1.4). The problem is illustrated in the present data (Fig. 1.16.3) in which *J. polycarpos* var. *turcomanica* is separated by 17 MEs from both *J. excelsa* and *J. procera*. Historically, *J. procera* has been recognized as a species, in spite of the paucity of morphological differences separating it from *J. excelsa*. Adams et al. (1993) found that DNA fingerprints (RAPDs) and terpenoids clearly separated *J. excelsa* and *J. procera*. Further confounding the situation is the fact that *J. p.* var. *turcomanica* has foliage and nrDNA that are similar to *J. excelsa*, but its cpDNA (tnrC-trnD) is like that of *J. polycarpos*. (Adams, 2011, Fig. 5.4.1). It may be that *J. p.* var. *turcomanica* is of hybrid origin between *J. excelsa* and *J. polycarpos* . At present, both *J. polycarpos* var. *polycarpos* and *J. p.* var. *turcomanica* are recognized (Table 1.4).

Plants from two populations of *J. seravschanica* (Kazakhstan, Pakistan) differed by no MEs (data not shown), but the taxon differs by 27 MEs from *J. p.* var. *polycarpos* (Fig. 1.16.3). It is also in a clade with *J. thurifera* not *J. polycarpos* or *J. excelsa* (Fig. 1.16.1). These data support the recognition of *J. seravschanica* Kom. (Table 1.4).

The *J. sabina - virginiana* clade (Fig. 1.16.1) is diverse. Of particular interest is the large difference (25 MEs, Fig. 1.16.4) between *J. sabina* (Europe) and the *J. davurica* group in Mongolia and Qinghai, China. Adams et al. (2006) reported considerable variation in the leaf terpenoids from Europe to Mongolia. RAPDs data indicated that *J. sabina* from Europe and the Tian Shan Mtns., China differed from *J. sabina* in Mongolia and Qinghai, China (Adams et al. 2007). The current DNA data indicate that the taxonomic separation in *J. sabina* is an even greater extent with 25 MEs between *J. sabina* (Europe) and the putative *sabina* varieties in Mongolia and Qinghai, China (Fig. 1.16.4). In fact, the data offer no support that *J. sabina* (*sensu stricto*) occurs in Mongolia and Qinghai, China, but rather that *J. davurica* Pall. should be recognized in this region (Table 1.4). It is apparent that *J. sabina* var. *arenaria* and *J. s.* var. *mongolensis* are not conspecific with *J. sabina*, but are varieties of *J. davurica* (Table 1.4, Fig. 1.16.4).

The *J. tsukusiensis* group (var. *tsukusiensis* - var. *taiwanensis*) is treated as *J. chinensis* var. *tsukusiensis* by Adams and Farjon (Table 1.4), but later treated as *J. tsukusiensis* (Adams et al., 2011), and is separated by 30 MEs from the *J. chinensis* group. The present data (Fig. 1.16.4) strongly support the recognition of *J. tsukusiensis* Masam. and *J. t.* var. *taiwanensis* (R. P. Adams and C-F. Hsieh) R. P. Adams (Table 1.4).

Juniperus erectopatens is an unusual taxon. I discovered the taxon at the margins (wasteland) of a cultivated field south of Songpan, Sichuan, China. It is a small tree, resembling *J. chinensis* (as recognized by Farjon, Table 1). However, its oils and RAPDs clearly separate it from *J. chinensis* and any other taxa (Adams, 1999). The sequence data separate *J. erectopatens* by 25 MEs from the nearest taxon (*J. microsperma*, Fig. 1.16.4) and 50 MEs from the *J. chinensis* group (Fig. 1.16.4). These data support the recognition of *J. erectopatens* (Table 1.4). More field work is needed to better understand this taxon. It may be a product of hybridization, such that these 'apparent mutations' might be DNA from another species.

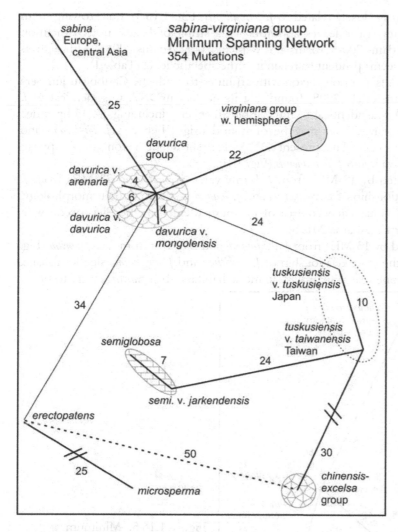

sabina-virginiana group
Minimum Spanning Network
354 Mutations

sabina
Europe,
central Asia

25

virginiana group
w. hemisphere

davurica
group

22

davurica v.
arenaria   4

6

davurica v.   4
davurica

davurica v.
mongolensis

24

tuskusiensis
v. tuskusiensis
Japan

10

34

tuskusiensis
v. taiwanensis
Taiwan

semiglobosa

7

24

semi. v. jarkendensis

erectopatens

50

30

25

chinensis-
excelsa
group

microsperma

Figure 1.16.4. Minimum spanning network for the *sabina-virginiana* group.
Numbers at branch points are the number of mutational events (MEs).
Nomenclature follows Adams and Schwarzbach (2013d).

The *J. virginiana* group is the least diverse and probably most recently diversified of the major groups (Figs. 1.16.1, 1.16.5). Overall, the group is divided into the *J. blancoi* group, the Caribbean group (*gracilior-barbadensis*) and outliers, *J. virginiana*, *J. horizontalis*, *J. maritima* and *J. scopulorum* (Fig. 1.16.5). The *J. blancoi* group is located in western and central Mexico and consists of 3 closely related varieties, differing by only 3-5 MEs (Fig. 1.16.5). Both Adams and Farjon (Table 1.4) accept these taxa.

Juniperus horizontalis is a cool-season species of northern United States and Canada (Adams 2011), thought to have been derived from *J. scopulorum* or *J. virginiana*, so it is surprising to find it linked to the Caribbean and *blancoi* groups (Fig. 1.16.5). It is interesting that it contains the cpDNA common in the Caribbean, but has nrDNA most like continental taxa: *J. blancoi*, *J. scopulorum* and *J. virginiana*.

The Caribbean group is closely related and difficult to separate by morphology. The Cuban *Juniperus saxicola* is a tree that has only juvenile leaves (neoteny), but is otherwise similar to *J. gracilior*. It differs from the shrubby *J. gracilior* var. *urbaniana* by only 3 MEs (Fig. 1.16.5). The DNA data suggest that *J. saxicola* is a conspecific member of the *J. gracilior* group. Adams and Schwarzbach (2013d) recognized *J. gracilior* var. *saxicola*. Thus, we are now left with only two species in the Caribbean: *J. barbadensis* and *J. gracilior* (Table 1.4).

Juniperus barbadensis (sensu stricto) is known only from Petit Piton, St. Lucia, BWI (Adams, 2011), whereas *J. b.* var. *lucayana* is widespread growing on Hispanola, Jamaica, Cuba and the Bahama Islands. The taxa are nearly indistinguishable; differing only by glands conspicuous on old whip leaves (Adams 2012). The DNA data show these taxa differ by only 1 ME (of 354), providing little support for the continued recognition of *J. b.* var. *lucayana* (Table 1.4).

Juniperus bermudiana is obviously closely related to *J. barbadensis* (Fig. 1.16.5), but growing in an isolated habitat it has evolved shorter scales leaves appearing as a string of beads, not to mention differences in its leaf essential oils (Adams 2000; Adams and Wingate 2008; Adams 2008). Although its DNA differences are only 8 MEs, it seems prudent to retain it at the specific level (Table 1).

Juniperus virginiana var. *silicicola* is a curious taxon with affinities to both the Caribbean junipers and the mainland, *J. virginiana*. Adams et al. (2008) found that its nuclear nrDNA was identical to *J. virginiana*, but its cpDNA (trnC-trnD) was identical to the Caribbean species including, a 245 bp indel. That data was suggestive that *J. v.* var. *silicicola* may be of hybrid origin between *J. virginiana* and Caribbean *J. barbadensis* (or an ancestor). The present DNA data show the taxon to occupy an intermediate position between *J. virginiana* and *J. barbadensis* (Fig. 1.16.5).

Juniperus scopulorum is separated by 12 MEs from *J. blancoi* var. *mucronata* and 23 MEs from *J. maritima* (Fig. 1.16.5). The close relationship of *J. scopulorum* and *J. blancoi* is shown in their morphology, and by likely hybridization in the Colonia Pacheco area of northern Mexico in the Pleistocene with nearby populations of *J. b.* var. *mucronata* (Adams 2011b).

Juniperus maritima is separated by 13 MEs from *J. virginiana*, and 23 MEs from *J. scopulorum* (Fig. 1.16.5). Several characters support the close relationship of *J. maritima* and *J. virginiana*: slender foliage, strong central axis, fruit maturing in one year, and growth in mesic habitats. If *J. maritima* were to be

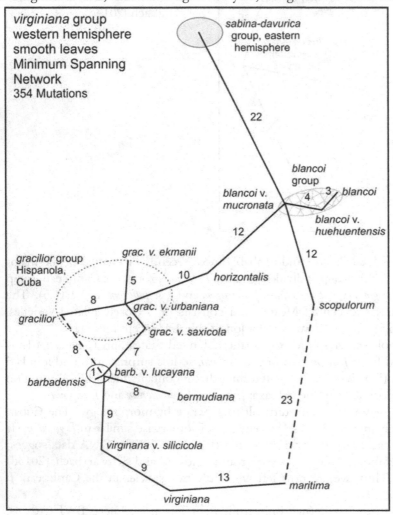

Figure 1.16.5. Minimum spanning network for the *virginiana* group. Numbers at branch points are the number of mutational events (MEs). Nomenclature follows Adams and Schwarzbach (2013d).

treated as a variety, it would have to be a variety of *J. virginiana*, not of *J. scopulorum*. However, DNA data give strong support to the continued recognition of *J. maritima* at the specific level (Table 1.4).

Table 1.4. Comparison of support using DNA sequencing data versus the taxonomies of Adams (2011) and Farjon (2010). Support levels: ++ = very strong, + = strong, +/- = not strong, - = not supported.

Adams(2011)	Farjon (2005, 2010)	Supported by DNA data
J. barbadensis	*J. barbadensis*	++ *J. barbadensis*
J. b. v. lucayana	*J. b. v. lucayana*	+ *J. b. v. lucayana* or *J. barbadensis*
J. bermudiana	*J. bermudiana*	++ *J. bermudiana*
J. blancoi	*J. blancoi*	++ *J. blancoi*
J. b. v. huehuentensis	*J. blancoi*	++ *J. blancoi* v. *huehuentensis*
J. b. v. mucronata	*J. b. v. mucronata*	++ *J. blancoi* v. *mucronata*
J. chinensis	*J. chinensis*	++ *J. chinensis*
J. c. v. sargentii	*J. c. v. sargentii*	+/- *J. chinensis* v. *sargentii*
J. c. v. tsukusiensis	*J. c. v. tsukusiensis*	++ *J. tsukusiensis*
J. c. v. taiwanensis	*J. c. v. tsukusiensis*	++ *J. tsukusiensis* v. *taiwanensis*
J. erectopatens	*J. chinensis*	++ *J. erectopatens*
J. excelsa	*J. excelsa* (in part)	++ *J. excelsa*
J. foetidissima	*J. foetidissima*	++ *J. foetidissima*
J. gracilior	*J. gracilior*	++ *J. gracilior*
J. g. v. ekmanii	*J. g. v. ekmanii*	++ *J. gracilior* v. *ekmanii*
J. g. v. urbaniana	*J. g. v. urbaniana*	++ *J. gracilior* v. *urbaniana*
J. horizontalis	*J. horizontalis*	++ *J. horizontalis*
J. jarkendensis	*J. semiglobosa*	++ *J. semiglobosa* v. *jarkendensis*
J. maritima	*J. scopulorum*	++ *J. maritima*
J. phoenicea	*J. phoenicea*	++ *J. phoenicea*
J. p. v. turbinata	*J. p.* subsp. *turbinata*	++ *J. turbinata*
J. procera	*J. procera*	++ *J. procera*
J. procumbens	*J. procumbens*	++ *J. chinensis* v. *procumbens*
J. polycarpos	*J. excelsa* subsp. *polycarpos*	++ *J. polycarpos*
J. p. v. seravschanica	*J. e.* subsp. *polycarpos*	++ *J. seravschanica*
J. p. v. turcomanica	*J. e.* subsp. *polycarpos*	+ *J. polycarpos* v. *turcomanica*
J. sabina	*J. sabina*	++ *J. sabina*
J. s. v. arenaria	*J. s. v. arenaria*	++ *J. davurica* v. *arenaria*
J. s. v. davurica	*J. s. v. davurica*	++ *J. davurica*
J. s. v. mongolensis	*J. s. v. arenaria*	++ *J. davurica* v. *mongolensis*
J. scopulorum	*J. scopulorum*	++ *J. scopulorum*
J. semiglobosa	*J. semiglobosa*	++ *J. semiglobosa*
J. saxicola	*J. saxicola*	++ *J. gracilior* v. *saxicola*
J. thurifera	*J. thurifera*	++ *J. thurifera*
J. t. v. africana	*J. thurifera*	+/- *J. thurifera* v. *africama*
J. virginiana	*J. virginiana*	++ *J. virginiana*
J. v. v. silicicola	*J. v. v. silicicola*	+ *J. virginiana* v. *silicicola*

There are three primary centers of *Juniperus* diversity:
1. the deserts of Mexico and the southwestern USA,
2. the Mediterranean region and
3. western China - central Asia.
    A minor area of *Juniperus* diversity is Japan and the far-eastern Asia.

The *Juniperus* species recognized in this treatment are listed in Table 1.2.

Table 1.2 Taxa of *Juniperus* - 76 species, 32 varieties, 8 formas

Section *Caryocedrus* (1 species)
   J. drupacea Labill.

Section *Juniperus* (=*Oxycedrus*): (14 species, 9 varieties)
   J. brevifolia (Seub.) Ant.
   J. cedrus Webb & Berthol.
   J. communis L. var. *communis*
      var. *charlottensis* R. P. Adams
      var. *depressa* Pursh
      var. *hemisphaerica* (J. & C. Presl) Parl.
      var. *kamchatkensis* R. P. Adams
      var. *kelleyi* R. P. Adams
      var. *megistocarpa* Fernald & H. St. John
      var. *nipponica* (Maxim.) E. H. Wilson
      var. *saxatilis* Pall. (only in eastern hemisphere)
   J. deltoides R. P. Adams
      var. *spilinanus* (Yalt., Elicin & Terzioglu) Terzioglu
   J. formosana Hayata
   J. jackii (Rehder) R. P. Adams
   J. macrocarpa Sibth. & Sm.
   J. maderensis (Menezes) R. P. Adams
   J. mairei Lemee & Lev.
   J. navicularis Gand.
   J. oxycedrus L. *(= var. badia)*
   J. rigida Mig. in Sieb. var. *rigida*
      var. *conferta* (Parl.) Patschka
   J. taxifolia Hook. & Arn.
   J. taxifolia var. *lutchuensis* (Koidz.) Satake

Section *Sabina*: (61 species, 23 vars., 8 forms)
   Serrate leaf margins, western hemisphere (22 species, 4 varieties, 7 forms)
      J. angosturana R. P. Adams
      J. arizonica (R. P. Adams) R. P. Adams
      J. ashei Buchholz
      J. californica Carriere
         f. *lutheyana* J. T. Howell & Twisselm.
      J. coahuilensis (Martinez) Gaussen ex R. P. Adams
      J. comitana Martinez
      J. deppeana Steudel var. *deppeana*
         forma *elongata* R. P. Adams
         forma *sperryi* (Correll) R. P. Adams
         forma *zacatacensis* (Mart.) R. P. Adams
         var. *gamboana* (Mart.) R. P. Adams
         var. *patoniana* (Martinez) Zanoni
         var. *robusta* Martinez
      J. durangensis Martinez
         var. *topiensis* R. P. Adams & S. Gonzalez
      J. flaccida Schlecht.
      J. grandis R. P. Adams
      J. jaliscana Martinez
      J. martinezii Perez de la Rosa
      J. monosperma (Engelm.) Sarg.

J. monticola Martinez forma *monticola*
    forma *compacta* Martinez
    forma *orizabensis* Martinez
J. occidentalis Hook.
    f. *corbetii* R. P. Adams
J. osteosperma (Torr.) Little
J. ovata (R. P. Adams) R. P. Adams
J. pinchotii Sudworth
J. poblana (Martinez) R. P. Adams
J. saltillensis M. T. Hall
J. standleyi Steyermark
J. zanonii R. P. Adams

Smooth leaf margins [including *J. phoenicea*, which has small teeth, but is atypical (see Fig. 1.5)]

One seed/ cone, turbinate or ellipsoidal shaped seed cones (16 species, 3 varieties, 1 form)
J. carinata (Y. K. Yu & L. K. Fu) R. P. Adams
J. convallium Rehder & Wilson
    f. *pendula* (Cheng & L. K. Fu) R. P. Adams
J. coxii A.B. Jacks
J. fargesii (Rehder & Wils.) Kom.
J. indica Bertol.
    var. *caespitosa* Farjon
J. komarovii Florin
J. morrisonicola Hayata
J. pingii Cheng & Ferre.
    var. *miehei* Farjon
J. przewalskii Kom.
J. pseudosabina Fisch., Mey. & Ave-Lall.
J. recurva Buch.-Ham. ex D. Don.
J. rushforthiana (R. P. Adams) R. P. Adams
J. saltuaria Rehder & Wils.
J. squamata D. Don in Lambert f. *squamata*
    var. *wilsonii* (Rehder) R. P. Adams
J. tibetica Kom.
J. uncinata (R. P. Adams) R. P. Adams

Multi-seeded ((1)2 - 6 seeds/ ovoid cone) Eastern Hemisphere (6 species, 2 varieties)
J. excelsa M.-Bieb.
J. foetidissima Willd.
J. polycarpos K. Koch var. *polycarpos*
    var. *turcomanica* (B. Fedtsch.) R. P. Adams
J. procera Hochst. ex. Endl.
J. seravschanica Kom.
J. thurifera L. var. *thurifera*
    var. *africana* Maire

One or more seeds / cone, both Eastern and Western Hemispheres
Eastern Hemisphere (9 species, 7 varieties)

J. chinensis L. var. *chinensis*
    var. *sargentii* Henry
    var. *procumbens* Sieb.ex Endl.
J. davurica Pall.
    var. *arenaria* (E. H. Wilson) R. P. Adams
    var. *mongolensis* (R. P. Adams) R. P. Adams
J. erectopatens (Cheng & L. K. Fu) R. P. Adams
J. microsperma (Cheng & L. K. Fu) R. P. Adams
J. phoenicea L.
J. sabina L.
J. semiglobosa Regel var. *semiglobosa*
    var. *jarkendensis* (Kom.) R. P. Adams
    var. *talassica* (Lipsky) Silba
J. tsukusiensis var. *tsukusiensis* Masam.
    var. *taiwanensis* (R. P. Adams & C-F. Hsieh) R. P. Adams
J. turbinata Guss.,

Western Hemisphere (8 species, 7 varieties)

J. barbadensis L.
    var. *lucayana* (Britton) R. P. Adams
J. bermudiana L.
J. blancoi Martinez var. *blancoi*
    var. *huehuentensis* R. P. Adams, S. Gonzalez, and M. G. Elizondo
    var. *mucronata* (R. P. Adams) Farjon
J. gracilior Pilger var. *gracilior*
    var. *ekmanii* (Florin) R. P. Adams
    var. *saxicola* (Britton & P. Wilson) R. P. Adams
    var. *urbaniana* (Pilger & Ekman) R .P. Adams
J. horizontalis Moench
J. maritima R. P. Adams
J. scopulorum Sarg.
J. virginiana L. var. *virginiana*
    var. *silicicola* (Small) E. Murray

Chapter 2: Geographic Variation

Pan-Arctic Variation in *Juniperus communis*

Juniperus communis has the largest distribution of any juniper species. It is the only species found in both the eastern and western hemispheres. Adams et al. (2003) examined Arctic populations of *J. communis* (Fig. 2.1).

Farjon (2001, 2005) recognized var. *communis* L. (n. Europe), var. *depressa* Pursh (North America), var. *megistocarpa* (e. Canada); and var. *saxatilis* Pall. (Europe, Siberia, central Asia, far east, Greenland, Iceland, and far western N. America). Farjon did not recognize var. *hemispherica* (J. & C. Presl.) Nyman (Sicily, Mediterranean) nor var. *oblonga* (M.-Bieb.) Parl. (Caucasus Mts.).   In addition, Adams et al. (2001a) have reported that in Japan, var. *saxatilis* and var. *nipponica* (Maxim.) E. H. Wilson are seemingly distinct from var. *saxatilis* from Europe.

Most of the northern portion of *J. communis'* range was glaciated during the Pleistocene (Flint, 1971; Graham, 1999).   Comparing the distribution of *J. communis* (Fig. 2.1) versus the ice cover (Fig. 2.2), only the Alaska and Kamchatka populations might have been ice free. However, whether *J. communis* survived in either location is uncertain.

Vast areas of the northern hemisphere that now have populations of *J. communis* have been colonized since the last glacial maximum (12,000-18,000 BP, Flint, 1971).   During the maximum of the Wisconsin glaciation, which occurred ca. 18,000 BP, the ice covered nearly all of Canada, except part of the Yukon, and the Arctic archipelago, reaching the Columbia Plateau in the west and the Ohio-Mississippi basin and New England in the east (Fig. 2.2). In Alaska, ice cover was restricted to southern Alaska

Figure 2.1. Populations of *J. communis* sampled by Adams et al. (2003), overlaid on a distribution map of *J. communis*.

Figure 2.2. Maximal Pleistocene ice cover.

and the Aleutians. Greenland was completely covered by ice. In North America, two large and confluent ice masses constituted the ice cover: the Cordilleran Ice Sheet, which occupied the Canadian Cordillera and the Laurentian Ice Sheet which covered the regions east of the Rocky Mountains. Smaller ice sheets occupied the High Arctic and Newfoundland. Mountain glaciers were also present in the western United States.   Sea levels lowered by as much as 200 m and large expanses of the continental shelf were colonized by vegetation. Between 14,000 and 9,000 BP, in response to the warming of the climate, the ice receded and ice sheets separated and broke into several domes. The

Cordilleran sheet thinned and fractionated into ice caps and then into families of valley glaciers. By the end of this period, much of the Arctic and coastal areas of Greenland were ice-free. The retreat of the Laurentide ice mass, north of the St. Lawrence Valley allowed marine transgression. Large post-glacial lakes, such as Agassiz and McConnell, existed along the western margins. Recession of the ice further accelerated and by 8,400 y BP, ice remained in the Canadian Cordillera and Arctic only in areas where glaciers persist today. The Laurentide sheet receded rapidly following the marine invasion of the Hudson's Bay region, separating into Labrador and Keewatin domes. By 8,000 y BP, the Tyrrel sea was transgressing the Hudson's Bay Lowlands and the Champlain sea had almost retreated from the St. Lawrence Valley. Ice gradually disappeared from Keewatin, New Quebec, and Labrador. In Greenland, the ice sheet reached a minimum about 5,000 BP, followed by a re-advancement to it present limits (Flint, 1971; Clague, 1989; Dyke et al., 1989; Occhietti, 1989).

Juniperus communis appeared soon after or at the end of the last glacial period in northern Britain (Bennett et al., 1992), in North America (Yu, 1997) and in Siberia (Pisaric et al., 2001). Whether *J. communis* existed in the Kamchatka peninsula during the last ice age is unknown to the present author. It is now generally thought that the Kamchatka peninsula probably lacked ice sheets and only local alpine glaciers occurred during the peak of the last glacial period (Savaskul, 1999). But severe conditions in western Beringia up to 12,800 BP (Lozhkin et al., 1993) probably made the area inhospitable for *J. communis* (or *J. jackii*). Of particular biogeographical interest is the colonization of Iceland and Greenland, and the question of the migration of *Juniperus* across the Bering land bridge between Siberia and Alaska.

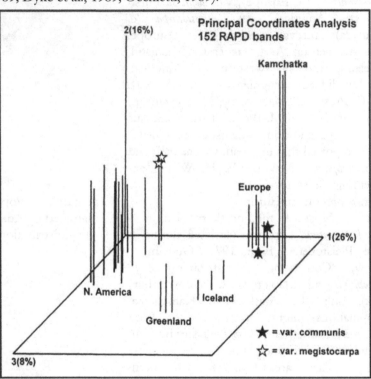

Figure 2.3. PCO of *J. communis* (from Adams et al., 2003).

Adams et al. (2003) found that these populations were separated into several groups (Fig. 2.3). Notice (Fig. 2.3) the trend from Kamchatka, Europe, Iceland, Greenland, to North America. It appears that the colonization of glaciated Greenland and Iceland was from Europe, not North America. The divergence of the Kamchatka population is also very obvious.

Several of the varieties of *J. communis* were included in this study. All of the plants in this study were small shrubs or prostrate plants except the small trees (2-3 m tall) of *J. communis* var. *communis* from Sweden. These var. *communis* plants (CC) link with var. *saxatilis* plants (UR) from the eastern hemisphere (Fig. 2.4). There appears to be no large distinction in the RAPDs of var. *communis* and var. *saxatilis* from Europe.

To better visualize the geography of these similarities, the minimum spanning network was plotted onto the distribution map of *J. communis* (Fig. 2.4.). Several important trends are now evident. The eastern North America plants are very similar (.920, .921). The linkage of MA to PE and to QB is suggestive of a colonization pathway. The colonization pathway to Iceland (IC) and Greenland (GR) appears to be from Europe (Sweden, SW) and not from North America (however the second largest link from Greenland is to Saskatchewan (SK), but at a much lower level of similarity). Several bird species stop in Iceland en route between Europe and Greenland: geese (*Anser brachyrhyncus* Baillon; *Branta leucopsis* Bechstein; *Anser albifrontus* Scop. and *Branta bernicla* L.); waders (sandpiper, *Calidris canutus* L.; sanderling, *Calidris alba* Pallas; turnstone, *Arenaria interpres* L.) and passerines (the Lapland bunting, *Calcarius lapponicus* L.) (Adams et al.

2003). Several bird species common in Greenland also occur in Iceland, such as the ptarmigan (*Lagopus mutus* Montin), snow bunting (*Plectrophenax nivalis* L.) and the Arctic redpoll (*Carduelis hornemanni* Holboll). As these species occur in both countries, it is very likely that they migrate between the Iceland and Greenland, and indeed snow buntings tagged in Iceland have been found in Greenland. The Greenland gyrfalcon (*Falco rusticolus* Linnaeus) is known to visit Iceland (where there is an indigenous population). It is possible that they eat birds that feed on juniper berries. Falcons often gulp the crop whole and its remains have been found in their vomit (Adams, 2003).

There is no evidence that the Alaska population (AL) came from across the Bering Strait, but likely from inland in North America. Of course, it did not come from Saskatchewan (SK), because SK was glaciated. All of the northern populations likely were colonized by

Figure 2.4. Linkage diagram showing the post-glacial re-colonization pathway from Europe to Iceland and Greenland.(from Adams et al. 2003).

spring bird migrations from more southerly regions. In North America, the dispersion of juniper seeds by birds has long been documented (Holthuijzenf and Sharik, 1985; McAtee, 1947; Phillips, 1910).

That study revealed that all of the populations (*J. communis* var. *depressa* and *J. communis* var. *megistocarpa*) from North America formed one group and all of the populations of the eastern hemisphere (including Greenland and Iceland), formed another group that included *J. communis* var. *communis* and *J. c.* var. *saxatilis*, except for the Kamchatka population that was quite dissimilar to any population examined. Most likely, the sites of all of the populations were covered with ice or uninhabitable to *J. communis* during the late Pleistocene (*ca.* 12,000 BP). Therefore, these populations are recent in origin. The path of re-colonization appears to have been northward in North America. Greenland appears to have been colonized from Iceland plants, which in turn came from northern Europe (Fig. 2.4). The Kamchatka population seems likely to have come from Japan.

Infra-specific variation in *Juniperus communis*

Adams and Pandey (2003) studied all the commonly recognized varieties of *J. communis* using DNA fingerprinting. As previously mentioned, *J. communis* is a highly variable taxon that is distributed around the northern hemisphere (Fig. 2.1). Ashworth, et al. (1999, 2001) used DNA fingerprinting to examine *J. communis* plants identified as *J. communis* var. *depressa*, *J. c.* var. *jackii* Rehder, *J. c.* var. *montana* Aiton (= *J .c.* var. *saxatilis* Pall. see Farjon, 2001) collected from California, Oregon, Nevada or Utah in the southwest and west coast of the United States. They did not get a clear pattern separating these taxa, and concluded that their samples represent a single taxon (variety). *Juniperus communis* var. *jackii* is quite distinctive in forming longer, more sparsely branched lateral branches and found only on serpentine soil. However, when

Figure 2.5. *Juniperus communis* var. *depressa* on granite in Canada.

J. c. var. *jackii* plants were transplanted to normal soil, the new growth reverted back to typical *J. c.* var. *montana* (*saxatilis*) growth (Adams, 1993). This led to the conclusion, at the time, that *J. c.* var. *jackii* is merely environmentally induced.

Figure 2.6 (modified from Adams and Pandey, 2003) shows a minimum spanning network. The largest cluster contains *J. c.* var. *communis*, var. *hemispherica*, var. *nipponica*, var. *oblonga* and var. *saxatilis*. This seems to imply that these varieties are not very distinct. *Juniperus communis* var. *oblonga* is relative distinct in having long needle-like leaves, but its DNA is not too different from var. *saxatilis* from the Urals and Greenland.

Surprisingly, *J. communis* var. *communis*, Sweden was most similar to *J. c.* var. *hemispherica*, Sicily (the type locality of var. *hemispherica*). The key character separating *J. c.* var. *communis* from other varieties is that it is an upright, small tree. It has been my field experience that in many locations (Hungary, Switzerland, Sweden) it is easy to find individuals that are more shrubby than tree-like. This character appears to be controlled by only a few genes throughout the species. In fact, near Amherst, Mass., USA, there are

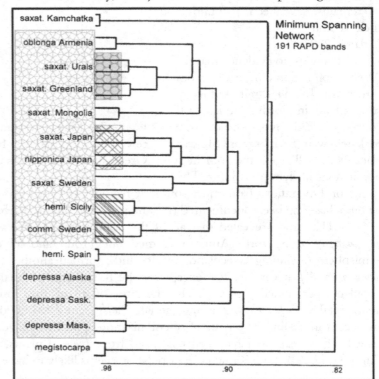

Figure 2.6. Minimum spanning network modified from Adams and Pandey, 2003.

individuals of the normally shrubby, *J. c.* var. *depressa*, that exhibit tree like growth, a further indication that the genes for apical dominance are not entirely well-behaved in this species. This lack of distinctness of *J. communis* var. *communis* from *J. c.* var. *saxatilis* has previously been reported (Adams et al., 2003).

The second large cluster (Fig. 2.6) is the *J. c.* var. *depressa* populations from North America. *Juniperus c.* var. *megistocarpa* appears very distinct in this analysis (Fig. 2.6), in contrast to a previous report (Adams et al., 2003). This taxon, with its very large female cones (9-13 mm, Adams, 1993) is endemic to sand dunes on islands in Quebec, Nova Scotia and Newfoundland and is the most distinct variety of *J. communis*.

The Kamchatka population (Fig. 2.6) is the quite distinct along with the Sierra Nevada, Spain population. Both fit loosely with *J. c.* var. *communis - saxatilis* populations (Fig. 2.6). Examination of the specimens did not reveal any major morphological differences in the Kamchatka plants, but the Sierra Nevada plants' leaves appeared a little broader than *J. communis* var. *communis* from Sweden.

Another view of the overall relationships is seen from Principal Coordinates Analysis (PCO, Fig. 2.7). The major trend in this data set is the separation of var. *depressa*, var. *megistocarpa*, Spain, Kamchatka and all the other *communis* var. *saxatilis* from Asia and Europe (Fig. 2.7). The first three coordinates accounted for only 44% of the variation among the 30 OTUs. This suggests that the large amount of variation seen among individuals and sample sites is not part of clearly defined patterns. The Kamchatka plants are placed near the var. *saxatilis - nipponica* from Japan (Fig. 2.7) concordant with the linkage map (not shown) that suggested the Kamchatka plants may have come from Japan.

Juniperus communis var. *depressa* is separated from *J. c.* var. *saxatilis* (= *J. c.* var. *montana* Aiton) in that the former has a stomatal band about as wide as the green leaf margin, whereas in var. *saxatilis*, the stomatal band is twice or more as wide as the green margin (Adams, 1993). Plants in North America near the west coast have stomatal bands indicative of *J. c.* var. *saxatilis*. However, Ashworth et al. (1999, 2002) studied plants from var. *saxatilis* and var. *depressa* from the western United States and found these varieties were not separated by their RAPDs data. Although specimens with broad stomatal bands have been found on the west coast, their DNA do not appear to be very similar to *J.c.* var. *saxatilis* as found in Europe and Asia.

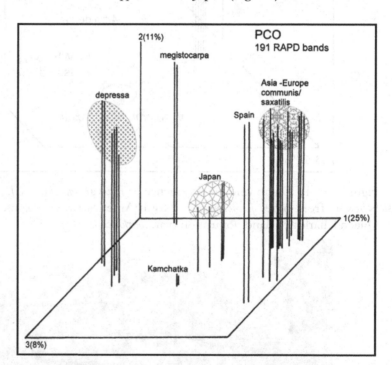

Figure 2.7. PCO of *J. communis* varieties based on RAPDs, from Adams and Pandey, 2003.

In a subsequent study, Adams (2008) utilized SNPs from nrDNA to examine populations in the Pacific Northwest of North America and found that *J. c.* var. *jackii* (now *J. jackii*) was quite distinct in its nrDNA (Fig. 2.7) as well as vars. *communis* and *saxatilis*, Norway. However, all of the samples of putative var. *saxatilis* and var. *depressa* from North America have essentially the same nrDNA (Fig. 2.8).

Adams (2008) removed var. *jackii* (*J. jackii*) from the data set and ran a new PCO (Fig. 2.9) that shows the lack of variation among of the North America samples of vars. *depressa* and *saxatilis* group.

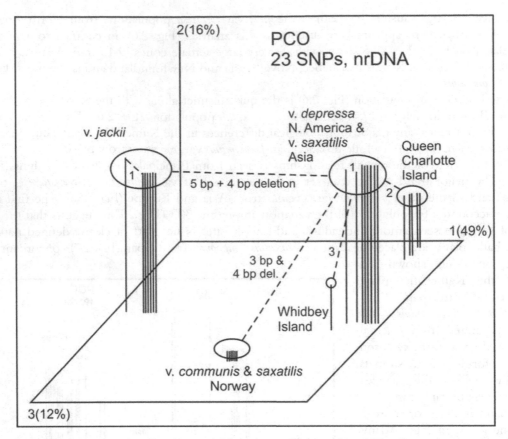

Figure 2.8. PCO showing the divergence of *J. c.* var. *jackii* (now *J. jackii*) and vars. *communis* and *saxatilis*, Norway, from other *J. communis* in North America (from Adams, 2008). Note the divergence of the Queen Charlotte Island population (var. *charlottensis*).

Habitat of *J. communis* var. *charlottensis* in a muskeg bog on Queen Charlotte Island.

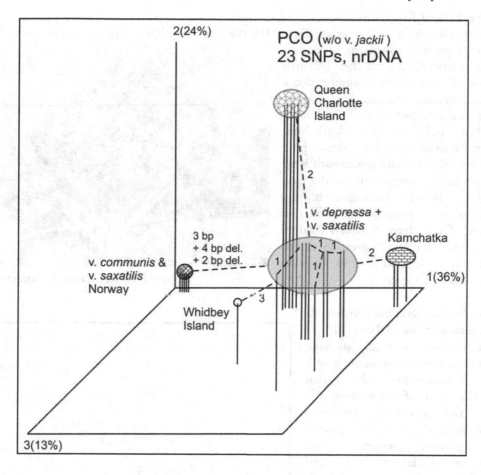

Figure 2.9. PCO without *J. jackii* (Adams, 2008) showing that the bulk of the North America samples of vars. *depressa* and *saxatilis* (now var. *kelleyi*) group with zero or 1 bp difference in their nrDNAs.

Based on DNA, morphology and physiological data, Adams (2008) recognized a new variety, *J. c.* var. *charlottensis*, discovered on Queen Charlotte Island, growing in muskeg bogs. In addition, I continued the recognition of var. *saxatilis* in North America based on leaf morphology, but noted that there appears to be little data to separate individuals of vars. *depressa* and *saxatilis* (now var. *kelleyi*) in North America except for stomatal band width and leaf shape.

In this treatment, I recognize the following varieties of *Juniperus communis*:

J. communis L. var. *communis* (= *v. oblonga*)

    var. *charlottensis* R. P. Adams

    var. *depressa* Pursh

    var. *hemisphaerica* (J. & C. Presl) Parl., see Fig. 1.11.2.

    var. *kamchatkensis* R. P. Adams (see Phytologia 94(2), 294, 2012.

    var. *kelleyi* R. P. Adams (prev. var. *saxatilis* in North America), see Fig. 1.11.2. (see Phytologia
        95(3), 215, 2013)

    var. *megistocarpa* Fernald & H. St. John

    var. *nipponica* (Maxim.) E. H. Wilson

    var. *saxatilis* Pall. (only in the eastern hemisphere), see Fig. 1.11.2.

In addition, I recognize *J. jackii* (Rehder) R. P. Adams (see Phytologia 94(2), 292, 2012) as species distinct from *J. communis*. (i.e., *J. communis* var. *jackii*).

Pleistocene relicts of *Juniperus ashei*

Juniperus ashei Buch. is a species (Fig. 2.10) that has a rather restricted range, occurring on limestone outcrops from northern Mexico to southern Missouri (USA). The Edwards Plateau region of central Texas support dense populations, covering millions of acres, whereas the disjunct populations (Lubbock-Post, Texarkana, Arbuckle Mts., Ozark Mts., and northern Mexico) often have almost pure stands of *J. ashei* that seldom cover such large areas. Being a fairly conspicuous conifer tree, one can be relatively confident in the taxonomic distribution records that imply that there are few, if any, trees between disjunct populations. Thus, *J. ashei* is a great species to test the hypothesis of gene flow vs. selection in the maintenance of a species.

Figure 2.10. *Juniperus ashei* tree native to limestone in Austin, Texas.

My study of geographic variation in *J. ashei* is reviewed in Adams (1977). Figure 2.11 presents a study showing contoured similarities of populations of *J. ashei* based on terpenoids. Notice that throughout the range of the species, it is very uniform, but Mexico, far west Texas and New Braunfels populations show a common divergence pattern. The populations in Oklahoma (4) and Arkansas (1, 2, 3) are completely discontinuous from the central Texas populations. Yet, there is no differentiation seen in their terpenoids.

This trend was further examined (Adams, 1977) in which three additional, disjunct populations were (24, Post, TX; 28, ne OK; 27, Texarkana) and the similarity matrices for both leaf terpenoids and morphology subjected to principal coordinate analyses (PCO). These coordinate scores were plotted onto maps (Fig. 2.12). For the terpenoids, the first and second coordinates accounted for for 50% and 9% of the terpenes variance.

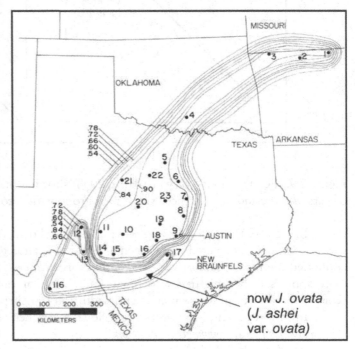

Figure 2.11. Contoured similarities based on terpenes. Note the large area of undifferentiated (ex Adams 1977).

For the morphology, the first and second coordinates accounted for 38% and 9% of the variance. Once again one sees the sharp divergence of the Mexico, west Texas populations. (popns. 12, 13, 26) and New Braunfels (popn. 2.12) from the central Texas and northward populations in both the terpenes and morphology (Fig. 2.12, A, B).

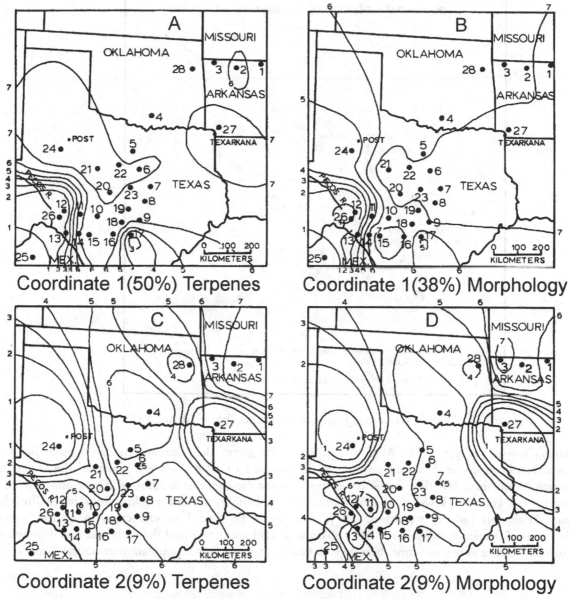

Figure 2.12. Principal coordinate analyses (PCO) of terpenoids and morphology of *J. ashei*. Adapted from Adams, 1977.

Interestingly, the second coordinates show the same pattern (Fig. 2.12, C, D) in which Texarkana and Post (24) are differentiated from the other populations.

Based on the trends in terpenoids and morphology, I examined the Pleistocene distribution of *J. ashei*. At the end of the Wisconsin glacial advance (10,000 - 15,000 yr bp), the north Texas, Oklahoma and Arkansas populations were likely extinct. It is postulated that *J. ashei* was pushed south into refugia in west Texas and Mexico with a relict stand near present day New Braunfels, TX (Fig. 2.13). In addition, a relict stand may have persisted in the central - southern part of the Edwards Plateau (Fig. 2.13). It may be that the original *J. ashei* was lower in camphor and had somewhat elongated leaf glands. The more recently derived populations (Figs. 2.11, 2.12) are higher in camphor and have the hemispherical oil glands that are unique to the genus, *Juniperus*. During the same period, *J. ashei* may have expanded south and west into the Chihuahuan desert, (Wells, 1966) but not as far south as

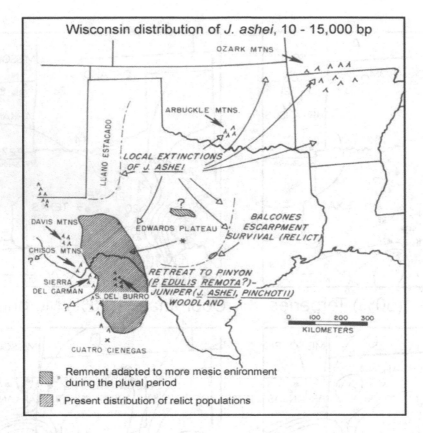

Figure 2.13. Possible Wisconsin *J. ashei* distribution. From Adams, 1977.

Cuatro Cienegas (Meyer, 1973). Migration west of the Sierra del Carman was also possible because *J. ashei* has been currently found growing at the top of La Cuesta Pass just south of the Sierra del Carman Mountains. Whether *J. ashei* could have crossed the high plateau around Alpine and Marfa (1, 500m) is not known. However, suitable habitat was probably available for colonization in the Presidio area. With this model, populations of *J. ashei* would be forced to extinction in central Texas, Oklahoma, Arkansas and Missouri. The subsequent recolonization could then take place as depicted in figure 2.14 over a very short time (hundreds of years?) from a relictual population in central Texas

Typical *Juniperus ashei* on limestone, Ozark Mtns., n Arkansas.

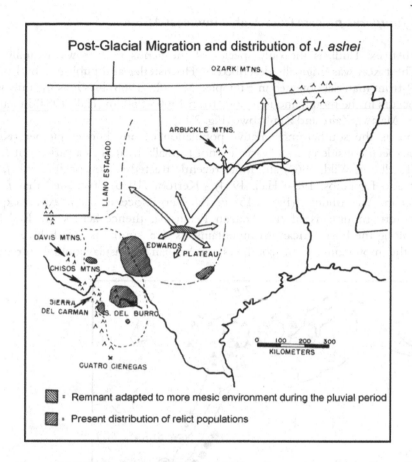

Figure 2.14. Hypothetical post-glacial migration and recolonization of *J. ashei* (adapted from Adams, 1977).

that may have gone through a selection 'bottleneck', perhaps coupled with migration and recolonization of *J. ashei* genetic drift. This 'relict' population would have had considerably more camphor in the oil (as a plant defense?), more hemispherical glands, larger female cones, fewer seeds (therefore a higher pulp to seed ratio as a reward for birds), and a more lax foliage which seems to be associated with more mesic Junipers. The rapid colonization of limestone outcrops (Fig. 2.12) could then lead to a uniform taxon from central Texas through the Ozarks. That study presented additional evidence that selection may be more important than gene flow in the maintenance of species. In *J. ashei*, one finds that populations with disjunctions of 200 - 300 km, and a small chance of gene exchange, were very similar to other populations covering 1,000 km in range (cf. Ozarks and central Texas populations). Yet, populations which are in close (that is, almost continuous populations) proximity have maintained either ancestral or modern patterns in spite of potentially large amounts of gene flow between these adjacent populations.

It should be noted that recent DNA sequencing data (Adams and Schwarzbach, 2013a, b) revealed that the divergent populations of *J. ashei* in the trans-Pecos, New Braunfels and Mexico are quite different in DNA, such that they were recognized as a new species, *J. ovata* (R. P. Adams) R. P. Adams (see also Figs., 1.14, 1.14.1).

Ancestral Migration of *Juniperus procera* from Arabia into east Africa

Juniperus procera Hochst. ex. Endl. is the only species of the genus that grows naturally in the southern hemisphere. The taxon was originally described by Hochstetler and published by Endlicher in 1847 (Endlicher, 1847) from material obtained in Ethiopia. Melville (1960) lists it as the only juniper in East Africa, where it occurs in the mountainous regions from 1500 - 2500m (Hall, 1984) in Ethiopia, Kenya, Tanzania, Uganda, Malawi, Zaire and Zimbabwe (Fig. 2.15).

The Zimbabwe site is the southernmost native population of any known juniper (Kerfoot, 1966). *Juniperus procera* has been considered as a "variant" (not formally treated as a variety) of *J. excelsa* M.-Bieb. from Europe (Exell and Wild, 1960) and more recently treated as conspecific with *J. excelsa* (Kerfoot, 1975; Kerfoot and Lavranos, 1984; Hall, 1984). Kerfoot (1975) postulated that *J. procera* originated from *J. excelsa* in Asia Minor   (Fig. 2.15) in the Mio-Pliocene as *J. excelsa* expanded southward along the western mountains of the Arabian Peninsula, thence across the Red Sea to Ethiopia and southward along the East African rift mountains as far south as Zimbabwe (see Fig. 2.15). The junipers in the mountains of the southwestern Arabian peninsula would then represent

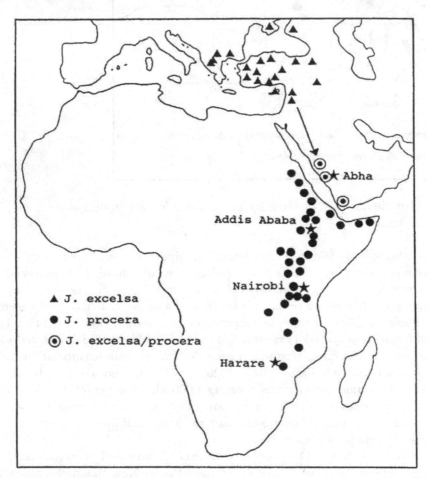

Figure 2.15. Distributions of *J. excelsa*, *J. procera* and *J. excelsa/ procera* populations.

a key link between *J. excelsa* in Asia Minor and the east African juniper (Fig. 2.15). So, we (Adams, Demeke and Abulfatih, 1993) examined *J. excelsa* from Greece, *J. procera* from Ethiopia and junipers from the Abha, Saudi Arabia population. Figure 2.16 shows typical RAPDs bands for individuals of *J. procera*, Ethiopia (lanes 1-3), Abha (4-6), and *J. excelsa*, Greece (7-9). Notice the major 500bp band shared by *J. procera* and the Abha junipers and the 390bp band characteristic of *J. excelsa*.

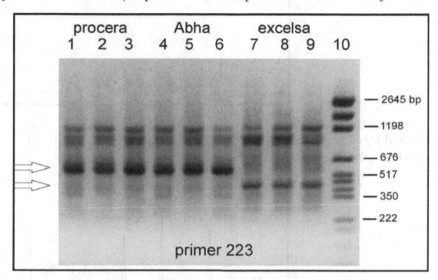

Figure 2.16. RAPDs for *J. procera*, Abha junipers and *J. excelsa*.
Modified from Adams and Demeke, 1993.

PCO ordination clearly reveals the Abha junipers are *J. procera* not *J. excelsa* (Fig. 2.17). *Juniperus procera* and the Abha junipers account for 54% of the variance among the samples.

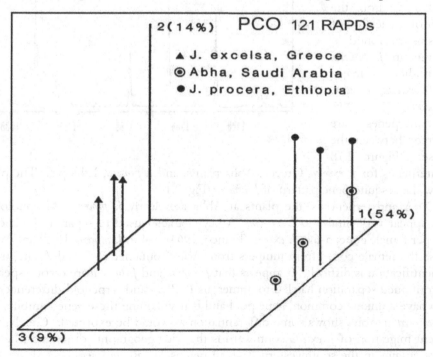

Figure 2.17. PCO based on 121 RAPD bands. Modified from Adams and Demeke, 1993.

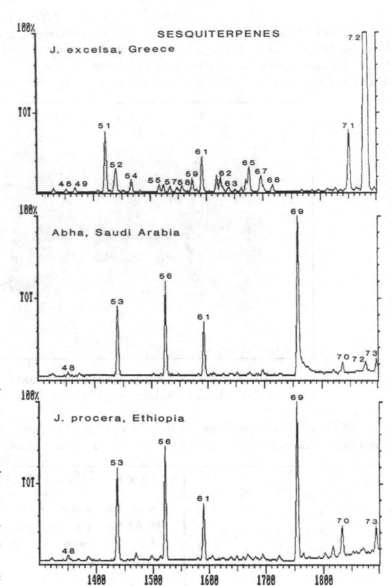

Figure 2.18. Sesquiterpenoids of *J. excelsa*, Abha juniper and *J. procera*. From Adams and Demeke, 1993.

It is noteworthy that *J. excelsa* has a great diversity of oxygenated terpenes and sesquiterpene hydrocarbons; these are absent in every population of *J. procera* examined. Cedrol, the major component of the leaf oil of *J. excelsa*, is missing or found in only trace amounts in *J. procera*. Limonene, another major component in *J. excelsa*, is also quite low in *J. procera*.

The sesquiterpenes are particularly different between the taxa in this case. Figure 2.18 shows the sesquiterpenes for *J. excelsa*, Greece, Abha plants, and *J. procera*, Ethiopia. The plants from Abha clearly show the sesquiterpene pattern of *J. procera* (Fig. 2.18).

By their DNA and terpenoids, the plants at Abha are clearly *J. procera*. Morphologically the junipers at Abha appear very similar to *J. excelsa*. A key character used to separate the species is the number of seeds per female cone: 5-6 in *J. excelsa* (Franco, 1964); 1-4 in *J. procera* (Exell and Wild, 1960). However, because the female cones from junipers from Abha contained 2-5 seeds/cone, unequivocal morphological identification is difficult. It appears that *J. excelsa* and *J. procera* are cryptic species. Their divergence and continued separation has led to numerous RAPDs and terpenoid differences. *Juniperus procera* appears to have a unique common, gene pool and is maintaining these gene combinations. The population of *J. procera* at Abha shows some differentiation as would be expected. Clearly, the barrier to past and present migration of *J. excelsa* southward is the wide geographic disjunction between Asia Minor and the mountains in the southwestern Arabian peninsula. Recent sequence data (Adams and Schwarzbach, 2013e) show (Fig. 1.14, Chpt. 1) that *J. procera* is in a separate clade from *J. excelsa*, supporting the continued recognition of *J. procera* as a distinct species.

Pleistocene Migrations of *Juniperus scopulorum*

Juniperus scopulorum is a widely occurring species in the mountains of western North America. Analysis of the leaf essential oils (Adams, 2011) revealed that *J. scopulorum* appears to be divided into 2 major groups: the central Rocky Mtn. portion and the northwest populations (Fig. 2.19).

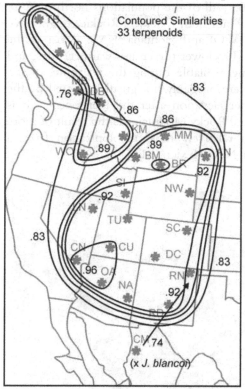

Figure 2.19. Contoured similarities, using 33 leaf terpenes. Modified from Adams, 2011.

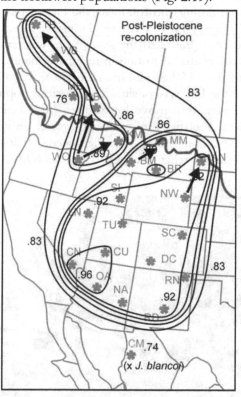

Figure 2.20. Hypothetical recolonization routes after the Wisconsin (Pleistocene). Modified from Adams, 2011.

The Mexico (Colonia Pacheco) population (CM, in Fig. 2.19) was later shown (Adams, 2011b) to be intermediate to *J. blancoi* in its oil (Fig. 2.20.1) and DNA (Fig. 2.20.2).

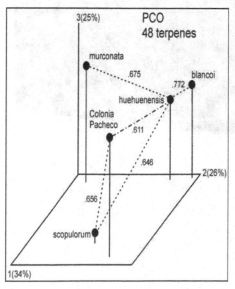

Fig. 2.20.1 (left). PCO of 48 terpenes shows Col. Pacheco intermediate between *J. scopulorum* and *J. blancoi* var. *huehuentensis*.

Fig. 2.20.2 (right) shows Col. Pacheco to be linked to *J. blancoi*.

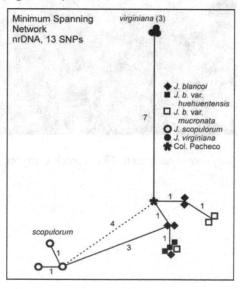

A population of putative *J. scopulorum* sampled in Palo Duro Canyon, Texas (PT) was found to be intermediate between *J. scopulorum* and *J. virginiana* (see Fig. 12 in Adams, 1983 and discussion in the chapter on hybridization later).

Based on these patterns it appears that there were several refugia and pathways of recolonization during and following the Pleistocene (Fig. 2.20). All of the populations north of WO, KM, BM, BR and AN were extinct during the Pleistocene. The northern Canadian Rockies might have been recolonized (Fig. 2.20) by seeds from western Oregon (WO) and Montana (KM). Most of the junipers in the central distribution may have merely descended to lower (dryer and warmer) sites, and then ascended to higher elevations as those habitats became available during the warming, drying period after the last Wisconsin. The Mexican populations were probably not greatly affected by the Pleistocene, except to move lower, then (re-)ascend to higher elevation after the Pleistocene. It is possible that the relative uniformity of *J. scopulorum* in the central Rocky Mountains is the result of more extensive extinction during the Pleistocene, followed by a relatively rapid recolonization from a genetically uniform relictual area [ex. Arizona (OA, NA) and s. New Mexico (RD)].

Juniperus scopulorum in Oak Creek Canyon, Arizona.

Infraspecific Variation in *Juniperus phoenicea*

Juniperus phoenicea L. is a small tree that is native to the Mediterranean area from Portugal to Israel and N. Africa in Algiers and Morocco as well as the Canary Islands and Madeira Island (Adams, Pandey, Rezzi and Casanova, 2002). Gaussen (1968) discussed several infraspecific taxa: var. *turbinata* (Guss.) Parl.(= var. *oophora* Kunze) with female cones elongated (turbinate) in littoral sites throughout the Mediterranean; var. *canariensis* Guyot on the Canary Islands; var. *lycia* (L.) Gaussen (*pro specie*) (= *J. phoenicea*), France littoral zone; var. *mollis* M & W., common in Morocco; and var. *megalocarpa* Maire, dunes near Mogador, Morocco. It is most likely that these varieties are not distinct taxa but reflect the lack of access to the type specimens by botanists who described local populations as new varieties of *J. phoenicea*. LeBreton and Thivend (1981), on the basis of total proanthocyanidins and the ratio of procyanidine to prodelphinidine, recognized *J. phoenicea* subsp. *eu-mediterranea* Lebr. & Thiv. as occurring on the Mediterranean islands, North Africa and southwestern Portugal. Later, LeBreton (1983) expanded his work to include more sample locations and showed all of the southwestern coastal populations of Portugal and Spain to have high proanthrocyanidines (implying *J. phoenicea* subsp. *eu-mediterranea*). He did not have samples from the Gibraltar (Tarifa) region of Spain and indicated uncertainty by use of a question mark (?) on his map (LeBreton, Fig. 1). Tarifa is an area where *J. phoenicea* var. *turbinata* occurs and LeBreton's population 70 (LeBreton, 1983), west of Setubal, is in the middle of his distribution for *J. phoenicea* subsp. *eu-mediterranea*.

Adams, Barrero and Lara (1996) sampled plants from the area of LeBreton's population 70, his pure *J. phoenicea* population (66-65) and *J. phoenicea* var. *turbinata* from Tarifa as well as a reference population of *J. phoenicea* in Greece. Based on leaf essential oils, Adams, Barrero and Lara (1996) concluded that *J. phoenicea* var. *turbinata* and *J. phoenicea* subsp. *eu-mediterranea* were conspecific and that the name *J. phoenicea* var. *turbinata* would take precedence over *J. p.* subsp. *eu-mediterranea*.

Recently, Rezzi, et al. (2001) reported on infraspecific variation in the leaf essential oils of *J. phoenicea* var. *turbinata* from Corsica. They found two chemical types: high α-pinene, low β-phellandrene, low α-terpinyl acetate (cluster I, 35 indvs.); and low α-pinene, high β-phellandrene, high α-terpinyl acetate (Cluster II, 15 indvs.). No morphological differences were found.

Adams, Pandey, Rezzi and Casanova (2002) reexamined *J. phoenicea* and its infraspecific taxa (var. *canariensis*, subsp. *eu-mediterranea* and var. *turbinata* using DNA fingerprints (RAPDs). There appears to be considerable infraspecific genetic variation within *J. phoenicea*. The Portugal population of *J. phoenicea* var. *turbinata* (previously recognized (LeBreton and Thivend, 1981) as *J. phoenicea* subsp. *eu-mediterranea*) is closely related to the Tarifa, Spain, *J. phoenicea* var. *turbinata* individuals.

The Adams, Pandey, Rezzi and Casanova (2002) study using RAPDs data confirmed the previous study based on leaf essential oils (Adams, Barrero and Lara, 1996) that *J. phoenicea* subsp. *eu-mediterranea* Lebr. & Thiv. from Portugal is clearly the same as *J. phoenicea* var. *turbinata*. So by nomenclatural priority, *J. phoenicea* subsp. *eu-mediterranea* Lebr. & Thiv. must be recognized as *J. phoenicea* var. *turbinata* (Guss.) Parl. Adams, Nguyen and Achak (2006) analyzed two populations of *J. phoenicea* from Morocco along with populations of *J. p.* var. *phoenicea*, Spain, *J. p.* var. *turbinata*, Tarifa, Spain and *J. phoenicea*, Canary Islands using RAPD markers.

Recently, Adams et al. (2013) examined DNA sequences (nrDNA and petN-psbM) from 19 populations of *J. phoenicea*. They found *J. p.* var. *phoenicea* only for the samples from Grazalema and El Penon, Spain; all other samples were all in a clade of *J. p.* var. turbinata (Fig. 2.21). Gaussen (1968) recognized var. *canariensis* on the Canary Islands, var. *megalocarpa* on sand dunes near Mogador (= Essaouira), Morocco, and var. *mollis*, as common in Morocco. Adams, et al. (2013) found these taxa to be in a clade (Fig. 2.21) with var. *turbinata* (Tarifa sands, Spain), and not sufficiently differentiated to warrant recognition of vars. *canariensis*, *megalocarpa* and *mollis* at this time.

A minimum spanning network based on 98 mutational events (SNPs plus indels) from nrDNA and petN-psbM sequences, showed that *J. p.* var. *phoenicea* from Spain was separated from the nearest *J. p.* var. *phoenicea* by 13 MEs (Fig. 2.22). This difference is comparable to differences found among recognized *Juniperus* species and led Adams et al. (2013) to conclude that differences in DNA, pollen shedding times (spring - var. *phoenicea*, fall - var. *turbinata*), morphology, and prodelphinidin content support the recognition of *J. turbinata* Guss.

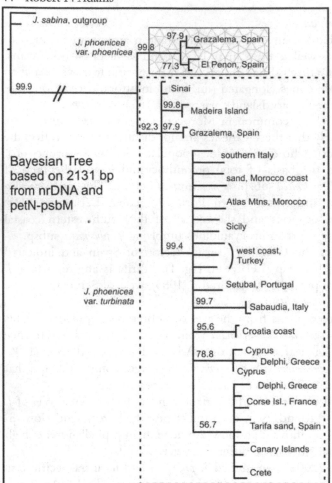

Figure 2.21. Bayesian tree of *J. phoenicea* and *J. p.* var. *turbinata* (now *J. turbinata*) (from Adams, et al., 2013).

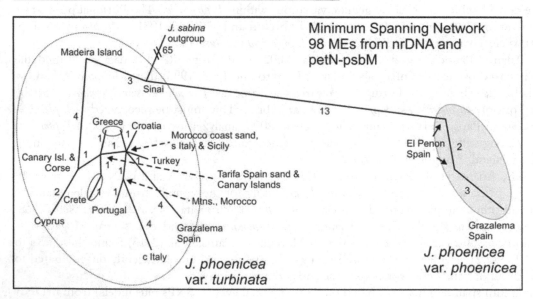

Figure 2.22. Minimum spanning network based on nrDNA and petN-psbM. Numbers on lines are MEs (mutational events = # base substitutions + # indels) for that link.

Variation in DNA sequence data in *J. turbinata* is considerable as seen in Fig. 2.23. Yet, the greatest differences between nearest neighbors is only 4 MEs Sicily - Grazalema, and Madeira - Tenerife). The central portion of the Mediterreanea is quite uniform, but the Sinai and Madeira populations share mutual divergence patterns.

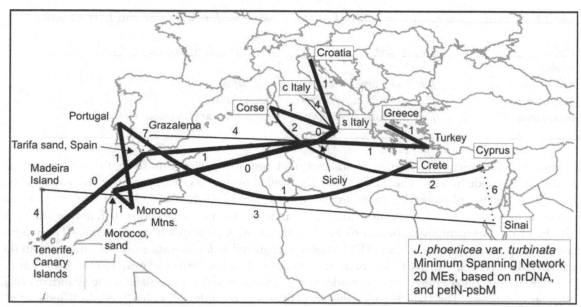

Figure 2.23. Minimum spanning network of *J. p.* var. *turbinata* (*J. turbinata*) sites. Line widths are proportional to number of ME differences between sites. Widest lines have 0 differences. The greatest difference between nearest neighbors is 4 MEs (Sicily - Grazalema, and Madeira - Tenerife).

In this treatment, *J. phoenicea* var. *phoenicea* and *J. turbinata* are recognized as distinct species.

Juniperus turbinata (center plant) on sand dunes at Tarifa, Spain.

Geographic variation in *Juniperus polycarpos*

There have been three recent taxonomic treatments of central Asian multi-seeded junipers (Table 2.1). I reported on the leaf oils and DNA fingerprinting for the multi-seeded junipers of the eastern hemisphere (Adams, 1999). However, more recent field work in

Table 2.1. Recent taxonomic treatments of *J. excelsa*, *J. polycarpos*, *J. seravschanica*, and *J. turcomanica*.

Initial Taxa	Farjon (2005)	Adams and Schwarzbach (2012c)
J. excelsa	*J. excelsa*	*J. excelsa*
J. excelsa var. *polycarpos*	*J. excelsa* subsp. *polycarpos*	*J. polycarpos*
J. seravschanica	*J. excelsa* subsp. *polycarpos*	*J. seravschanica*
J. turcomanica	*J. excelsa* subsp. *polycarpos*	*J. polycarpos* var. *turcomanica*

Armenia and Turkmenistan seemed to indicate that previous collections from the Tbilisi Botanic Garden were misidentified. The Tbilisi plants, identified as *J. excelsa* var. *polycarpos* (K. Koch) Silba, appeared to be, morphologically, more similar to *J. excelsa* M.-Bieb. from Greece than to the new collections of *J. polycarpos* K. Koch in Armenia. In addition, the new collections of *J. turcomanica* B. A. Fedtsch. from Turkmenistan appeared to be very similar to *J. polycarpos* from Armenia. The juniper from the Kopet Mts., Turkmenistan (TK) has been recognized as *J. turcomanica* and the juniper from the Talasskiy Mts., Kazakhstan (KT) has been recognized as *J. seravschanica* (Adams, 1999). Both Farjon (1992, 2005) and Silba (1986, 1990) considered *J. seravschanica* and *J. turcomanica* to be synonyms of *J. excelsa* var. *polycarpos*. The Balochistan, Pakistan juniper was recently still called *J. excelsa* (Ciesla, et al., 1998).

Figure 2.24 shows the clustering based on presence/absence matching for *J. excelsa*, *J. procera* and putative *J. polycarpos* populations using 106 terpenes. Notice that *J. excelsa* from the Tbilisi Botanic Garden (ET) clusters closely with *J. excelsa* from Greece (EG). The plants from Kazakhstan (identified in the field as *J. seravschanica*), Pakistan (*J. seravschanica*), and Turkmenistan (*J. turcomanica*) all cluster with *J. polycarpos* from Armenia (Fig. 2.24) and not with *J. excelsa*. This work seemed to indicate that *J. polycarpos* was a variable species ranging from Armenia to Pakistan. In addition, *J. procera* is very distinct as is *J. excelsa*.

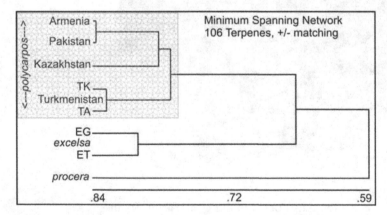

Figure 2.24. Minimum spanning network based on terpenes (from Adams, 2001). Note the large difference in the clustering of *J. polycarpos* away from *J. excelsa* (EG = Greece, ET = Tbilisi Bot. Gard.) and *J. procera*. From Adams, 2001.

However, recent work using DNA sequencing (Adams and Schwarzbach, 2012c) gives a different interpretation (Fig. 2.25). Notice that *J. seravschanica* has 100% support as a separate clade. *Juniperus polycarpos* and *J. p.* var. *turcomanica* are supported at 91% from *J. excelsa* (Fig. 2.25).

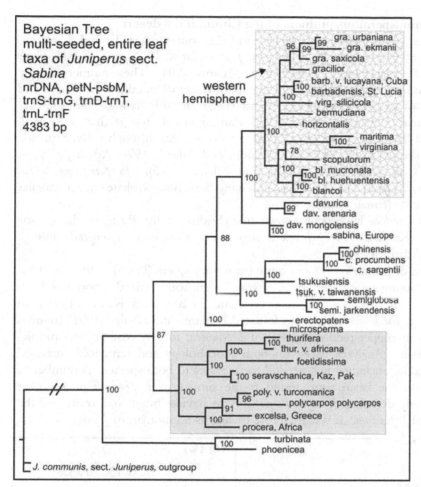

Figure 2.25. Bayesian tree with *J. excelsa - J. polycarpos - J. seravschanica* clades shaded.

Recent DNA sequencing of nrDNA and four cpDNA regions (Adams and Schwarzbach (2013d) gives additional insight into the relationship of *J. polycarpos* to *J. excelsa* (Fig. 2.26). *Juniperus polycarpos* var. *polycarpos* (Armenia) is separated by 8 mutations from *J. polycarpos* var. *turcomanica* (Kopet Mtns., Turkmenistan, *J. p.* var. *turcomanica* is separated by 17 MEs from *J. excelsa*, and by 17 MEs from *J. procera*. The entire *J. excelsa - J. polycarpos* complex is separated by 27 MEs from *J. seravschanica* (Fig. 2.26), giving concurrent support with the Bayesian tree (Fig. 2-25) for the recognition of *J. seravschanica*.

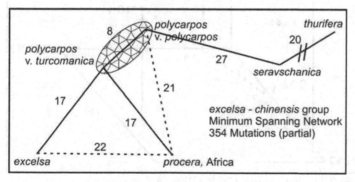

Figure 2.26. Minimum spanning network based on 354 DNA mutations (Adams and Schwarzbach (2013d).

These DNA data support the following taxa:

J. excelsa M.-Bieb.

J. polycarpos K. Koch var. *polycarpos*

J. polycarpos K. Koch var. *turcomanica* (B. Fedtsch) R. P. Adams

J. seravschanica Kom.

Geographic variation in the monospermous junipers of the Chihuahuan desert

The one-seeded, elongated gland junipers of Mexico and the southwestern United States are comprised of *J. coahuilensis* (Mart.) Gaussen ex R. P. Adams, *J. arizonica* R. P. Adams, *J. monosperma* (Engelm.) Sarg., *J. angosturana* R. P. Adams and *J. pinchotii* Sudw. (Adams, 1994). These junipers occupy mostly allopatric sites distributed in northern Mexico and the southwestern United States. Vasek and Scora (1967) reported two races of *Juniperus monosperma* ('A' and 'B') based on terpenoids, but upon re-examination of plants from their populations, Adams and Zanoni (1979) found that 'A' was *J. erythrocarpa* (now *J. arizonica*) and 'B' was *J. monosperma*. *Juniperus monosperma*, although often reported from Mexico, has not been confirmed using terpenoids (Zanoni and Adams, 1976; Adams, Zanoni, von Rudloff and Hogge, 1981) or morphology (Zanoni and Adams, 1975). *Juniperus angosturana* (previously *J. monosperma* var. *gracilis*) from Mexico differs in having bark that exfoliates in rectangular plates, thin foliage, and smaller fruits (female cones).

Juniperus coahuilensis and *J. pinchotii* have been shown to hybridize in the Basin of the Chisos Mountains, Texas (Adams and Kistler, 1991). *Juniperus angosturana* appears to intergrade into *J. coahuilensis* in Coahuila, Mexico (Adams, 1994).

Because the type specimen of *J. erythrocarpa* Cory came from the Laguna (Basin) at the west base of Mt. Emery in the Chisos Mountains of western Texas and significant hybridization has been reported between *J. erythrocarpa* and *J. pinchotii* at that very site (Adams and Kistler, 1991), an examination of the type specimen for *J. erythrocarpa* was initiated. Adams and Kistler (1991) found a range of fruit colors, varying from copper-reddish (typical of *J. pinchotii*) to very rose (almost purple) colored fruits in the Chisos basin. Analyses based on both morphology and terpenoids revealed essentially a continuum of individuals including hybrids and backcrosses to both species. A number of 'bright red' fruited individuals in the basin were found to be similar to *J. pinchotii* in all other characteristics. Cory (1936) clearly described *J. erythrocarpa* Cory as having 'bright red' fruits and the type itself is clearly a segregate of *J. pinchotii*. It was therefore treated as a synonym of *J. pinchotii*, along with another 'red' fruited taxon, *J. texensis* van Melle (Adams, 1994).

The description for *J. erythrocarpa* var. *coahuilensis* Martinez reveals this to be a rose-fruited taxon, such fruits being typical of the taxon throughout its range in Mexico and Texas. Martinez (1946) mentions, in detail, that his var. *coahuilensis* had a blue-gray waxy bloom on the fruits. *Juniperus erythrocarpa* sensu Cory lacks bloom on its fruit. These facts, along with additional data led Adams (1994) to recognize *J. erythrocarpa*. var. *coahuilensis* at the specific level as proposed by Gaussen, as *Juniperus coahuilensis* (Mart.) Gaussen ex R. P. Adams.

Analysis of *J. angosturana, J. arizonica, J. coahuilensis, J. monosperma* and *J. pinchotii* (Adams, 1994) using RAPDs revealed that the four taxa are well separated (Figs. 2.31, 2.32).

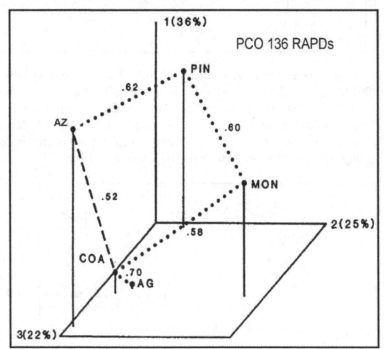

Figure 2.31. PCO using 136 RAPDs. AZ = *J. arizonica*, COA = *J. coahuilensis*, MON = *J. monosperma*, PIN = *J. pinchotii*. AG = *J. angosturana*, Modified from Adams (1994).

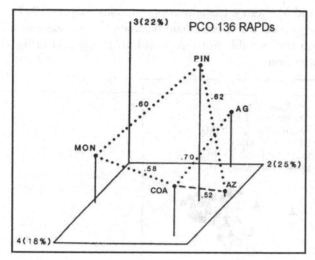

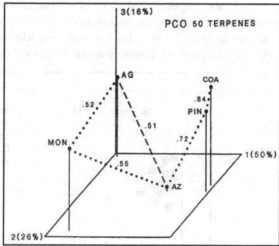

Figure 2.32. PCO, axes 2, 3, 4. Note that AG (*J. angosturana*) is clearly separated from COA(AT) (*J. coahuilensis*) on axis 4. From Adams (1994).

Figure 2.33. PCO 50 terpenes. Note that *J. arizonica* and *J. coahuilensis* are well separated on the first 3 axes. From Adams (1994).

Juniperus arizonica (AZ) and *J. coahuilensis* from Texas (COA) are quite different. This was also seen in the leaf terpenoids (Fig. 2.34). The composite sample from Alvaro Obregon (OM) is not included within the *J. coahuilensis* ellipse as it appears to be intermediate to *J. angosturana*.

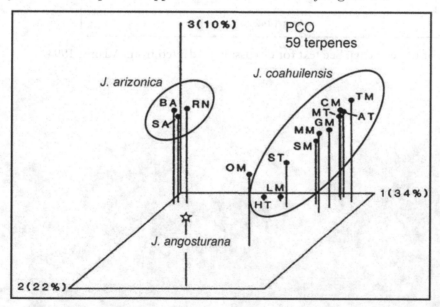

Figure 2.34. PCO, terpenes. Note the separation of *J. angosturana* (AG), *J. arizonica* and *J. coahuilensis*. Modified from Adams (1994).

PCO Analysis of terpenes for *J. angosturana*, *J. arizonica*, and *J. coahuilensis* show the clearly differentiated nature of *J. angosturana* (AG) and the populations of *J. arizonica* from Arizona and New Mexico (BA, SA, RN) from *J. coahuilensis* of Mexico (Fig. 2.34). There appears to be a trend of Mexico populations of *J. coahuilensis* towards *J. angosturana* (AG, Fig. 2.34).

Using terpenes and PCO of *J. angosturana* and Mexican *J. coahuilensis* populations Adams (1994) found that *J. coahuilensis* populations nearest to *J. angosturana* are intermediate between *J. angosturana* and *J. coahuilensis* (Fig. 2.35). This suggests hybridization and possible introgression into *J. coahuilensis* (Fig. 2.35). (see Chpt. 6, Hybridization for additional discussion).

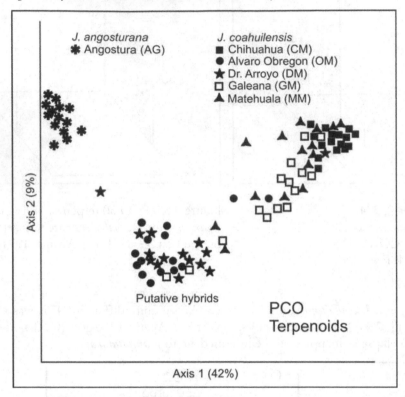

Figure 2.35. PCO of terpene data. See text for discussion. Adapted from Adams, 1994.

Juniperus angosturana in its habitat near Angostura, Mexico, south of Dr. Arroyo (site of population AG in Fig. 2.35, above).

Geographic Variation in _Juniperus thurifera_

    Juniperus thurifera from north Africa has been the subject of several nomenclatural changes having been treated as a variety [(_J. thurifera_ var. _africana_ Maire, Bull. Soc. Hist. Nat. Afrique N. 17: 125, (1926)], then as a distinct species [_J. africana_ (Maire) Villar, Types Sols Afrique N. 1:91 (1947)].

    More recently, a study (Gauquelin, Hassani and LeBreton, 1988) using proanthocyanidins and the number of seeds per cone, resulted in the naming of a new subspecies, _J. thurifera_ subsp. _africana_ (Maire) Gauquelin, Hassani et LeBreton, and 3 chemivars, _hispanica_ (Spain), _gallica_ (France) and _corsicana_ (Corse).   Gauquelin, Hassani and LeBreton (1988) found that the number of seeds per cone was significantly fewer (1.21 seeds/cone) in Morocco than in Europe where the number of seeds/cone ranged from 2.9 to 3.1 (Fig. 2.36).   Barbero et al. (1994) reported additional morphological measurements that seemed to support the recognition of the subspecies.   Barbero et al. (1994) discussed the chemivars as varieties, but they did not appear to actually name the chemivars as varieties in a nomenclatural sense.   In a recent treatment of the conifers, Farjon (1992) recognized _J. thurifera_ and treated _J. africana_ and _J. thurifera_ var. _africana_ as synonyms.

    Adams et al. (2003) examined both the leaf oils and DNA fingerprints for populations of _Juniperus thurifera_ from Morocco, Spain and France to aid in understanding the variation in this species.

    The leaf oils were found to be polymorphic in the Tizi-n' Ait-Imi, Morocco, Ruidera, Spain and Pyrenees population samples.   In spite of the variation among individuals, there is a clear pattern in the leaf oils. All the European populations are low in sabinene (0.3 to 9.7%) and the Moroccan composite sample is high (33.9%) (Fig. 2.37).   A reverse trend is seen with limonene (Fig. 2.38) in high European populations (28.8 to 61.8%) and low Moroccan composite (2.6%).

    The δ-2-carene variation had a similar pattern (not shown) with 2.85 to 4.4% in the European populations and none in the Moroccan composite. Each of these three compounds shows the same trend: divergence of the Moroccan plants, clinal variation from Morocco to Spain to France, and polymorphisms.

    Analysis using the DNA fingerprinting data (Adams et al., 2003) resulted in 142 RAPD bands that showed fidelity within populations (bands that only occurred once or randomly among   populations   were not   scored).

    A minimum spanning network (Fig. 2.39) includes two _J. foetidissima_ plants from Greece as an outgroup. This network shows that most plants from

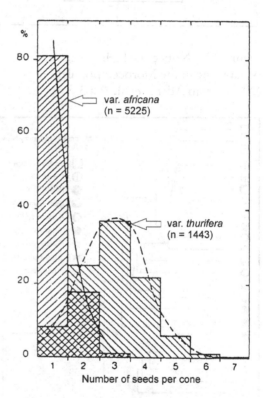

Figure 2.36. Histogram, number seeds/cone. From Adams et al. (2003).

each population cluster together (Fig. 2.39).   The most distinct cluster is composed of the northernmost plants from France.   Notice that one of the Moroccan plants clusters with the Ruidera, Spain individuals (Fig. 2.39).   The Oukaimeden and Tizi-n-Tichka, Morocco plants form a distinct cluster (Fig. 2.39), even though they are separated from the Tizi-n' Ait-Imi population by only about 100 km.   To better visualize the relationships among these _J. thurifera_ populations, the _J. foetidissima_ plants were removed from the data set and a new similarity matrix was computed and factored.   An ordination using the first three coordinates (Fig. 2.40) shows separation between the Moroccan and European plants, with the plants from Spain being somewhat intermediate.

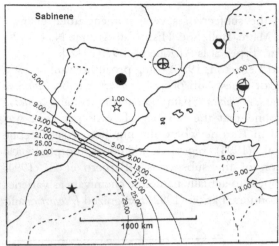

Figure 2.37. Notice the high concentration of sabinene in the Morocco population (29%). From Adams et al., 2003.

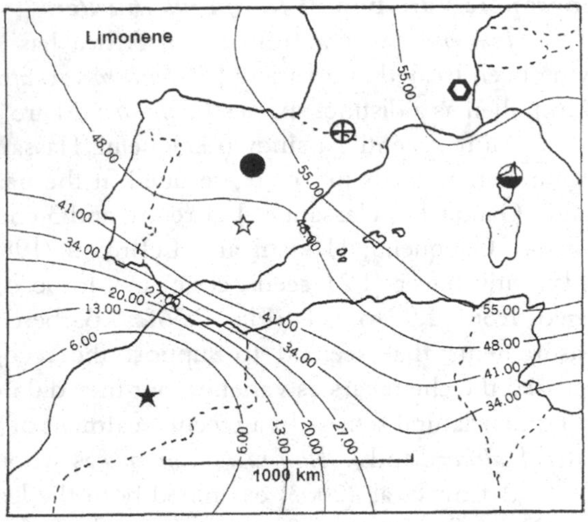

Figure 2.38. Contour of high levels of limonene in the European populations. From Adams, et al., 2003.

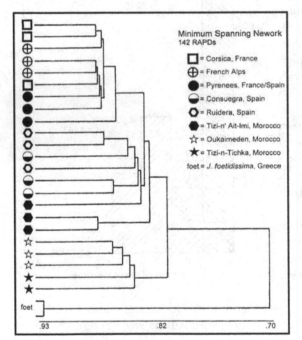

Figure 2.39. Minimum spanning network based on 142 RAPDs. foet is *J. foetidissima*, added as an outgroup. From Adams et al., 2003.

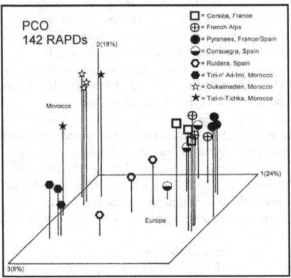

Figure 2.40. PCO based on 142 RAPDs. The trend from left to right is from Morocco to Spain, thence to Europe. From Adams et al., 2003.

This is exactly the order that would be expected if colonization occurred from the Laurasian land mass to the African continent. It is also possible that the African populations are of more recent origin. Seeds could easily have been carried from the Iberian peninsula to Algeria and Morocco by birds. Birds (thrushes) are primary dispensers of *J. thurifera* seeds (Santos, et al., 1999) and this would also be a route that birds might take. This is most certainly the case for the origin of *J. phoenicea* on the Canary Islands, which is more distant from Europe than North Africa is from Europe. In addition, it is likely that birds continue to carry seeds between Morocco and Spain, providing gene flow.

The previous studies using seeds per cone and proanthocyanidins (Gauquelin, Hassani and LeBreton, 1988; Barbero, et al., 1994) indicated that the Moroccan population was quite differentiated from the European populations and that study is supported by the essential oils and to some extent by the RAPDs data in this study. However, although the major oil components show this pattern, most of the minor components do not show much differentiation between the Moroccan and European populations.

The Adams et al. (2003) data correlates with previous studies on the seeds per cone, proanthrocyanidines. In a study of other *Juniperus* species (Adams, Schwarzbach and Pandey, 2003) reported that the high correlations between classifications based on essential oils, RAPDS, ISSRs, and ITS sequence data. However, in this case, the classifications based on RAPDS, ISSRs and ITS sequence data were highly correlated but none of these data sets were correlated with the essential oil classification. It is possible that the divergence in sabinene, etc. is controlled by only a few genes with major effects. Likewise, perhaps the number of seeds/cone is influenced by the more arid conditions in Morocco, where the fewest number of seeds per cone have been reported (Gauquelin, Hassani and LeBreton, 1988).

Adams et al. (2003) concluded that evolutionary divergence of the Moroccan populations in their seeds per cone, proanthocyanidins, and leaf oils appear to support the recognition of *J. thurifera* var. *africana* Maire [syn: *J. africana* (Maire); *J. thurifera* subsp. *africana* (Maire) Gauquelin, Hassani et LeBreton] in North Africa until additional data can be gathered that is pertinent to this question. It is important to conserve these north African populations (Gauquelin, et al., 1999) as they may represent significant gene combinations.

Terrab et al. (2008), using AFLPs data in an exhaustive study, confirmed that the Moroccan *J. thurifera* deserve recognition as a variety. In this treatment *J. thurifera* L. var. *thurifera* and *J. thurifera* L. var. *africana* Maire are recognized.

Recently, Adams and Schwarzbach (2012c, 2013d), using nrDNA and four cp DNA regions, found 100% support for *J. thurifera* and *J. t.* var. *africana* (Fig. 1.16.1). A minimum spanning network of these and related taxa shows (Fig. 2.40.1) that *J. thurifera* and *J. t.*

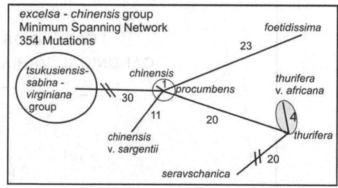

Fig. 2.40.1. Min. spanning network.

var. *africana* differ by 4 MEs. This is certainly a smaller difference than in most verities of *Juniperus*, but when one considers differences in seed number, proanthocyanidins and terpenoids, it seems reasonable to maintain *J. thurifera* var. *africana*.

Geographic Variation in *Juniperus virginiana*

There has been disagreement as to whether *J. silicicola* (Small) Bailey is a distinct species apart from *J. virginiana* in the southeastern region of the United States. Adams (1986) reviewed the literature and presented data on morphology and leaf terpenoids in an examination of geographic variation in both taxa.   Hall (1954) refers to the southeastern coastal juniper as the Florida race of *J. virginiana*. Flake and Turner (1973), using volatile leaf oils, analyzed 24 populations of *J. virginiana* and found 3 major groups: populations east of the Mississippi River to Washington, D. C.; an Ozark group; and an East Texas group extending to Bastrop, Texas.   They did not find evidence of a Florida race. However, Flake and Turner's southeastern most populations were in Burlington, NC; Columbia, SC; Macon, GA; and Birmingham, AL, which are outside of the recorded range of *J. silicicola*.

Canonical variate analysis of the leaf terpenoids resulted in 9 significant eigenroots.   The first five canonical variate axes accounted for 37.7, 26.5, 12.2, 7.0 and 4.9% of the variance among populations.   A contour map of canonical variate one (Fig. 2.41) shows the differentiation of the southeastern coastal (cf. *J. silicicola*) populations (note contours 4 - 7).

Canonical variate two shows (Fig. 2.42) the trans-Mississippi (Texas) grouping and differentiation of the Washington, D. C. (DC) population.

The first three canonical variates accounted for 76.4% of the variance among populations in their leaf terpenoids.   This overall clustering is shown in figure 2.38.   All the southeastern coastal populations (cf. *J. silicicola*) cluster except the WT population.   The WT (West Columbia, Texas) population (open hexagon, Fig. 46) has traditionally been called *J. silicicola* but it is clearly closely allied to the Bastrop, TX (BT) population, not with the southeastern coastal populations (*J. silicicola*, open hexagons, Fig. 2.38).   It should be noted that the West Columbia population is not on coastal foredunes, but is inland in old fields along the Bernard River, approximately 44 km from the ocean dunes.   This population was called to my attention by the report of their having flat-topped crowns (L. Gilbert, pers. comm.).   Because they are in the range of the distribution of *J. silicicola* reported by Little (1971), they were included in this study.

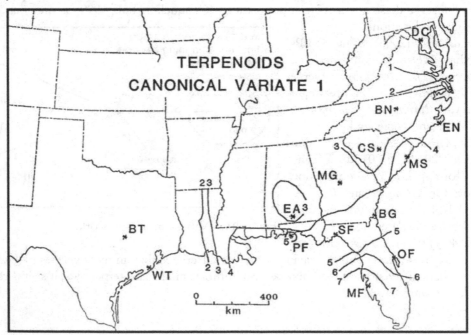

Figure 2.41. The first canonical variate shows the differentiation of var. *silicicola* in Florida and coastal populations. From Adams (1986).

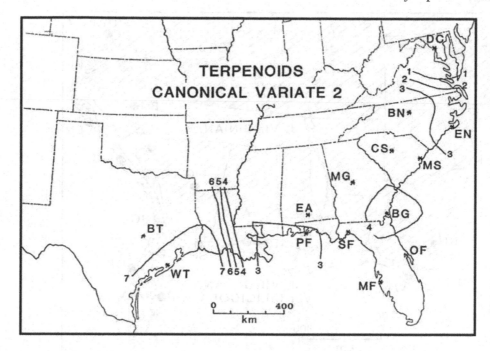

Figure 2.42. The second canonical variate depicts divergence of the trans-Mississippi populations and some differentiation of the DC population. From Adams (1986).

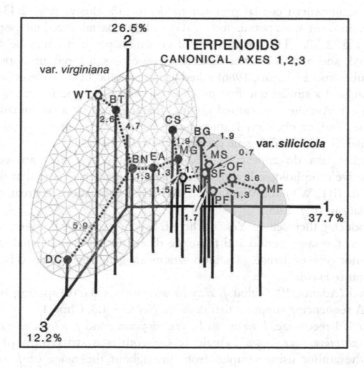

Figure 2.43. Ordination of populations based on canonical distances (dotted lines). Shaded circles are *J. v.* var. *virginiana*. Open hexagons are putative *J. v.* var. *silicicola*. From Adams (1986).

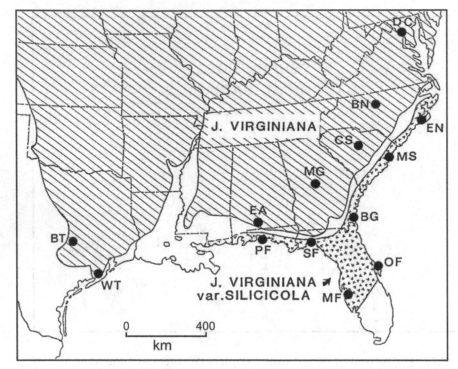

Figure 2.44. Distribution of *J. virginiana* var. *virginiana* and *J. v.* var. *silicicola*, based on terpenoids and morphology. From Adams (1986).

Although the southeastern coastal populations do form a cluster (Fig. 2.43), there is only a minor gap between inland *J. virginiana* populations (MG) and the coastal, foredune populations (cf. *J. v.* var. *silicicola*, EN, SF, Fig. 2.38). The quantitation of crown shape (if it could be shown to not be environmentally induced) and bark color would likely separate coastal foredune populations from the inland *J. virginiana* populations. I (Adams, 1986) asked if these should be considered a separate species. Ownbey (1950) encountered a similar question in *Tragopogon* and listed the following criteria (which I rephrased as questions): 1. Are the taxa natural groups, characterized by a combination of distinctive morphological features? (and/or chemical features, *my addition*); 2. Are they reproducing themselves under natural conditions?; 3. Is there free gene exchange between the taxa?

The questionable taxa do appear to be natural groups, but they are certainly not well characterized by distinctive morphological (and/or terpenoid) characters. Note that the differentiation of the Texas populations (BT, WT) and Washington, DC (DC) populations is greater than that of the coastal *J. silicicola* populations (Fig. 2.43).

Are they reproducing themselves? Yes. Is there free gene exchange between the taxa? Because they both shed pollen in the same period and these are dioecious, wind pollinated taxa, we have no assurance that there is not gene exchange. The populations are generally separated by the coastal plain (Fig. 2.44), but this distance is small.

So, I concluded (Adams, 1986) that *J. silicicola* was not a separate species, but a variety of *J. virginiana*. Recent DNA sequencing supports that decision (see Fig. 1.3, Chpt. 1).

In this treatment I recognize, *J. virginiana* L. var. *virginiana*, and *J. virginiana* var. *silicicola* (Small) E. Murray. *Juniperus virginiana* var. *crebra* Fernald & Griscom is a strict form of *J. virginiana*. A geographic study by the author using samples from throughout the range of *J. virginiana* is under investigation using DNA sequencing.

Chapter 3: Speciation Concepts in *Juniperus*

Species Concepts

As one discusses speciation, it is important to discuss "What is a species?" There are several major species concepts. Stuessy (1990) gives a good review of species concepts.

In considering the recognition of species status, Ownbey (1950) encountered a similar situation in *Tragopogon* and listed three criteria: 1. The taxa are natural groups, characterized by a combination of distinctive morphological features (and/or chemical/DNA features, my addition); 2. The taxa are reproducing themselves under natural conditions; 3. There is no free gene exchange between the taxa.

This seems to be near the "genetic species concept" (Stuessy, 1990) and is the species concept that I have tried to apply to *Juniperus*.

How is "Ownbey species concept" applied to *Juniperus*? Let us examine each of the three points.

1. The taxa are natural groups, characterized by a combination of distinctive morphological features (and/or chemical/DNA features, my addition).

Over the past 41 years, I have gathered data on morphology, ecology, leaf terpenoids, RAPDs and DNA sequences. By comparing the individuals of putative species with closely related, recognized *Juniperus* species I have attempted to be consistent in recognizing species that have about the same level of distinctness (in their genetic characters, e. g., morphology, terpenoids, RAPDs and DNA sequences) as recognized *Juniperus* species.

Issakainen (1999) wrote "We easily forget that different parts of a single organism's genome may have a different evolutionary history." We might modify his statement to read "different parts of the genome may be under differential selection pressure". We, as taxonomists, have relied on morphology as the deciding data for the recognition of species, varieties, and indeed most of our nomenclatural taxa. This is only natural, as the morphology is "what you see". The morphology is a product of the plant's genes plus the environment. The genes are composed of DNA. In tomato, the genome size is about 700,000,000 base pairs (bp) versus 4,000,000 bp in *E. coli* and 230,000,000,000 bp in man (Brown, 1986) and these may represent 20,000 to 30,000 genes (Somerville and Somerville, 1999). The amount of the genome that we see in the morphology is not known precisely. But, in an interesting study of two species of goldenrod (*Solidago*), Charles and Goodwin (1953) made the following estimates for the minimum number of genes for several key taxonomic characters:

Character	Minimum number of genes
leaf margins: entire vs. serrate	7
leaf surface: glabrous to pubescent	6
leaf thickness	6
basal leaves: length	8
leaf cuticle: degree of sculpturing	5
stomatal apparatus: length	3

For these 6 key characters used to separate *Solidago sempervirens* from *S. rugosa*, they estimated that the species differed by a minimum of 35 genes. How many DNA base pairs this represents is unknown.

Irving and Adams (1973) applied these methods to estimate the minimum number of genes controlling monoterpenes in *Hedeoma*. They found that 20 monoterpenoids were inherited by from 1 to 7 genes, with an average of 1.95 genes per compound. Thus, these 20 monoterpenoids appeared to be inherited by a minimum of 39 genes. Again, a small sample of the total genome.

If *Solidago* and *Hedeoma* have 20,000 to 30,000 genes as commonly expected in plants (Somerville and Somerville, 1999), then the *Solidago* morphology and *Hedeoma* monoterpenes are small samples of these genomes. Somerville and Somerville (1999) show that, in *Arabidopsis*, 54% of the genes can be assigned a known function. Although they did not show morphology *per se*, they did show that of the genes with known function, approximately 5% control cell structure and 6% code for secondary metabolism in *Arabidopsis*.

The situation for natural plant products is further complicated by the possible effects of a single gene on an entire class of compounds (ex., acetylation of terpene alcohols). One gene might acetylate more than one terpene alcohol. If that gene mutates to become non-functional, then all those acetates would be missing from the plant extract. A second complicating situation is shown the following schematic synthesis diagram: A > B > C > D > E. If the enzyme that converts A to B has mutated (and is not functional), then the plant loses not only compound B, but also compounds C, D and E. This is likely the case for chemical polymorphisms that are seen in many mints (and some *Juniperus* leaf oils). Thus, when one is computing a similarity measure, a single gene difference could be counted as four differences in this example. Central to the above discussion are two questions: How much of the genome are we using as data? And, is the portion of the genome that we are using as data representative of the entire genome?

Adams, Schwarzbach and Pandey (2003) examined classifications based on terpenoids, RAPDs, ISSRs, and nrDNA (ITS) sequence data. They found the classifications based on three kinds of DNA data to be highly correlated (Table 3.1), but not correlated with terpene classification. It appears that the terpenoids (using presence/absence data) may not be too useful for classifications above the species level. However, recall the previous cited study of *J. excelsa* and *J. procera* (Adams, Demeke and Abulfatih, 1993) in which there was a perfect correlation between the RAPDs and terpenoids data.

Table 3.1. Correlation among the similarity matrices of the four data sets.

	ITS	RAPDs	ISSRs	Terpenes
ITS	---	0.95	0.83	0.03
RAPDs	0.95	---	0.84	0.38
ISSRs	0.83	0.84	---	-0.04
Terpenes	0.30	0.38	-0.04	---

It is clear, however, that evolution does not proceed in all characters at the same rate. We have found, for example (Adams, Schwarzbach and Pandey, 2003) that *J. deltoides* is quite differentiated from the sibling species, *J. oxycedrus*, in its terpenes, DNA fingerprints and nrDNA (ITS) sequence, yet its morphology is very similar to *J. oxycedrus*.

Figure 3.1 shows two hypothetical species, A and B. Their morphology is essentially identical (as noted by the + symbol, Fig. 3.1). Their terpenoids are very similar (as noted by the square and octagon, Fig. 3.1). However, they are very different in some DNA sequences (denoted by the circle and bow tie, Fig. 3.1). Combining these characters to obtain an overview (lower portion, Fig. 3.1) gives us the character differences. So I believe that one cannot rely only on the similarity of morphology (or chemistry or any single character set) in making the determination that "these taxa are natural groups, characterized by a combination of distinctive features". In *Juniperus*, I have tried to utilize a number of kinds of data in answering: Are the taxa natural groups, characterized by a combination of distinctive morphological features? However, it must be admitted that the recognition of "natural groups" is still somewhat subjective.

It is particularly difficult to ascertain if a taxon is a variety or a new species. In these cases, I have tried to be consistent in trying to judge the level of differentiation of a putative species from a recognized species to be comparable with differences among recognized *Juniperus* species.

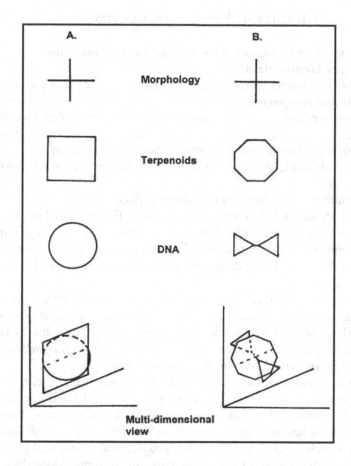

Figure 3.1. Multi-dimensional viewing of taxa. A multi-dimensional view is more than the sum of A and B. See text for discussion.

2. The taxa are reproducing themselves under natural conditions. This aspect is determined in the field. If the "new" taxon is disjunct (allopatric) to the "other" taxon, it is easy to note if a range of age classes are observed in the population(s). This is strong evidence that the taxon is reproducing under natural conditions. If the "new" taxon is growing interspersed with the "other" taxon, then one needs to look for hybridization as well as age classes. If the "new" taxon is involved in hybrid swarms, it is unlikely that it will reproduce itself.

3. There is no free gene exchange between the taxa. This is the most difficult factor to determine. If the taxa are widely allopatric (as in the case of *J. squamata* on mainland China and *J. morrisonicola*, endemic to Taiwan), then it is easy to say that they are not freely exchanging genes. But if the taxa are sympatric or nearly sympatric as with *J. coahuilensis* and *J. pinchotii*, and their pollen shedding times overlap, then one can only examine plants to see there is any sign of hybridization. Lacking any evidence of hybridization, then one assumes that there is not free gene exchange.

Although there are numerous species concepts, the species concept of Ownbey (1950) seems to be one of the most practical and it is applied in making decisions about species in *Juniperus* in this treatment. Of course, biology is an always changing discipline and new data (and ways to view old data) will always occur. No doubt the taxonomy of *Juniperus* will continue to change long after I am gone.

Chapter 4: Keys to *Juniperus*

In order to make the keys easier to use, several geographic keys are presented.

Key to *Juniperus* of Eastern Hemisphere

1a. All leaves acicular (subulate), jointed at the base, in whorls of 3, cones borne axillary (sections **Caryocedrus** and **Juniperus**)

  2a Seed cones 20-25 mm, seeds united to form a stone; male cones in axillary fascicles ....*J. drupacea*

  2b. Seed cones 6-15 mm, seeds free (not united), male cones solitary, axillary

    3a. Seed cones blue, bluish-black, black when mature (2 -3 yrs.), may be reddish when immature (*J. formosana*)

      4a. Leaves with 1 white stomatal band on the adaxial surface

        5a. Leaves 4 - 10 (-15) mm x 1 mm or more wide, nearly flat to curved in cross section, white stomatal band wider than green margins..........................*J. communis*

        5b. Leaves 7 - 25 mm x 0.5 - 0.9 mm wide, 'U' or 'V' shaped in cross section, with a keel (ridge) on the abaxial side, deeply grooved, white stomatal band narrower than green margins

          6a. Low procumbent shrub on sands at seashores; cones 8-12 mm; leaves 10-22 mm. long ....*J. rigida* var. *conferta*

          6b. Upright shrub or tree; dry areas in mountains; cones 6-9 mm; leaves 12-25 mm. long..*J. rigida* var. *rigida*

      4b. Leaves with 2 white stomatal bands on the adaxial surface

        7a. Leaves with mucronate tips, flat to U shaped, 12 - 20 mm; Taiwan, c. & e. China

        7.1a. Persistent branch (brown) leaf sheath with raised, elongate to linear gland, seed cones reddish-brown when immature, turning black when mature, Taiwan and possibly on the China mainland across from Taiwan......................*J. formosana*

        7.1b. Persistent branch (brown) leaf sheath with raised, oval to elliptical gland, seed cones green with glaucous bloom when immature, turning reddish-blue or reddish under bloom when mature, central & eastern China...............................*J. mairei*

        7b. Leaves with acute to obtuse tips, boat shaped, 7 - 12 mm, Japan

          8a. Leaves with acute to apiculate tips; boat-shaped, 7 - 10 mm and wide (1.1 - 1.5 mm), endemic to Ryukyu Islands and s. coast of Shizuoka Pref., Oshima (Island), Japan..*J. taxifolia* var. *lutchuensis*

          8b. Leaves with acute to obtuse tips, boat-shaped, 8 - 12 (-14) mm; endemic to Bonin Islands..(some leaves may have 2 stomatal bands)....................*J. taxifolia*

    3b. Seed cones reddish-brown, red, copper-red, reddish purple when mature

      9a. Fastigiate shrubs to 2 m; seed cones red when mature, 7 - 10 mm; leaves about as long as seed cone diam., Pliocene sands near the sea coast of w. Portugal ....*J. navicularis*

      9b. Shrubs or trees; seed cones reddish-brown, to dark purple when mature; leaves up to 25 mm.

        10a.Mature seed cones 12 - 15 mm, purple, with 3 distinct leaf impressions (from fused cone scales); leaves 2 - 2.5 mm wide, near the sea coast.......*J. macrocarpa*

        10b.Mature seed cones 8 - 10 mm, reddish- brown to red, globose, without cone scale impressions; leaves 1 - 2 mm wide.

          11a. Leaves curved, shorter than mature seed cones; shrubs; endemic to Azores...*J. brevifolia*

          11b. Leaves straight, longer than mature seed cones

            12a. Mature seed cones copper-red, scarcely glaucous; leaves closely spaced; endemic to Canary and Madeira Islands, trees to 30 m, shrubs at high elevations.

12a.1 Stomatal bands about as wide as midrib, mature seed cones about as long as leaves, endemic to Madeira Island...............*J. maderensis*

12a.2 Stomatal bands wider than midrib, mature seed cones longer that leaves, endemic to the Canary Islands................................*J. cedrus*

12b.Mature seed cones reddish-brown, dark red, reddish-purple, glaucous; leaves loosely spaced, shrubs or trees to 6 m

13a.Leaves narrowing at the base of attachment, stomatal bands sunken; seed cones without raised cone scale tips, seed cones globose, shrubs and trees with round crowns; France, Spain, Portugal, Algeria, Morocco ....*J. oxycedrus*

13b.Leaves with base of attachment nearly as wide as the blade, stomatal bands scarcely sunken; seed cones with raised cone scale tips, with shrub and trees with pyramidal crown, Italy, Greece, Turkey and eastward...*J. deltoides*

1b. Leaves decurrent (not subulate, not jointed at the base), both scale-like and whip (decurrent with free tips) leaves, only whip-leaves on juvenile plants and at the tips of adult foliage, leaves in pairs or whorls of 3; cones borne terminal (section *Sabina*)

14a. (1-) 2 - 9 seeds per globose cone or 2-3 (-4) seeds per bi-lobed cone

15a. All leaves of one kind, decurrent, with a blade (that is free) and a sheath (that the stem), procumbent shrubs, 2 - 3 seeded, sea shores, Japan, Korea............. ....*J. chinensis* var. *procumbens*

15b. Leaves of 2 kinds: scale like and whip (decurrent with free blade tips) leaves

16a.Whip-leaves found only on juvenile plants or at the ends of rapidly growing (juvenile) branchlets, otherwise, all leaves are scale-like

17a. Seed cones bi-lobed (occasionally globose on the same tree), 6 - 9 mm, 2 - 3 (-4) seeds, trees with strong central axis

18a.Scale-leaves mostly flat, scarcely overlapping, giving the ultimate branchlets a smooth appearance, glands on whip-leaves not conspicuous, not generally raised...*J. semiglobosa*

18b.Scale-leaves beaked, overlapping by about 1/4, giving the ultimate branchlets a rough appearance, glands on whip-leaves conspicuous, generally raised..*J. semiglobosa* var. *jarkendensis*

17b. Seed cones globose to subglobose

19a.Seed cones small, 4 - 6 mm, irregularly globose, (1-) 2 seeds per cone, shrubs or trees

20a. Trees, seed cones 4 - 5 mm, very fine foliage (0.6 - 0.9 mm in diam.). ....*J. erectopatens*

20b. Shrubs, seed cones 5 - 6 mm, fine foliage (0.8 - 1 mm in diam.)..*J. sabina*

19b. Seed cones large 7 - 14 mm diam.

21a.Twigs distichously arranged; seed cones dark purple, glaucous; 2 - 4 seeds per cone...*J. thurifera*

21b.Twigs irregularly arranged; seed cones reddish brown, dark blue, purplish-black; black, 1 - 8 seeds per cone

22a.1-2 (-3) seeds per cone; scale-leaf tips often free, scale-leaves 2 - 2.7 mm; leaves with a fetid odor, ultimate branchlets 1.2 - 1.7 mm diam., seed cones almost black...*J. foetidissima*

22b. 3 - 6 (-8) seeds per cone

23a.Seed cones copper-red to dark red when mature, 3 - 9 seeds per cone, leaf margins serrate (at 40X), Mediterranean, Madeira, Canary Islands.

23.1a. Seed cones globose (even when immature), pollen shed in late winter-
spring, branchlets bark brown, seed cones 5 - 9 mm long, crown usually
rounded, in mountainous areas on dolomite soils.........*J. phoenicea*

23.1b. Seed cones turbinate (especially when immature) or globose when
mature, pollen shed in Oct.-Nov., branchlets reddish, seed cones 7 - 11
mm long, crowns more conical, on coastal sand dunes and limestone
(in mountains)..*J. turbinata*

23b. Seed cones reddish-brown, bluish-red, blue, purple or black when
mature, 2 - 8 seeds per cone, leaf margins smooth (at 40X)

24a. Ultimate branchlets 0.6 - 1 mm diam.; scale-leaves 0.5 - 1 mm; 2 - 5
seeds per cone, seed cones reddish-brown to purple-black; trees
with pendulous foliage; endemic to e. Africa, Arabian Peninsula......
....*J. procera*

24b. Ultimate branchlets 0.7 - 1 mm, scale-leaves 0.6 - 1.6 mm long, 3 - 6
(-8) seeds per cone, reddish-brown to purple black, trees and
shrubs, foliage erect to pendulous

25a. Ultimate branchlets 0.7 - 1 mm diam.; scale-leaves very small, 0.6 -
1.1 mm long, appressed; seed cones 6 - 11 mm diam, globose.
....*J. excelsa*

25b. Ultimate branchlets 1 - 1.3 mm diam.; scale-leaves coarse, 1.2 - 1.6
mm long, appressed or apex free; seed cones 8 - 14 mm diam, globose

25.1a. Foliage slender, 0.7- 0.8 mm in cross section on ultimate branchlets,
seed cones 7-9(-10) mm, scale leaves tightly appressed, giving a smooth
branchlet, (1-)2-3(-4) seeds/cone..*J. polycarpos* var. *turcomanica*

25.1b. Foliage stout, 0.9-1 mm in cross section on ultimate branchlets, seed
cones 9-11 mm or more, scale leaves with a beak or keel so branchlet
appears as a string of beads, 3-6 seeds/cone

25.2a. Seed cones 9-11mm, at least some scale leaves with very narrow,
elongated, brown glands, not ruptured.
....*J. polycarpos* var. *polycarpos*

25.2b. Seed cones 8-10 mm, scale leaves with clear, ellipsoid glands, often
ruptured, with a clear exudate....................................*J. seravschanica*

16b. Both scale and whip-leaves found on mature plants, interspersed on a single
branchlet, not just at the ends of rapidly growing branchlets. Occasionally, one
type leaves predominates.

26a. Both scale and whip-leaves interspersed in mature foliage; decumbent shrubs,
glands conspicuous on whip-leaves and some scale-leaves, n. Mongolia, far e.
Russia..*J. davurica*

26b. Whip-leaves in clumps next to clumps of scale-leaves in mature foliage;
shrubs, trees (very variable taxon), glands on scale and whips leaves usually
not conspicuous, or almost all leaves scale-like and a decumbent shrub(Japan and
Taiwan).

26.1a. Both scale-like and whip (decurrent with free tips) leaves present; whip-leaves
interspersed among sections of scale-leaves in mature branchlets, trees................
e. China, Korea, Japan, ...*J. chinensis*

26.1b. Almost all leaves scale-like and a decumbent shrub (Japan and Taiwan).........
....*J. tsukusiensis*

14b. Seed cones 1 seeded, subglobose to turbinate, often with a pointed tip

27a. All leaves whip (decurrent with free tips), no scale-leaves (or scale-leaves appearing
decurrent in *J. carinata*).

28a.Leaf tips (blades) appressed to stem giving the ultimate branchlets a smooth
    rope appearance
  29a.Foliage pendulous and recurved, blades long and narrow, and very tightly
      recurved or appressed to the stem.
    30a. Single white stomatal band on adaxial side of leaves.
      30.1a. Terminal branchlets lax, but not hooked, whip leaves 35-50 mm, trees and
          shrubs with strong central axis, even when a shrub..........................*J. recurva*
      30.1b. Terminal branchlets hooked, whip leaves 20-28 mm, multi-stemmed shrubs.
          ....*J. uncinata*
      30b. Two white stomatal bands on adaxial side of leaves..................................*J. coxii*
    29b. Foliage not pendulous, blades shorter and wider, not so appressed
      31a.Trees with strong central axis in boreal forests, leaves with abaxial keel and
          narrow grooves next to the keel (or a raised oil gland in some cases)......*J. pingii*
      31b.Shrubs at timberline, leaves without abaxial keel and without narrow
          grooves..*J. carinata*
28b.Leaf tips (blades) free and divergent from stem giving the ultimate branchlets a
    thorny appearance
  32a. Plants of mainland China
    32.1a. Leaf blades 6-7 mm long, seed cones 8-9 mm long..........................*J. squamata*
    32.1b. Leaf blades 3-5 mm long, seed cones 4-5 mm long
      32.2a. Leaf blades 3 mm long, seed cones 5 mm long...............................*J. fargesii*
      32.2a. Leaf blades 4-6 mm, seed cones 4-5 mm long......*J. squamata* var. *wilsonii*
  32b. Plants endemic to Taiwan..*J. morrisonicola*
27b.Leaves of two kinds: scale-like and whip (decurrent) with free tips, whip-leaves
    only on juvenile plants or on fast growing branchlets
  33a. Gland at the base of scale-leaves on abaxial surface
    34a.Branchlets quadrangular (leaves opposite), curved, seed cones 4 - 8 mm............
        ....*J. saltuaria*
    34b.Branchlets terete, straight, rarely curved, seed cones 8 - 13 mm.
      35a.Branchlets tapering, ultimate branchlets gradually shorter toward the apex of
          branch; leaves without noticeable cuticular wax (glaucous) covering, foliage
          weeping (pendulous), seed cones 8 - 10 mm...................................*J. komarovii*
      35b.Branchlets not tapering, ultimate branchlets of equal length or irregularly
          equal; leaves with noticeable cuticular wax (glaucous) covering, foliage erect
          (except for var. *pendula*), seed cones 8 - 13 mm diam., scale-leaves with
          hyaline margins..*J. przewalskii*
  33b.Gland centrally positioned on scale-leaves on abaxial surface
    36a.Ultimate branchlets thin, less than 1 mm diam.; seed cones small, 5 -8 mm,
        globose
      37a. Seed cones 5 mm diam..*J. microsperma*
      37b. Seed cones 7 - 8 mm diam...*J. convallium*
    36b.Ultimate branchlets stout, more than 1 mm diam.; seed cones large, 8 - 16 mm,
        turbinate
      38a.Branchlets terete with a few branchlets quadrangular (leaves opposite), limbs
          on trees upturned at the tips reminiscent of the roofs of Tibetan temples..
          ....*J. tibetica*
      38b.Branchlets quadrangular (leaves opposite), a few branchlets terete, limbs not
          upturned at the tips
        39a.Scale-leaves overlapping by 1/4, with scarious border, whip-leaves with
            oval to elliptical glands, and gland ray usually absent or extending less than
            1/2 the distance to the leaf tip, green foliage..........................*J. pseudosabina*

39b.Scale-leaves not overlapping, without scarious border, whip-leaves with oval to elliptical glands with a gland ray (narrow groove) that extends more than 1/2 the distance to the leaf tip, dark green foliage

    39.1a. Low shrubs 50 -100 cm tall, foliage very dense, seed cones 5-8 mm long ....*J. indica* var. *caespitosa*

    39.1b.Shrubs or tree, foliage loosely set, seed cones 9-12 mm long

        39.2a. Monecious, shrubby trees, pollen: late winter-early spring..*J. indica*

        39.2b. Dioecious, trees with strong central axis, pollen: April - May..... ....*J. rushforthiana*

Key to *Juniperus* of Europe, Canary Islands, Azores, Asia Minor and Africa

1a. All leaves acicular (subulate), jointed at the base, in whorls of 3; cones borne axillary (sections ***Caryocedrus*** & ***Juniperus***)

  2a. Seed cones 20-25 mm, seeds united to form a stone; male cones in axillary fascicles.. ....*J. drupacea*

  2b. Seed cones 6-15 mm, seeds free (not united), male cones solitary, axillary

    3a. Leaves with one (rarely divided by a faint midrib) broad stomatiferous band above (adaxial, closest to the stem), seed cones 6 - 9 mm, black when mature (2 - 3 yrs.).. ....*J. communis*

    3b. Leaves with two stomatiferous bands above (adaxially); seed cones 6 -15 mm, reddish-brown to dark purple when mature (1, 2 yrs.)

      4a. Fastigiate shrubs to 2 m; seed cones red when mature, 7 - 10 mm; Pliocene sands near the sea coast of w. Portugal.................................*J. navicularis*

      4b. Shrubs or trees; seed cones reddish-brown, to dark purple when mature; leaves up to 25 mm.

        5a. Mature seed cones 12 - 15 mm, purple, with 3 distinct leaf impressions (from fused cone scales); leaves 2 - 2.5 mm wide.......................*J. macrocarpa*

        5b. Mature seed cones 8 - 10 mm, reddish- brown to red, globose, without leaf impressions; leaves 1 - 2 mm wide

          6a. Leaves curved, shorter than mature seed cones; shrubs; endemic to Azores..... ....*J. brevifolia*

          6b. Leaves straight, longer than mature seed cones

            7a.1 Stomatal bands about as wide as midrib, mature seed cones about as long as leaves, endemic to Madeira Island.................................*J. maderensis*

            7a.2 Stomatal bands wider than midrib, mature seed cones longer that leaves, endemic to the Canary Islands…...................................*J. cedrus*

          7b. Mature seed cones reddish-brown, dark red, reddish-purple, glaucous; leaves closely spaced, shrubs or trees to 6 m

            8a. Leaves narrowing at the base of attachment, stomatal bands sunken; seed cones globose, shrubs and trees with round crowns; France, Spain, Portugal, Algeria, Morocco..*J. oxycedrus*

            8b. Leaves with base of attachment nearly as wide as the blade, stomatal bands scarcely sunken; seed cones with raised cone scale tips, shrub and trees with pyramidal crown, Italy, Greece, Turkey and eastward........*J. deltoides*

1b. Leaves decurrent (not subulate, not jointed at the base), scale-like and whip-leaves, whip leaves only on juvenile plants and at the tips of adult foliage, leaves in pairs or whorls of 3; cones borne terminal (section ***Sabina***)

  9a. Seed cones small, 5 - 6 mm, irregularly globose, (1-) 2 seeds per cone, shrubs, fine foliage (0.8 - 1 mm in diam.)...*J. sabina*

  9b. Seed cones large 7 - 14 mm

10a. Twigs distichously arranged; seed cones dark purple, glaucous; 2 - 4 seeds per cone..
....***J. thurifera***

10b. Twigs irregularly arranged; seed cones red, reddish brown, dark blue, purplish-black; black, 1 - 8 seeds per cone

11a. 1-2 (-3) seeds per cone; scale-leaf tips often free, scale-leaves 2 - 2.7 mm; leaves with a fetid odor, ultimate branchlets 1.2 - 1.7 mm diam. seed cones almost black..............
....***J. foetidissima***

11b. 3 - 6 (-8) seeds per cone

12a. Seed cones copper-red to dark red (mature), 3 - 9 seeds per cone, leaf margins serrate (at 40X), Mediterranean, Madeira, Canary Islands.

12.1a. Seed cones globose (even when immature), pollen shed in late winter-spring, branchlets bark brown, seed cones 5 - 9 mm long, crown usually rounded, in mountainous areas on dolomite soils.............................***J. phoenicea***

12.1b. Seed cones turbinate (especially when immature) or globose when mature, pollen shed in Oct.-Nov., branchlets reddish, seed cones 7 - 11 mm long, crowns more conical, on coastal sand dunes and limestone (in mountains)..***J. turbinata***

12b. Seed cones reddish-brown, bluish-red, blue, purple or black when mature, 2 - 8 seeds per cone, leaf margins smooth (at 40X)

13a. Ultimate branchlets 0.6 - 1 mm diam.; scale-leaves 0.5 - 1 mm; 2 - 5 seeds per cone, seed cones seed cones reddish-brown to purple-black; trees with pendulous foliage; endemic to e. Africa, Arabian Peninsula.....................***J. procera***

13b. Ultimate branchlets 0.7 - 1 mm, scale-leaves 0.6 - 1.6 mm long, 3 - 6 (-8) seeds per cone, reddish-brown to purple black, trees and shrubs, foliage erect to pendulous

14a. Ultimate branchlets 0.7 - 1 mm diam.; scale-leaves very small, 0.6 - 1.1 mm long, appressed; seed cones 6 - 11 mm diam., globose...........................***J. excelsa***

14b. Ultimate branchlets 1 - 1.3 mm diam.; scale-leaves coarse, 1.2 - 1.6 mm long, appressed or apex free; seed cones 8 - 14 mm diam., globose..

14.1a. Foliage slender, 0.7- 0.8 mm in cross section on ultimate branchlets, seed cones 7-9(-10) mm, scale leaves tightly appressed, giving a smooth branchlet, (1-)2-3(-4) seeds/cone...***J. polycarpos*** var. ***turcomanica***

14.1b. Foliage stout, 0.9-1 mm in cross section on ultimate branchlets, seed cones 9-11 mm or more, scale leaves with a beak or keel so branchlet appears as a string of beads, 3-6 seeds/cone

14.2a. Seed cones 9-11mm, at least some scale leaves with very narrow, elongated, brown glands, not ruptured..............***J. polycarpos*** var. ***polycarpos***

14.2b. Seed cones 8-10 mm, scale leaves with clear, ellipsoid glands, often ruptured, with a clear exudate...***J. seravschanica***

Key to *Juniperus* of Central Asia (from Turkmenistan/Iran through the former Soviet Union, Pakistan and the Western Himalayas, including Nepal)

1a. All leaves acicular (subulate, jointed at the base), in whorls of 3, cones borne axillary (section ***Juniperus***) seed cones 6-15 mm, blue to bluish-black when mature (2 - 3 yrs), seeds free, male cones solitary, axillary, leaves with one (rarely divided by a faint midrib) broad stomatiferous band above (adaxial, closest to the stem)...***J. communis***

1b. Leaves decurrent (not subulate, not jointed at the base), both scale-like and whip (decurrent with free tips) leaves, only whip-leaves on juvenile plants and at the tips of adult foliage, leaves in pairs or whorls of 3, cones borne terminal (section ***Sabina***)

  2a. 2 - 6 (-8) seeds per cone, seed cones globose or bi-lobed (multi-seeded junipers)

    3a. (1-) 2 seeds per globose cone or 2-3 (-4) seeds per bi-lobed cone

      4a. Seed cones 4 - 6 mm when mature, irregularly globose, (1-) 2 seeds per cone, shrubs...***J. sabina***

      4b. Seed cones 6 - 9 mm, bi-lobed (occasionally globose on the same tree), 2 - 3 (-4) seeds, trees with strong central axis

        5a. Scale-leaves mostly flat, scarcely overlapping, giving the ultimate branchlets a smooth appearance, glands on whip-leaves not conspicuous, not generally raised ....***J. semiglobosa***

        5b. Scale-leaves beaked, overlapping by about 1/4, giving the ultimate branchlets a rough appearance, glands on whip-leaves conspicuous, generally raised................ ....***J. semiglobosa*** var. ***jarkendensis***

    3b. (2-) 3 - 6 (-8) seeds per globose cone

      6a. Ultimate branchlets 0.7 - 1 mm diam., scale-leaves very small, 0.6 - 1.1 mm long, appressed, seed cones 6 - 11 mm diam, globose...................................***J. excelsa***

      6b. Ultimate branchlets 1 - 1.3 mm diam., scale-leaves coarse, 1.2 - 1.6 mm long, appressed or apex free, seed cones 8 - 14 mm diam, globose

        6.1a. Foliage slender, 0.7- 0.8 mm in cross section on ultimate branchlets, seed cones 7-9(-10) mm, scale leaves tightly appressed, giving a smooth branchlet, (1-)2-3(-4) seeds/cone................................***J. polycarpos*** var. ***turcomanica***

        6.1b. Foliage stout, 0.9-1 mm in cross section on ultimate branchlets, seed cones 9-11 mm or more, scale leaves with a beak or keel so branchlet appears as a string of beads, 3-6 seeds/cone

          6.2a. Seed cones 9-11mm, at least some scale leaves with very narrow, elongated, brown glands, not ruptured.............***J. polycarpos*** var. ***polycarpos***

          6.2b. Seed cones 8-10 mm, scale leaves with clear, ellipsoid glands, often ruptured, with a clear exudate..***J. seravschanica***

  2b. 1 seed per cone, seed cones ovoid to turbinate, if turbinate then with pointed tip (1 seeded junipers)

    7a. All leaves of one kind, decurrent, with a blade that is free and sheath that clasps the stem.

    8a. Foliage pendulous and recurved, blades long and narrow, and very tightly recurved or appressed to the stem.

      8.1a. Terminal branchlets lax, but not hooked, whip leaves 35-50 mm, trees and shrubs with strong central axis, even when a shrub....................................***J. recurva***

      8.1b. Terminal branchlets hooked, whip leaves 20-28 mm, multi-stemmed shrubs... ....***J. uncinata***

    8b. Foliage erect and ascending, leaf tips free and divergent from stem giving the ultimate branchlets a thorny appearance...***J. squamata***

    7b. Leaves of two kinds: scale-like and decurrent with free tips (only on juvenile plants or on fast growing tips)

9a. Scale-leaves overlapping by 1/4, with scarious border, whip-leaves with oval to elliptical glands, but gland ray usually absent or extending less 1/2 the distance to the leaf tip...***J. pseudosabina***

9b.Scale-leaves not overlapping, without scarious border, whip-leaves with oval to elliptical glands with a gland ray (narrow groove) that extends more than 1/2 the distance to the leaf tip, dark green foliage

   9.1a. Low shrubs 50 -100 cm tall, foliage very dense, seed cones 5-8 mm long ....***J. indica* var. *caespitosa***

   9.1b. Shrubs or tree, foliage loosely set, seed cones 9-12 mm long

     39.2a. Monecious, shrubby trees, pollen: late winter-early spring.....................***J. indica***

     39.2b.Dioecious, trees with strong central axis, pollen: April - May....***J. rushforthiana***

Key to *Juniperus* of China, e. Himalayas, Mongolia, e. Russia, Korea.

1a. All leaves acicular (subulate, jointed at the base), in whorls of 3, cones borne axillary (sect. ***Juniperus***)

  2a. Leaves with 2 white stomatal bands on the adaxial surface, seed cones reddish when immature, turning reddish-blue to blue/black at maturity

    2.1a. Persistent branch (brown) leaf sheath with raised, elongate to linear gland, seed cones reddish-brown when immature, turning black when mature, Taiwan and possibly on the China mainland across from Taiwan...................................***J. formosana***

    2.1b. Persistent branch (brown) leaf sheath with raised, oval to elliptical gland, seed cones green with glaucous bloom when immature, turning reddish-blue or reddish under bloom when mature, central & eastern China..***J. mairei***

  2b. Leaves with 1 white stomatal band on the adaxial surface

    3a. Leaves 'V' shaped in cross section, 12 - 22 mm x 0.5 - 0.9 mm wide, with a keel (ridge) on the abaxial side, deeply grooved, white stomatal band narrower that green margins, seed cones reddish when immature, turning bluish-black when mature...............***J. rigida***

    3b. Leaves nearly flat to curved in cross section, 4 - 10 (-12) mm x 1 mm or more wide, white stomatal band wider than green margins, seed cones blue to blue-black.....***J. communis***

1b. Leaves decurrent (not subulate, not jointed at the base), both scale-like and whip (decurrent with free tips) leaves, only whip-leaves on juvenile plants and at the tips of adult foliage, leaves in pairs or whorls of 3, cones borne terminal (section ***Sabina***)

  4a. All leaves of one kind, decurrent, with a blade (that is free) and a sheath (that clasps the stem)

    5a. Seed cones 2 - 3 seeded...***J. chinensis* var. *procumbens***

    5b. Seed cones 1 seeded

      6a. Leaf tips (blades) appressed to stem giving the ultimate branchlets a smooth rope appearance

        7a. Foliage pendulous and recurved, blades long and narrow, and very tightly recurved or appressed to the stem.

        8a. Single white stomatal band on adaxial side of leaves.

          8.1a. Terminal branchlets lax, but not hooked, whip leaves 35-50 mm, trees and shrubs with strong central axis, even when a shrub........................***J. recurva***

          8.1b. Terminal branchlets hooked, whip leaves 20-28 mm, multi-stemmed shrubs... ....***J. uncinata***

        8b. Two white stomatal bands on adaxial side of leaves....................................***J. coxii***

        7b. Foliage not pendulous, blades shorter and wider, not so appressed

         9a. Trees with strong central axis in boreal forest, leaves with abaxial keel and narrow grooves next to the keel (or a raised oil gland in some cases)......***J. pingii***

         9b. Shrubs at timberline, leaves without abaxial keel and without narrow grooves.... ....***J. carinata***

    6b.Leaf tips (blades) free and divergent from stem giving the ultimate branchlets a thorny appearance

6.1a.   Leaf blades 6-7 mm long, seed cones 8-9 mm long...............................*J. squamata*

6.1b.   Leaf blades 3-5 mm long, seed cones 4-5 mm long

  6.2a.   Leaf blades 3 mm long, seed cones 5 mm long......................................*J. fargesii*

  6.2a.   Leaf blades 4-6 mm long, seed cones 4-5 mm long..*J. squamata* var. *wilsonii*

4b. Leaves of two kinds: scale-like and whip (decurrent) with free tips, whip-leaves only on juvenile plants or on fast growing branchlets or interspersed with sections of scale-like leaves on mature branchlets.

  10a. (1-) 2 seeds per globose cone or 2-3 (-4) seeds per bi-lobed cone, seed without pointed end

   11a.Whip-leaves found only on juvenile plants or at the ends of rapidly growing (juvenile) branchlets, otherwise, all leaves are scale-like

    12a.Seed cones 4 - 6 mm when mature, irregularly globose, (1-) 2 seeds per cone, shrubs or trees

     13a.Trees, seed cones 4 - 5 mm, very fine foliage (0.6 - 0.9 mm in diam.)...

      ....*J. erectopatens*

     13b.Shrubs, seed cones 5 - 6 mm, fine foliage (0.8 - 1 mm in diam.)............*J. sabina*

    12b.Seed cones 6 - 9 mm, bi-lobed (occasionally globose on the same tree), 2 - 3 (-4) seeds, trees with strong central axis

     14a.Scale-leaves mostly flat, scarcely overlapping, giving the ultimate branchlets a smooth appearance, glands on whip-leaves not conspicuous, not generally raised...*J. semiglobosa*

     14b.Scale-leaves beaked, overlapping by about 1/4, giving the ultimate branchlets a rough appearance, glands on whip-leaves conspicuous, generally raised............

      ....*J. semiglobosa* var. *jarkendensis*

   11b.Both scale and whip-leaves found on mature plants, interspersed on a single branchlet, not just at the ends of rapidly growing branchlets. Occasionally, one type of leaves predominates.

    15a.Both scale- and whip-leaves interspersed in mature foliage; decumbent shrubs, glands conspicuous on whip-leaves and some scale-leaves, n. Mongolia, far e. Russia...*J. davurica*

    15b.Whip-leaves in clumps next to clumps of scale-leaves in mature foliage; shrubs, trees (a very variable taxon), glands on scale and whips leaves usually not conspicuous, e. China, Korea, Japan, Taiwan......................................*J. chinensis*

  10b.One seed per cone, seed cones ovoid to turbinate, if turbinate, then with pointed tip, seeds with pointed end.

   16a. Gland at the base of scale-leaves on abaxial surface

    17a.Branchlets quadrangular (leaves opposite), curved, seed cones 4 - 8 mm...

     ....*J. saltuaria*

    17b.Branchlets terete, straight, rarely curved.

     18a.Branchlets tapering, ultimate branchlets gradually shorter toward the apex of branch; leaves without noticeable cuticular wax (glaucous) covering, seed cones 8 - 10 mm, foliage weeping (pendulous)...*J. komarovii*

     18b.Branchlets not tapering, ultimate branchlets or equal length or irregularly equal; leaves with noticeable cuticular wax (glaucous) covering, seed cones 8 - 13 mm, , foliage erect (except pendulous in f. *pendula*).......................*J. przewalskii*

   16b. Gland centrally positioned on scale-leaves on abaxial surface

    19a.Ultimate branchlets thin, less than 1 mm diam.; seed cones small, 5 - 8 mm, globose

     20a.   Seed cones 5 mm diam...*J. microsperma*

     20b.   Seed cones 7 - 8 mm diam..*J. convallium*

    19b.Ultimate branchlets stout, more than 1 mm diam.; seed cones large, 8 - 16 mm, turbinate

21a. Branchlets terete with a few branchlets quadrangular (leaves opposite), limbs on trees upturned at the tips reminiscent of the roofs of Tibetan temples... ....*J. tibetica*

21b. Branchlets quadrangular (leaves opposite), a few branchlets terete, limbs not upturned at the tips

    22a.Scale-leaves overlapping by 1/4, with scarious border, whip-leaves with oval to elliptical glands, and gland ray usually absent or extending less 1/2 the distance to the leaf tip, foliage green......................................*J. pseudosabina*

    22b.Scale-leaves not overlapping, without scarious border, whip-leaves with oval to elliptical glands with a gland ray (narrow groove) that extends more than 1/2 the distance to the leaf tip, dark green foliage

        22.1a. Low shrubs 50 -100 cm tall, foliage very dense, seed cones 5-8 mm long ....*J. indica* var. *caespitosa*

        22.1b.Shrubs or tree, foliage loosely set, seed cones 9-12 mm long

        22.2a. Monecious, shrubby trees, pollen: late winter-early spring........*J. indica*

        22.2b.Dioecious, trees, strong cent axis, pollen: April-May...*J. rushforthiana*

Key to *Juniperus* of the Far East (Japan, Taiwan, Sakhalin Island)

1a. Leaves all acicular (subulate, joined the base), in whorls of 3, spreading, articulate at base; flowers axillary

  2a. Leaves with 1 white stomatal band on the adaxial surface

    3a. Leaves 4 - 10 (-12) mm x 1 mm or more wide, nearly flat to curved in cross section, white stomatal band wider than green margins...................................*J. communis*

    3b. Leaves 7 - 25 mm x 0.5 - 0.9 mm wide, 'U' or 'V' shaped in cross section, with a keel (ridge) on the abaxial side, deeply grooved, white stomatal band narrower that green margins

      4a.Upright shrub or tree; dry areas in mountains; cones 6-9 mm; leaves 12-25 mm. long ....*J. rigida* var. *rigida*

      4b.Low procumbent shrub on sands at seashores; cones 8-12 mm; leaves 10-22 mm. long ....*J. rigida* var. *conferta*

  2b. Leaves with 2 white stomatal bands on the adaxial surface

    5a. Leaves with mucronate tips, flat to U shaped, 12 - 20 mm; Taiwan, e. China... ....*J. formosana*

    5b. Leaves with acute to obtuse, boat shaped, 7 - 12 mm, Japan

      6a. Leaves with one stomatal band or if 2, then midrib faint and not full length of leaf, leaves with acute to obtuse tips, boat-shaped, 8 - 12 (-14) mm; endemic to Bonin Islands...*J. taxifolia* var. *taxifolia*

      6b. Leaves with 2 distinct stomatal bands, with midrib the runs full length of leaf, leaves acute to apiculate tips; boat-shaped, 7 - 10 mm and wide (1.1 - 1.5 mm), endemic to Ryukyu Islands and s. coast of Shizuoka Pref., Oshima (Island), Japan.. ....*J. taxifolia* var. *lutchuensis*

1b. Leaves not acicular (not jointed at the base), but decurrent (with a sheath and blade), scale-like and whip (decurrent with tips free), flowers terminal

  7a All leaves whip type (decurrent with tips free), no scale-like leaves

    8a. Procumbent shrubs, (1-) 2 - 3 seeds per cone, sea shores, Japan and Korea.. ....*J. chinensis* var. *procumbens*

    8b. Erect shrub or tree, 1 seed per cone; rocky cliffs, endemic to Taiwan...*J. morrisonicola*

  7b. Both scale-like and whip (decurrent with free tips) leaves present; whip-leaves interspersed among sections of scale-leaves in mature branchlets, (1-) 2 seeds per cone or almost all leaves scale-like, and a decumbent shrub.

    7.1a. Both scale-like and whip (decurrent with free tips) leaves present; whip-leaves interspersed among sections of scale-leaves in mature branchlets, trees............ *J. chinensis*

    7.1b. Almost all leaves scale-like and a decumbent shrub (Japan and Taiwan).*J. tsukusiensis*

Key to *Juniperus* of Continental North America

1a. Leaves all acicular (subulate, jointed at the base) and spreading; seed cones sessile, axillary; decumbent or rarely upright shrubs (in the western hemisphere)

    1.1a. Seed cones globose, shorter, or about equal leaf length, shrubs, spreading to upright shrubs (or shrubby tree in New England), leaves straight to curved, flat or V shaped, not boat shaped, free from stem (25 - 90°)...*J. communis*

    1.1b. Seed cones elongated oval (ellipsoid), especially on immature cones, shorter as long as or longer than leaf length, shrubs, low to prostrate, leaves curved, boat shaped, usually appressed to stem, gland on brown sheath long, narrow, raised, on serpentine or volcanic rocks..*J. jackii*

1b. Leaves decurrent (not jointed at the base), both whip- and scale-like; seed cones terminal; trees or decumbent to upright shrubs.

2a. Whip- and scale-leaf margins entire (20 X) or with irregular teeth (40 X) and then with scale-leaves with acuminate to mucronate tips and seed cones tan-brown to brownish purple

3a. Whip- and scale-leaf margins with irregular teeth (40 X), scale-leaves acuminate; tan-brown to brownish-purple; branches pendulous to slightly erect

    3.1a. Seed cones 9-20 mm diam., (4) 6 - 10 (-13) seeds per cone, terminal branch tips drooping (hanging)

        3.2a. Branching radially, seed cones tan to brownish/purple........................*J. flaccida*

        3.2b. Branching planate, seed cones bluish-brown..*J. poblana*

    3.1b. Seed cones (5-) 6 (-9) mm diam., 1-2 (3) seeds per cone, terminal branch tips erect ....*J. martinezii*

3b. Whip- and scale-leaf margins smooth (entire) (40 X), scale-leaves obtuse to acute to apiculate; seed cones 1-2(3) seeded, blue-black to brownish-blue when mature; branches not drooping (but ultimate branchlets are often flaccid).

4a. Prostrate to decumbent shrub; scale-leaves apiculate; both whip- and scale-leaves growing along the branchlets; peduncles generally curved..........................*J. horizontalis*

4b. Tree with 1(2-3) stems and rounded, flattened, pyramidal, or strict crowns; scale-leaves obtuse to acute; whip-leaves growing only at branchlet tips (on mature trees); peduncles generally straight.

    5a. Seed cones mostly bi-lobed (reniform), a few may be globose, seed cones maturing in 1 year.

      6a. Scale-leaves with acute to acuminate tips

        6.1a. Tree or shrubby tree, females not on understand of planate branches, foliage not dense, often along streams......................*J. blancoi* var. *blancoi*

        6.1b. Prostrate shrub forming patches, one to several meters in diameter, female cones on underside of planate branches, foliage dense, sub-alpine.......... ....*J. blancoi* var. *huehuentensis*

      6b. Scale-leaves with mucronate tips...................................*J. blancoi* var. *mucronata*

    5b. Seed cones globose, or occasionally a few on a tree are bi-lobed, seed cones maturing in 1 or 2 years.

      7a. Scale-leaves not overlapping, or, if so, not by more than 1/5 the length, leaf tips obtuse to acute; seed cones globose to reniform

        7.1a. Twigs (3-5 mm dia.) with smooth bark, twigs (6-15 mm dia.) with bark exfoliating in plates, reddish-copper beneath; seed cones maturing in 2 yrs, most seed cones normal, rarely with exserted seeds..*J. scopulorum*

        7.1b. Twigs (3-5 mm dia.) with persistent dead whip-leaves, twigs (6-15 mm dia.) reddish-brown beneath; seed cones maturing in 1 yr. (14-16 mos.), often the seed cones with exserted (exposed) seeds; Pacific northwest near the seaside in Georgia Straits and Puget Sound, rarely on mountains (in Olympic National Park)...........*J. maritima*

7b. Scale-leaves overlapping (more than 1/4 length), leaf tips acute; twigs (3-5 mm dia.) with persistent dead whip-leaves, twigs (6-15 mm dia.) with bark not exfoliating in plates, or, if so, brownish beneath; seed cones maturing in 1 yr..***J. virginiana***

2b. Whip- and scale-leaf margins denticulate (20 X)

8a. Seed cones (1) 2 - 11 seeded, if 1-2 seeded, then bark exfoliates in quadrangular plates.

9a. Seed cones (1-3 -) 4 -5 (- 6) seeded, (1-2 seeded in var. *gamboana*) fibrous to obscurely woody; trunk bark exfoliating in square or quadrangular plates or strips (if in strips on the trunk, then plates on the upper limbs (*J. d. f. sperryi*).......................***J. deppeana***

9b. Seed cones 2- 9 seeded

10a.Mature seed cones reddish-brown to brown, with a light coat of bloom; seed cone peduncles straight.

11a.Mature seed cones irregular and gibbous; seeds (2-)4-9(-11) per cone; terminal whip branches strict, foliage very loosely spaced; scale-leaves do not appear as a string-of-beads on the ultimate twig..***J. jaliscana***

11b.Mature seed cones bluish-red, very glaucous, globose; seeds 1-4 per cone; terminal whip branches recurved (nodding hook) at tip; foliage tightly-spaced; scale-leaves appear as a string-of- beads..***J. durangensis***

10b.Mature seed cones dark blue to bluish-black, with light coat of bloom, cone peduncles usually curved

12a.Twigs not distichous, or sometimes sub-distichous, scale-leaves thick but not with a hood or beak, ultimate branchlets 1.3 - 1.4 mm wide, central and n. Mexico

12.1a. Whip leaf glands oval to elongate, usually raised, branchlets (3-5mm diam.) rough with persistent old leaves, mature twigs 5-10 mm. long, angle of branching of ultimate twig 50-60 degrees, foliage tightly compacted, prostrate shrubs (to 1 m.), sub-alpine habitat, on limestone..........***J. zanonii***

12.1b. Whip leaf glands elongate, often sunken, branchlets (3-5mm diam.) smooth, mature twigs 10-40 mm. long, angle of branching of ultimate twig 40-60 degrees, foliage loose to tightly compacted, low shrub to tree (to 10 m), sub-alpine to montane habitat, especially on volcanic soils.................***J. monticola***

12b.Twigs distichous, scale-leaves thick, usually a hood or beak, ultimate branchlets 1.4 - 1.8 mm wide, Chiapas, Mexico and n. Guatemala...........................***J. standleyi***

8b. Seed cones 1-2 (-3) seeded, fleshy to fibrous (when mature and fresh), bark exfoliating in strips (if bark exfoliating in quadrangular plates, then seed cones 5-6 mm long)

13a. Scale-leaves with a raised hemispherical gland, whip leaf gland raised, oval or hemispherical

14a.Scale leaves on ultimate twig do not appear as a string-of-beads, scale-leaf glands protruding like a bead, hemispherical, dark brownish-green on dark grayish-green mature leaf; angle of branching of ultimate twig 25-40 degrees

14.1a. Whip-leaf gland hemispherical and raised, scale leaf glands hemispherical, 1 seed/cone (rarely 2, avg. 1.01), seed cones (8) 9 (10) mm diam..............***J. ashei***

14.1b. Whip-leaf glands oval to elliptical and raised, 2 seeds/cone (avg. 1.7), seed cones (5) 6 (-8) mm in diam.............................. ***J. ovata*** (***J. ashei*** var. ***ovata***)

14b.Scale-leaves on ultimate twig appear as a string of beads, whip-leaf glands elongate, not hemispherical, scale-leaf gland protruding, oval to elliptical, same color as light gray-green mature leaf; twig bark grayish brown, smooth; angle of branching of ultimate twig about 60 degrees...***J. saltillensis***

13b. Scale-leaves without a raised hemispherical or oval gland

15a.Mature seed cones orange, reddish-orange, red, bronze, or reddish-brown, appearing pink or rose-color if covered with bloom.

16a.Stem bark exfoliating in quadrangular plates; cones reddish-brown; terminal whip branches tips straight, not curved; seeds subglobose or broadly ovoid...

....***J. deppeana*** var. ***gamboana***

16b. Stem bark exfoliating in longitudinal strips, often interconnected; seed cones rose, orange to reddish-brown or bronze-color; seed ovoid or if bark in quadrangular plates, then seed cones bluish-rose.

    17a. Mature seed cones orange to red, with light bloom appearing pink or rose-colored; whip-leaf ventral side white-glaucous, glands on whip-leaves visible, raised, elongated and divided (often 3 glands); often single stemmed shrub-trees, with stocky, clumpy foliage

        17.1a. Shorter whip-leaf glands, half or less as long as the associated sheath ....*J. arizonica*

        17.1b. Longer whip-leaf glands, more that half as long as the associated sheath ....*J. coahuilensis*

    17b. Mature seed cones copper to reddish-brown, with no bloom; whip-leaf ventral side not white-glaucous, glands on whip-leaves visible, raised, oval, not divided; shrubs with elongated terminal whips.......................*J. pinchotii*

15b. Mature seed cone dark blue, dark bluish-black to bluish-brown, with a light to heavy coat of bloom appearing light blue.

    18a. Glands on scale-leaves visible (conspicuous) and ruptured.

        19a. Seed cones maturing in 2 yrs., 2(3) seeded, 7-10 mm long; bark on twigs (5-10 mm dia.) reddish and exfoliating in scales or flakes; single stemmed tree to 20 (-30)m, dioecious or monoecious

            19.1a. Trunk bark red-brown to cinnamon; seeds cones avg. 7.7 mm (5-9); approx. 95% of the plants dioecious, leaf glands usually not ruptured, if ruptured with clear to pale yellow (or whitish) exudate................................*J. grandis*

            19.1b. Trunk bark brown; seeds cones avg. 8.5 mm (7-10); approx. 50% of the plants dioecious, leaf glands ruptured with yellow exudate turning dark brown to black..*J. occidentalis*

        19b. Seed cones maturing in 1 yr., 1 (-2) seeded, 6-10 mm long, bark on trigs brown to ash, not exfoliating in scales or flakes, shrubs to small trees, mostly dioecious.

            20a. Seed cones fibrous to woody pericarp, (7-) 9-10 (-13) mm. diam., bluish-brown under glaucous, 1 (-2) seeded; dioecious (1.9% monoecious), branchlets approx. as wide as scale-leaf length; scale-leaves closely appressed and generally flattened, branchlets terete.......................*J. californica*

            20b. Seed cones with soft, juicy pericarp, 6-8 mm diam., reddish-blue to brownish-blue, globose to ovoid, 1 seeded, dioecious, scale-leaf glands barely visible not conspicuous, few (less than 1/5) of the whip-leaf glands with a white crystalline exudate (visible without a lens), ultimate twigs 1.3-1.5 mm wide...*J. monosperma*

    18b. Glands on scale-leaves scarcely visible, not conspicuous

        21a. Plants monoecious, ultimate twigs 1.3 - 1.5 mm wide, seed cones bluish brown, glaucous, 8-9 mm diam., 1 seeded, leaf glands not conspicuous (embedded in the leaf, therefore not visible), not ruptured......*J. osteosperma*

        21b. Plants dioecious, ultimate twigs 0.7 - 1.1 mm wide, seed cones blue.

            22a. Ultimate twigs very slender (0.7- 0.8 mm. wide); stem bark of fibrous-ragged; longitudinal strips; mature seed cones globose, bluish, 5-8 mm. diam., 1 (2) seeds per cone; seeds 4-6 mm. long, 3-4 mm. wide; strong central axis, usually single-stemmed tree (to 10 m.), with a globose or broad-conic crown; foliage with faint odor of terpenoid compounds, scale-leaf glands not conspicuous, visible on whip-leaves but seldom ruptured..*J. comitana*

            22b. Ultimate twigs slender (0.8-1.1 mm. wide), trunk bark exfoliating in quadrangular plates on the lower part, then grading into longitudinal strips

on the upper branches, occasionally trunk exfoliating only in strips; mature
seed cones bluish and very glaucous, 5-7 mm. diam., 1 (2) seeds per cone;
seeds globose to ovoid, 5-6 mm. long, 4-5 mm. wide; shrubs branched at
base or short trees (to 10 m.) with horizontally broadened crowns; foliage
with strong odor of terpenoid cpds., scale-leaf glands not conspicuous,
whip-leaf glands visible but not usually ruptured..................*J. angosturana*

Key to *Juniperus* of Canada and the United States

1a. Leaves all acicular (subulate, jointed at the base) and spreading; seed cones sessile, axillary;
decumbent or rarely upright shrubs (in the western hemisphere)

  1.1a. Seed cones globose, shorter, or about equal leaf length, shrubs, spreading to upright
shrubs (or shrubby tree in New England), leaves straight to curved, flat or V shaped,
not boat shaped, free from stem (25 - 90°).....................................*J. communis*

  1.1b. Seed cones elongated oval (ellipsoid), especially on immature cones, shorter as long as
or longer than leaf length, shrubs, low to prostrate, leaves curved, boat shaped,
usually appressed to stem, on serpentine or volcanic rocks..................*J. jackii*

1b. Leaves decurrent (not jointed at the base), both whip- and scale-like; seed cones terminal;
trees or decumbent to upright shrubs.

  2a. Whip- and scale-leaf margins entire (20 X) or with irregular teeth (40 X) and then with
scale-leaves with acuminate to mucronate tips and tan-brown to brownish purple seed
cones.

   3a. Whip- and scale-leaf margins with irregular teeth (40 X), scale-leaves acuminate; seed
cones (4-) 6-10 (-13) seeded, and tan-brown to brownish-purple; branches
pendulous (weeping)..*J. flaccida*

   3b. Whip- and scale-leaf margins smooth (entire) (40 X), scale-leaves obtuse to acute to
apiculate; seed cones 1-2(3) seeded, blue-black to brownish-blue when mature;
branches not drooping (but ultimate branchlets are often flaccid).

    4a. Prostrate to decumbent shrub; scale-leaves apiculate; both whip- and scale-leaves
growing along the branchlets; peduncles generally curved.....................*J. horizontalis*

    4b. Tree with 1(2-3) stems and rounded, flattened, pyramidal, or strict crowns;
scale-leaves obtuse to acute; whip-leaves growing only at branchlet tips (on mature
trees); peduncles generally straight.

     5a. Scale-leaves not overlapping, or, if so, not by more than 1/5 the length, obtuse to
acute; seed cones globose to reniform

      5.1a. Twigs (3-5 mm dia.) with smooth bark, twigs (6-15 mm dia.) with bark
exfoliating in plates, reddish-copper beneath; seed cones maturing in 2 yrs,
most seed cones normal, rarely with exserted seeds.....................*J. scopulorum*

      5.1b. Twigs (3-5 mm dia.) with persistent dead whip-leaves, twigs (6-15 mm dia.)
reddish-brown beneath; seed cones maturing in 1 yr. (14-16 mos.), often the
seed cones with exserted (naked) seeds; Pacific northwest near the seaside or
nearby dry mountains in Georgia Straits and Puget Sound...............*J. maritima*

     5b. Scale-leaves overlapping (more than 1/4 length) acute; twigs (3-5 mm dia.) with
persistent dead whip-leaves, twigs (6-15 mm dia.) with bark not exfoliating
in plates, or, if so, brownish beneath; seed cones maturing in 1 yr...
....*J. virginiana*

  2b. Whip- and scale-leaf margins denticulate (20 X).

   6a. Seed cones with (3-) 4 - 5 (-6) seed, fibrous to obscurely woody, trunk bark exfoliating in
square or quadrangular plates (except. in f. *sperryi* with bark that exfoliates in strips)
....*J. deppeana*

   6b. Seed cones with 1-2 (3) seeds, fleshy to fibrous (when mature and fresh)

    7a. Scale-leaves with a raised hemispherical gland, whip leaf gland raised, oval or hemispherical

     7.1a. Whip-leaf gland hemispherical and raised, scale leaf glands hemispherical,

1 seed/cone (rarely 2, avg. 1.01),seed cones (8) 9 (10) mm diam....................*J. ashei*

7.1b. Whip-leaf glands oval to elliptical and raised, scale leaf glands hemispherical, 2 seeds/cone (avg. 1.7) (5) 6 (-8) mm in diam...............***J. ovata*** (***J. ashei*** var. ***ovata***)

7b. Scale-leaves without a raised hemispherical or oval gland

8a. Mature seed cones orange, reddish-orange, red, bronze, or reddish-brown, appearing pink or rose-color if covered with bloom.

9a. Mature seed cones orange to red, with light bloom appearing pink or rose-colored; whip-leaf ventral side white-glaucous, glands on whip-leaves visible, raised, elongated and divided (often 3 glands); often single stemmed shrub-trees with stocky, clumpy foliage

9.1a. Shorter whip-leaf glands, half or less as long as the associated sheath ....*J. arizonica*

9.1b. Longer whip-leaf glands, more that half as long as the associated sheath ....*J. coahuilensis*

9b. Mature seed cones copper to reddish-brown, with no bloom; whip-leaf ventral side not white-glaucous, glands on whip-leaves visible, raised, oval, not divided; shrubs with elongated terminal whips ....*J. pinchotii*

8b. Mature seed cone dark blue, dark bluish-black to bluish-brown, with a light to heavy coat of bloom (waxy glaucous) appearing light blue.

10a.Glands on scale-leaves visible (conspicuous) and ruptured.

11a.Seed cones 7-10 mm long; maturing in 2 yrs., 2(3) seeded; bark on twigs (5-10 mm diam.) reddish and exfoliating in scales or flakes; single stemmed tree to 20 (-30)m, dioecious or monoecious.

11.1a. Trunk bark red-brown to cinnamon; seeds cones avg. 7.7 mm (5-9); approx. 95% of the plants dioecious, leaf glands usually not ruptured, if ruptured with clear to pale yellow (or whitish) exudate..*J. grandis*

11.1b. Trunk bark brown; seeds cones avg. 8.5 mm (7-10); approx. 50% of the plants dioecious, leaf glands ruptured with yellow exudate turning dark brown to black..*J. occidentalis*

11b.Seed cones 6-10 mm long, maturing in 1 yr., 1 (-2) seeded, bark on twigs brown to ash, not exfoliating in scales or flakes, shrubs to small trees, mostly dioecious.

12a.Seed cones with a fibrous to woody pericarp, (7-) 9-10 (-13) mm. diam., bluish-brown under glaucous, 1 (-2) seeded; dioecious (1.9% monoecious), branchlets approx. as wide as scale-leaf length; scale-leaves closely appressed and generally flattened, branchlets terete..*J. californica*

12b.Seed cones with a soft, juicy pericarp, 6-8 mm diam., reddish-blue to brownish-blue, globose to ovoid, 1 seeded, dioecious, scale-leaf glands barely visible not conspicuous, few (less than 1/5) of the whip-leaf glands with a white crystalline exudate (visible without a lens), ultimate twigs 1.3-1.5 mm wide...*J. monosperma*

10b.Glands on scale-leaves not conspicuous (embedded in the leaf, therefore not visible), not ruptured, plants monoecious, ultimate twigs 1.3-1.5 mm wide, seed cones bluish brown, very glaucous, 8-9 mm diam., 1-seeded.......................*J. osteosperma*

Key to *Juniperus* of Mexico and Guatemala

1a.Whip- and scale-leaf margins entire (20 X) or with irregular teeth (40 X) and then with scale-leaves with acuminate to mucronate tips.

2a. Whip- and scale-leaf with irregular teeth (40 X), scale-leaves acuminate; seed cones 1(2), (4-) 6-10 (-13) seeded, tan-brown to brownish-purple; branches pendulous to erect.

2.1a. Seed cones 9-20 mm diam., (4) 6 - 10 (-13) seeds per cone, terminal branch tips drooping (hanging)

3a. Branching radially, seed cones tan to brownish/purple............................*J. flaccida*

3b.   Branching planate, seed cones bluish-brown...*J. poblana*

2.1b. Seed cones (5-) 6 (-9) mm diam., 1-2 (3) seeds per cone, terminal branch tips erect
....*J. martinezii*

2b. Whip- and scale-leaf margins smooth (entire) (40 X), scale-leaves obtuse to acute to
   apiculate; seed cones 1-2(3) seeded, blue-black to brownish-blue when mature; branches
   not drooping (but ultimate branchlets are often flaccid).

4a. Seed cones mostly bi-lobed (reniform), a few may be globose, seed cones maturing in 1 year

5a. Scale-leaves with acute to acuminate tips

5.1a. Tree or shrubby tree, females not on understand of planate branches,
     foliage not dense, often along streams...................................*J. blancoi* var. *blancoi*

5.1b. Prostrate shrub forming patches, one to several meters in diameter, female cones
     in underside of planate branches, foliage dense, sub-alpine..........
     ....*J. blancoi* var. **huehuentensis**

5b. Scale-leaves with mucronate tips..*J. blancoi* var. **mucronata**

4b.   Seed cones globose, rarely bi-lobed, maturing 1 or 2 years

4.1a. Scale-leaves not overlapping, or, if so, not by more than 1/5 the length, leaf tips
     obtuse to acute; seed cones maturing in 2 yrs, globose but rarely reniform (bi-lobed),
     twigs (3-5 mm dia.) with smooth bark, twigs (6-15 mm dia.) with bark exfoliating in
     plates, reddish-copper beneath...*J. scopulorum*

4.1b. Scale-leaves overlapping (more than 1/4 length), leaf tips acute; cones maturing
     in 1 yr, twigs (3-5 mm dia.) with persistent dead whip-leaves, twig (6-15 mm dia.)
     bark not exfoliating in plates, or, if so, brownish beneath......................*J. virginiana*

1b. Whip- and scale-leaf margins denticulate (20 X)

6a. Seed cones (1) 2 - 11 seeded, if 1 seeded, then bark exfoliating in quadrangular plates

7a. Seed cones (1-3 -) 4 -5 (6) seeded (except var. *gamboana* with 1-2 seeds/cone), fibrous
   to obscurely woody; trunk bark exfoliating in quadrangular plates or if trunk
   exfoliating in strips, then upper branches exfoliating in quadrangular plates
   (f. *sperryi*)..*J. deppeana*

7b. Seed cones 2- 9 seeded, bark exfoliating in strips (if exfoliating in quadrangular plates,
   then seed cones small, 5-6 mm long).

8a. Mature seed cones reddish-brown to brown, with a light coat of bloom; seed cone
   peduncles straight.

9a. Mature seed cones irregular and gibbous; seeds (2-)4-9(-11) per cone; terminal whip
   branches strict; foliage very loosely spaced; scale-leaves do not appear as a string-of-
   beads on the ultimate twig..*J. jaliscana*

9b. Mature seed cones bluish-red, very glaucous, globose; seeds 1-4 per cone; terminal
   whip branches erect to strict, recurved (hooked) at tip; foliage tightly-spaced; scale
   leaves appear as a string-of-beads on the ultimate twig.........................*J. durangensis*

8b. Mature seed cones dark blue to bluish-black, with light coat of bloom, cone peduncles
   usually curved

10a. Twigs not distichous, or sometimes sub-distichous, scale-leaves thick but not with a
   hood or beak, ultimate branchlets 1.3 - 1.4 mm wide, central and n. Mexico.

10.1a. Whip leaf glands oval to elongate, usu. raised, branchlets (3-5mm diam.)
     rough with persistent old leaves, mature twigs 5-10 mm. long, angle of
     branching of ultimate twig 50-60 degrees, foliage tightly compacted,
     prostrate shrubs (to 1 m.), sub-alpine habitat, on limestone........*J. zanonii*

10.1b. Whip leaf glands elongate, often sunken, branchlets (3-5mm diam.) smooth,
     mature twigs 10-40 mm. long, angle of branching of ultimate twig 40-60
     degrees, foliage loose to tightly compacted, low shrub to tree (to 10 m), sub-
     alpine to montane habitat, especially on volcanic soils.................*J. monticola*

10b. Twigs distichous, scale-leaves thick, usually a hood or beak, ultimate branchlets 1.4 -
   1.8 mm wide, Chiapas, Mexico and n. Guatemala......................................*J. standleyi*

6b. Seed cones 1-2 (-3) seeded, fleshy to fibrous (when mature and fresh)
  11a. Scale-leaves with a raised hemispherical gland, whip leaf gland raised, oval or hemispherical
    12a.Scale-leaf glands protruding like a bead, hemispherical, whip-leaf glands
        oval (var. *ovata*), dark brownish-green on dark grayish-green mature leaf; angle
        of branching of ultimate twig 25-40 degrees; scales leaves on ultimate twig do
        not appear as a string-of-beads..*J. ovata (J. ashei* var. *ovata)*
    12b.Whip-leaf glands elongate, not hemispherical, scale-leaf gland protruding, oval to
        elliptical, same color as light gray-green mature leaf; twig bark grayish brown,
        smooth; angle of branching of ultimate twig about 60 degrees; scale-leaves on
        ultimate twig appear as a string of beads..*J. saltillensis*
  11b. Scale-leaves without a raised hemispherical gland
    13a.Seed cones (7-) 9-10 (-13) mm. diam., fibrous to woody pericarp, bluish-brown under
        glaucous, 1 (-2) seeded; dioecious (1.9% monoecious), branchlets approx. as wide as
        scale-leaf length; scale-leaves closely appressed and generally flattened, branchlets
        terete, Baja California...*J. californica*
    13b.Seed cones 5-8 mm diam., fleshy when ripe and fresh.
      13.1a. Mature seed cones orange, reddish-orange, red, bronze, or reddish-brown,
          appearing pink or rose-color if covered with bloom.
        14a.Stem bark exfoliating in quadrangular plates; cones reddish-brown; terminal whip
            branches tips straight, not curved; seeds subglobose or broadly ovoid...................
            ....*J. deppeana* vat. *gamboana*
        14b.Stem bark exfoliating in longitudinal strips, often interconnected; seed cones
            rose, orange to reddish-brown or bronze-color; seed ovoid or if bark in
            quadrangular plates, then seed cones bluish-rose.
          15a.Mature seed cones orange to red, with light bloom appearing pink or rose-
              colored; whip-leaf ventral side white-glaucous, glands on whip-leaves visible,
              raised, elongated and divided (often 3 glands); often single stemmed shrub-
              trees, with stocky, clumpy foliage
            15.1a. Shorter whip-leaf glands, half or less as long as the associated sheath
                (not yet confirmed in Mexico, but likely grows in n. Sonora on the
                Arizona border)..*J. arizonica*
            15.1b. Longer whip-leaf glands, more that half as long as the associated sheath
                ....*J. coahuilensis*
          15b.Mature seed cones copper to reddish-brown, with no bloom; whip-leaf
              ventral side not white-glaucous, glands on whip-leaves visible, raised, oval,
              not divided; shrubs with elongated terminal whips..........................*J. pinchotii*
      13.1b.Mature seed cone dark blue to bluish-brown with a light to heavy coat of bloom
          appearing light blue.
        16a.Ultimate twigs very slender (0.7- 0.8 mm. wide); stem bark fibrous-ragged;
            narrow strips; seed cones globose, bluish, 5-8 mm. diam., 1 (2) per cone; seeds 4-
            6 mm. long, 3-4 mm. wide; strong central axis, usually single-stemmed tree (to 10
            m.), with a globose or broad-conic crown; foliage with faint odor of aromatic
            (terpenoid) compounds, scale-leaf glands not conspicuous, visible on whip-leaves
            but seldom ruptured..*J. comitana*
        16b.Ultimate twigs slender (0.8-1.1 mm. wide), trunk bark exfoliating in quadrangular
            plates on the lower part, then grading into longitudinal strips on the upper
            branches, occasionally trunk exfoliating only strips; mature seed cones bluish
            and very glaucous, 5-7 mm. diam., 1 (2) seeds per cone; seeds globose to ovoid,
            5-6 mm. long, 4-5 mm. wide; shrubs branched at base or short trees (to 10 m.)
            with horizontally broadened crowns; foliage with strong odor of aromatic
            (terpenoid) compounds, scale-leaf glands not conspicuous, whip-leaf glands
            visible but not usually ruptured ....*J. angosturana*

Key to Caribbean Junipers (3 species) (plus *J. v.* var. *virginiana* and *J. v.* var. *silicicola* from se USA to aid in identification)

1a. All leaves whip or decurrent (composed of a blade with free tip and a sheath that clasps the stem), no scale-leaves; endemic to Cuba...*J. gracilior* var. *saxicola*

1b. Plants with both scale-like and decurrent (at least on juvenile growth) leaves

  2a. Ultimate leafy branchlets 1.3-1.6 mm. wide, and 2-4 cm. long, scale-leaves overlapping by almost 1/2 their length; leaves opposite, branchlets square, persistent brown scale-leaves on branches, decurrent leaves rarely present, endemic to Bermuda..............*J. bermudiana*

  2b. Ultimate leafy branchlets 0.7-1.0 mm. wide and 0.5-2.0 cm. long, scale-leaves overlapping by about 1/4 their length

    3a. Scale-leaf tips acuminate to mucronate; 1-2 seeds per seed cone, single-seeded cones, globose, but 2- seeded cones nearly reniform (bi-lobed)

      4a. Branchlets pendulous, ultimate leafy branching angle 20-30 degrees; glands on old, brown, persistent whip-leaves not conspicuous: endemic to Hispaniola........................ ....*J. gracilior* var. *gracilior*

      4b. Branchlets erect or procumbent, ultimate leafy branching angle 35-40 degrees; glands on old, brown, persistent whip-leaves conspicuous

        5a. Erect trees; whip-leaf glands oval (to twice as long as wide); endemic to Haiti ....*J. gracilior* var. *ekmanii*

        5b. Prostrate shrubs; whip-leaf glands elongated (3-4 times as long as wide); endemic to Haiti ....*J. gracilior* var. *urbaniana*

    3b. Scale-leaf tips obtuse to acute; 1-4 seeds per seed cone, seed cones reniform or globose

      6a. 1-(2-3) seeds per cone, seed cones globose, continental eastern USA and Canada, crowns pyramidal to flattened

        7a. Seed cones 4-6 mm; crowns strict, pyramidal to round; bark reddish-brown; scales leaves acute; pollen cones 3-4 mm; inland and in old fields, inland eastern USA and Canada, not on coastal fore dunes......................*J. virginiana* var. *virginiana*

        7b. Seed cones 3-4 mm; crowns flattened; bark cinnamon-reddish; scale leaves bluntly obtuse to acute; pollen cones 4-5 mm; on sand on coastal fore-dunes, se USA..... ....*J. virginiana* var. *silicicola*

      6b. 2-4 seeds per seed cone, seed cones reniform (bi-lobed), on coppice rocks or rocky summits, crowns pyramidal to round, Caribbean

        8a. Glands on old brown persistent whip-leaves conspicuous, sunken and extending almost to the whip-leaf tip; known only from the summit of Petite Piton, St. Lucia, BWI...*J. barbadensis* var. *barbadensis*

        8b. Glands on old brown persistent whip-leaves not conspicuous, if visible then neither sunken nor extending almost to the whip-leaf tip; Cuba, Bahamas, Jamaica...*J. barbadensis* var. *lucayana*

Chapter 5: Species Descriptions, Distribution Maps and Photos

Note on Status: The status is taken from the "Status survey and conservation action plan: Conifers", A. Farjon and C. N. Page, IUCN, Cambridge, UK. 1999. See pp. 118-120 for codes explanations.   Listed below are the basic categories:

EX - Extinct

EW - Extinct in the wild.

CR - Critically endangered

EN - Endangered

VU - Vulnerable

LR - Lower risk

DD - data deficient.

As additional information is gathered, the status of the taxa will be revised (see www.juniperus.org) for the most recent status.

Juniperus angosturana R.P. Adams, Biochem. Syst. Ecol. 22 (7):704 (1994). Cedro, slender one-seed juniper. Type: Mexico: San Luis Potosi: Hacienda de Angostura. *C. G. Pringle 3771* (holotype: not found at MEXU: Lectotype: VT, San Luis Potosi: Hacienda de Angostura.; designated by Zanoni & Adams, Bol. Soc. Bot. Mex. 38: 105. 1980. Isolectotypes: ARIZ!, F!, GH!, MO!, NY!, UC!)

> *J. monosperma* (Engelm.) Sarg. var. *gracilis* Martinez, Anales Inst. Biol. Univ. Nac. Mexico 17(1):111 (1946)

Dioecious. Trees to 8 m. **Trunk bark** brown, exfoliating in square plates (checkered). **Branches** erect, ultimate twigs, 6-12 mm long, 1-13 mm diam. usually very slender often with broader angle of branching of ultimate twig, about (50-)60-70 degrees. **Leaves** scale and decurrent (whip), glands visible but sunken, whip leaves with elongated glands (0.5 x 3 mm). **Seed cones** globose, bluish, with glaucous bloom, 4-6 mm. **Seeds** 1 (2) per cone. **Pollen shed** Dec.-Jan. **Habitat** In grasslands, *Acacia* scrub, *Agave-Yucca-Opuntia-Juniperus* scrub, or pine woods, on gravelly or rocky limestone soils in the w. foothills of the Sierra Madre Oriental at 1050-2800 m elevation. **Uses** posts. **Dist.:** e. Coahuila, s. Nuevo Leon, se . Tamaulipas, ne. Queretaro and n. Hidalgo, Mexico. **Status:** restricted to a small area, it may become threatened in the future.

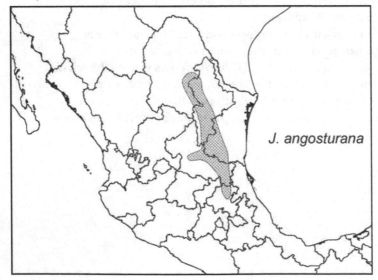

J. angosturana, bark exfoliating in quadrangular plates

J. angosturana
leaves and
seed cones.

J. angosturana, near Angostura,
Mexico with Glenn Harris (right)
and local rancher. Santo Domingo (e
of Villa Juarez), San Luis Potosi,
Mexico. cf. *Adams 8713-8715*

J. angosturana habitat, near Angostura, Mexico (type locality). cf. *Adams 8713-8715*

Juniperus arizonica (R. P. Adams) R. P. Adams, Phytologia 88(3): 306 (2006).
 Arizona juniper. Type: USA, Arizona, Yavapai Co., Camp Verde, 72 km south of Flagstaff,
1160 m. *R. P. Adams 2132* (holotype BAYLU!)
   Juniperus coahuilensis (M. Martinez) Gaussen ex. R.P. Adams var. *arizonica* R.P. Adams,
   Biochem. Syst. Ecol. 22 (7): 708 (1994).
   J. erythrocarpa Cory (in part: New Mexico, Arizona)

Dioecious. **Trees** large shrub to small tree, 3-8 m, often with a single stem to 1 m, when shrubs branched at the base, with flattened-globular or irregular crowns. **Trunk bark** brown, thin, exfoliating in long ragged strips. **Branches** ascending to erect in shrubs, spreading in trees. Branch bark scaly, ashy-gray. Stumps sprouting after burning or cutting. **Leaves** decurrent (whip) and scale. Whip- and scale-leaf margins denticulate (20 X), white glaucous on adaxial leaf surface. At least ¼ or more of the whip leaf glands with a white crystalline exudate. **Seed cones** rose to pinkish but yellow-orange, orange or dark red beneath the white-blue glaucous, soft and juicy, globose to ovate, 6-7 mm, 1(-2) seeded, the hilum scar pale brown, approx. ½ as long as seed. **Seeds** 4-5 mm long. **Pollen shed** Jan. - March. **Habitat** *Bouteloua* grasslands and adjacent rocky slopes; 980-1600 (-2200) m. **Uses** fence posts. Sprouts from cut stumps and is thus a pest in grasslands. **Dist.**: Arizona, South of the Mogollon Rim; and in southwestern New Mexico and northeastern Sonora, Mexico. **Status**: abundant and weedy in areas.

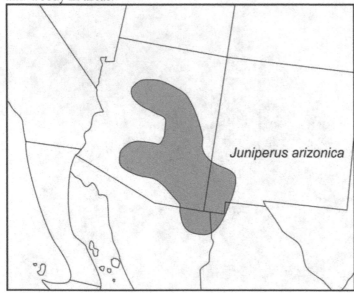

J. arizonica, bark

J. arizonica, with
Glenn Harris
Rock Hound State Park,
Deming, New Mexico
cf. *Adams 2203-2217.*

J. arizonica, seed cone and leaves.

Juniperus arizonica, s of Cottonwood, AZ with Judy Thornburg. This might to be the National Big Tree (USA) for *J. arizonica*.

Juniperus ashei

The divergent populations of *J. ashei* in the Trans-Pecos, TX, n Mexico and around New Braunfels were treated as *J. ashei* var. *ovata* in the previous edition (Adams, 2011). However, the recent DNA sequence data is so strong (see Chapter 1), that *J. ashei* var. *ovata* is treated as *J. ovata* in this edition. No doubt additional research will clarify this difficult taxonomic situation.

As the two taxa are difficult to recognize, a key is provided:

1. Whip-leaf glands hemispherical, scale-leaf glands hemispherical; female cones (8) 9 (10) mm in diameter; seeds 16-27 mm^2, 1 (avg. 1.01, rarely 2,) per cone......................***J. ashei* var. *ashei***

1. Whip leaf glands oval to elliptical; scale-leaf glands hemispherical, female cones (5) 6 (8) mm in diameter; seeds 13-16 mm^2, 2 (avg. 1.7), per cone...........***J. ovata* (*J. ashei* var. *ovata*)**

Juniperus ashei J. Buchholz. var. ***ashei***, Bot. Gaz. (Crawfordsville) 90(3):329 (1930). Mountain cedar, rock cedar, post cedar, Mexican Junipers, Ashe juniper. Type: USA, Arkansas, Sylamore. *W. W. Ashe s. n. 1923-1925*. NCU! Named in honor of W. W. Ashe (1872-1932), a prominent botanist. The Ashe herbarium is located at NCU. duplicates at MO.

> *Cupressus sabinoides* Kunth in Humboldt et al., Nov. Gen. Sp. Pl. 2: 3 (1817)
> *J. sabinoides* (Kunth) Nees, Linnaea 19: 706(1847), *non* Griseb. (1846)
> *J. sabionoides* Sarg., Silva N. Amer. 10:91 (1896), *non* Griseb. (1846)
> *J. sabinoides* (H.B.K.) sensu Sargent *non* Nees
> *J. occidentalis* Hook. var. *texana* Vasey, (Cat. Forest Trees U.S. 37) Rep. U.S. Dept. Agric. 1875: 185 (1876)
> *J. occidentalis* Hook. var. *conjugens* Engelm., Trans. St. Louis Acad. Sci. 3:590 (1878)
> *J. tetragona* Moench var. *oligosperma* Engelm., Trans. St. Louis Acad. Sci. 3:590 (1878)
> *Sabina sabinoides* Small, Fl. S.E. United States: 33 (1903)
> *J. mexicana* Sprengel in part, see Zanoni, 1978

Dioecious. **Trees** with broad, bushy rounded or irregularly open crown, to 15 m, with a single trunk branching at 1-3 m or occasionally branching at the base. **Trunk bark** exfoliating in thin brown strips. **Branches** brown but usually with a grey-white fungus. **Leaves** both whip and scale-like. Whip leaves with a raised, hemispherical glands (not prominent on scale leaves). Whip- and scale-leaf margins denticulate (20 X). **Seed cones** ovoid to subglobose, maturing in one year, dark blue and glaucous, 6-9 mm in diam., 1(2-3) seeded. **Seeds** 4-6 mm long. 2n = 22 (Irving, 1980). **Pollen shed** Dec. -Feb. **Habitat** Limestone glades and bluffs, 150-600 m. **Uses** source of Texas cedar wood oil (Adams, 1987), fence posts. **Dist**.: Ark., Okla., Tex. Maps: Adams, 1977 (amended Little, 1971, 21-E, W). **Status**: abundant on limestone in central/ west Texas. The range is expanding, and it is regarded as a weed in Texas.

The type for *J. ashei* Buch. consisted of one male and three female specimens (Hall, 1954). To resolve this problem, Hall (1954) selected a female specimen (acc. number 22520, dated Sept. 16, 1923, UNC) and designated it as the lectotype. All of the material cited by Buchholz (1930) was collected on limestone bluffs, above the White River, near Sylamore, Arkansas. It is clear in Buchholz (1930) that his illustration (Fig. 1) is of post-Pleistocene *J. ashei* with the hemispherical glands on the whip leaves.

The whip leaf glands are illustrated in figure 5.1. Notice hemispherical glands on *J. ashei* (Fig. 5.1, left) and the raised, oval to elongated glands on *J. ovata* (Fig. 5.1, right). It should be noted that a few nearly hemispherical glands are present on whip leaves of *J. ovata*. This is informative, as these characters can be used to distinguish *J. ovata* from *J. ashei*, yet exclude other nearby juniper species such as *J. monosperma* (Englem.) Sarg., *J. pinchotii* Sudw. and *J. coahuilensis*. (Mart.) Gaussen ex R. P. Adams.

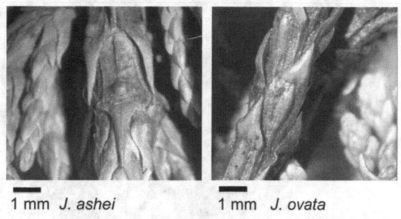

1 mm *J. ashei*   1 mm *J. ovata*

Figure 5.1. Comparison of whip leaf glands for *J. ashei* and *J. ovata*.

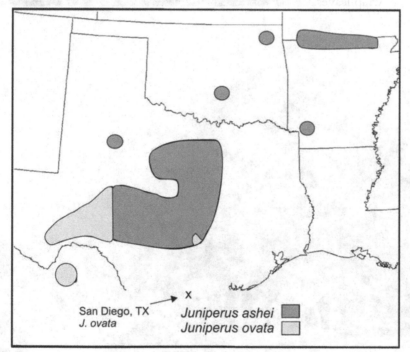

Distributions of *J. ashei* var. *ashei* and *J. ovata*.

The area of possible sympatry in west Texas and around New Braunfels is not well understood and additional field collections are needed to define better their distributions in these areas (see Adams, 2008).

J. ashei bark.

J. ashei with hemispherical glands on whip leaves.

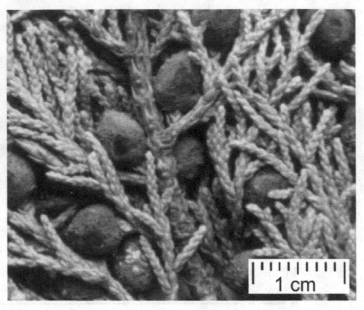

J. ashei, typical habit, branching at or 1 - 2 m above the ground, Austin, Texas, USA.

J. ashei invading an old field near Austin, Texas.

Truck load of *J. ashei* wood near Leakey, Texas bound for a steam distillation plant to produce commercial Texas cedarwood oil.

Earlier reports of hybridization between *J. ashei* and *J. virginiana* (Hall, 1954) have been negated in later studies (Adams, 1977; Flake, von Rudloff and Turner, 1969). Hybridization with *J. pinchotii* (Hall, McCormick and Fogg, 1961) could not be substantiated using numerous chemical and morphological characters (Adams, 1969; 1977).

Author (RPA) with US National Champion tree for *J. ovata* tree, New Braunfels, TX, 2007. cf. *Adams 11309*.

This champion is incorrectly entered into the records as *J. ashei*, but should be changed to *J. ovata* in view of the recent DNA data.

Juniperus barbadensis

1a. Glands on old brown persistent whip-leaves conspicuous, sunken and extending
almost to the whip-leaf tip; known only from the summit of Petit Piton,
St. Lucia, BWI..*J. barbadensis* var. *barbadensis*

1b. Glands on old brown persistent whip-leaves not conspicuous, if visible then neither
sunken nor extending almost to the whip-leaf tip; Cuba, Bahamas, Jamaica
....*J. barbadensis* var. *lucayana*

Juniperus barbadensis L. var. *barbadensis* Sp. Pl. 2:1039 (1753). Barbados juniper. Type: LINN!,
Barbados, (now extinct there), *Linnaeus 1198.1* (lectotype LINN!, see Adams et al., 1987)

    J. virginiana humilis Lodd., Cat. (1836)
    J. virginiana var. *barbadensis* (L.) Gordon, Pinetum: 114 (1858)
    Sabina barbadensis (L.) Small, Fl. S.E. United States: 33 (1903)

Dioecious. Trees to 10 m. **Trunk bark,** thin and bark separating in strips. **Branches,** branchlets
slender, leaves usually opposite, thus branchlets quadraform. **Leaves** light green, decurrent on young
plants and on rapidly growing shoots but otherwise scale-like, about 1 mm long, with obtuse to acute
tips, rounded on the back and tightly appressed. Glands on scale-like leaves inconspicuous, but oval to
ovate, glands on old, brown persistent whip leaves conspicuous, sunken and elongated extending
almost to the whip leaf tip. **Seed cones** reddish-blue with bloom, reniform (occasionally subglobose)
and somewhat flattened, 4 - 5 mm long, 6-8 mm wide when reniform. **Seeds** 2 - 3 seeds/cone.
Pollen shed Jan.-Mar. **Habitat** rocky areas at the summit of Petit Piton. **Uses** none known.
Dist.: endemic to Petit Piton, St. Lucia, BWI. **Status**: CR, D. Now known only from the island of St.
Lucia, BWI, near the summit of Petit Piton, 700-730 m. The taxon is now extinct on Barbados, having
been cut out before 1700 (Adams et al. 1987a,b; Adams, 1989). A visit to Barbados revealed that the
habitat was converted to sugar cane fields over 280 years ago. Very endangered species.

J. barbadensis, bark

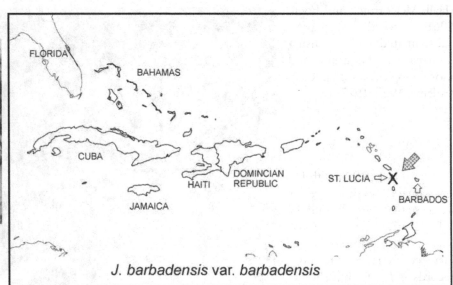

J. barbadensis var. *barbadensis*

Leaves and seed cones, *J. barbadensis* var. *barbadensis*

J. barbadensis, habit, top of Petit Peton, St. Lucia, BWI, cf. *Adams 5367-5372*

The only known population of *Juniperus barbadensis* var. *barbadensis* grows at the summit of Petit Piton. In 1986, the author collected from 5 individuals (*Adams 5367-5373*) and observed: approximately 25 trees ranging from the top of Petit Piton to about 30 m down the east escarpment; 2 trees on the north side about 10 m below the summit; 4 trees on the west side, about 5 m below the summit. The trees on the east side are subjected to prevailing winds and have round crowns that lean downwind. Trees on the north and west sides are protected in the forest and have a strong central axis with a pyramidal shape.

Juniperus barbadensis grows at the top of Petit Piton (on the right, 2610 ft.). *Juniperus barbadensis* has not been reported from Gros Piton (on the left, 2619 ft).

Juniperus barbadensis L. var. **lucayana** (Britton) R. P. Adams, Phytologia 78: 145 (1995) Lucayan juniper (Lucayan is another name for the Arawak Indians of the Bahamas), Bahaman juniper. Type: New Providence, Southwest Bay, Coast coppice, *Britton & Brace 497* (holotype: not published; lectotype: Britton & Brace 497 NY).

    J. virginiana L. var. *australis* Endl. Syn. Conif. 28 (1847)

    Sabina virginiana (L.) Antoine var. *australis* (Endl.) Antoine, Cupress.-Gatt.: 62 (1857)

    Juniperus lucayana Britton, North Amer. Trees 121 (1908).

    J. australis (Endl.) Pilg. in Urban, Symb. Antill. 7: 479 (1913)

    J. barbadensis L. var. *jamaicensis* Silba, J. Int. Conifer Preserv. Soc. 7(1):24 (2000)

Dioecious. **Trees** to 12 m. **Trunk bark** thin and separating in strips. **Branches** branchlets slender, 0.5 - 2 cm long, width 0.8 - 0.9 mm. Leaves usually opposite, thus branches quadraform. **Leaves** decurrent on young plants and on rapidly growing shoots but otherwise scale-like, about 1 mm long, with obtuse to acute tips, rounded on the back and tightly appressed. Glands on scale-like leaves inconspicuous, but oval to ovate, glands on old, brown persistent whip leaves not conspicuous, but if visible, then not sunken and extending almost to the whip leaf tip. **Seed cones** reddish blue to blue with bloom (glaucous), usually reniform (occasionally subglobose) and somewhat flattened, 4 - 5 mm long by 5 - 7 mm wide when reniform. **Seeds** 2 - 4 seeds/cone. **Pollen shed** in Jan-Feb. **Habitat** swamps to limestone coppice areas near the shore (in the Bahamas), or to 1600m in Mts. (Jamaica, Cuba). **Uses** wood carving. **Dist.**: Cuba: Sierra de Nipe region, Holguin Province, swamps in the south-central portion of the Isle of Pines (= Isla de Pinos). Jamaica. 1,100 - 1,200 m near Clydesdale in St. Andrew Parish. Bahamas. On coppice limestone near sea level on Andros, Grand Bahama and Great Abaco Islands. **Status**: VU, B1+2c. Due to the increased population pressure and clearing of land for resorts, coupled with its limited distribution, it should be considered threatened.

Reports of *J. b.* var. *lucayana* on New Providence could not be confirmed in 1980. The taxon is also reported from Cat Cay. Specimens have been examined from Haiti (St. Michel de l'Atalaye; Gros Morne near Pendu; Bassin Bleu; and Crete-Sale), but field trips to the areas failed to discover any plants in recent years. *Juniperus b.* var. *lucayana* is now presumed to be extinct on Hispaniola.

    Juniperus b. var. *lucayana* and *J. barbadensis* are very closely related. Based on terpenoids (Adams, 2000a), *J. b.* var. *lucayana* is most similar to *J. barbadensis*. And this is also shown in the DNA data (Figs. 1.16.1, 1,16,5) *Juniperus b.* var. *lucayana* appears to hold a central role in the island hopping of *Juniperus* in the Caribbean Islands and Bermuda.

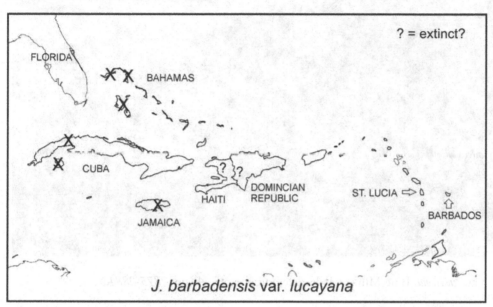

J. barbadensis var. *lucayana* leaves and seed cones

J. b. var. *lucayana* bark exfoliating in thin, wide strips.

Juniperus barbadensis var. *lucayana*, Blue Mtns., Clydesdale, Jamaica, cf. *Adams 2875-2884*

J. barbadensis var. *lucayana*, cultivated from native plants, Abaco Island, Bahamas. Michael Wheeler, Crossing Rocks, Abaco, the owner, calls it the 'Bahamian Red Cedar' The top was damaged by hurricane Floyd, leading to the shrubby form, rather than a strong central axis.
cf. *Adams 11404*

Kenneth Major with a monecious (very rare) *J. barbadensis* var. *lucayana* plant about 1 km west of Coopers Town, Abaco. The top was damaged in hurricane Floyd.
cf. *Adams 11408*

Juniperus bermudiana L., Sp. Pl.2: 1039 (1753). Bermuda juniper. Type: Bermuda , *Linnaeus 1198.2* (holotype LINN!, see Adams et al., 1987).

    J. oppositifolia Moench, Methodus: 699. (1794)

    J. pyramidalis Salisb., Prodr. Chap. Allerton: 397 (1796)

    Sabina bermudiana (L.) Antoine, Cupress. -Gatt.: 65. (1857)

    J. virginiana L. var. *bermudiana* (L.) Vasey, Rep. (U.S.) Comm. Agric. 1875:185 (1876)

Dioecious. Trees to 15 m. **Trunk bark** thin bark exfoliating in strips. **Branches** erect, leafy branchlets 2 – 4 cm long, and 1.3-1.6 mm wide, branching angle 30 - 35°. **Leaves** scale leaves opposite, thus branchlets strongly quadraform (four sided), overlapping by about 0.4 - 0.5 times their length, green, decurrent on young plants and on rapidly growing shoots but otherwise scale-like, about 1 mm long, with obtuse to acute tips, rounded on the back and tightly appressed. Glands on scale-like leaves inconspicuous, elongated and sunken, but extending almost to the tip of whip leaves. **Seed cones** dark blue with bloom, subglobose to reniform, 4 - 5 mm long, 6-8 mm wide when reniform, mature in 1 yr. **Seeds** 1-2(3) seeds/cone. **Pollen shed** spring. **Habitat** limestone area in Bermuda. **Uses** previously used for ship repair and masts, but no present use. **Dist.**: Endemic to Bermuda Island. **Status**: CR, B1+2c. very endangered (see below).

    Juniperus bermudiana has been subject to attack by two scale insects, *Lepidosaphes newsteadi* and *Carulaspis minima*, that were apparently introduced from the U.S. mainland prior to 1942 (Bennett & Hughes 1959; Groves 1955). These insects cause defoliation and death. Groves (1955) estimated that 90% of the trees were dead by 1955. In 1978, William E. Sterrer, Bermuda Biological Station, (pers. comm.) estimated that perhaps 99% of the original trees were dead. To make the situation even more precarious, *J. virginiana* (1950s) and *J. v.* var. *silicicola* (1940s) were introduced into Bermuda and these junipers are resistant to the scale insects (Adams and Wingate, 2008). In addition, they are hybridizing with *J. bermudiana*. Thus, the unique *J. bermudiana* germplasm is gradually being depleted through hybridization and introgression. It seems likely that pure *J. bermudiana* germplasm will eventually be lost on Bermuda. However, it should be noted that *J. bermudiana* was introduced to St. Helena Island, south Atlantic Ocean before 1859 (Phillip Ashmole and Andrew Darlow, pers. comm.) so this would be the pre-insect invasion, *J. bermudiana* germplasm. It was later planted on nearby Ascension Island.

J. bermudiana, bark

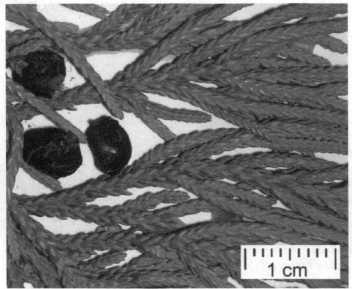

J. bermudiana, leaves and seed cones.

J. bermudiana, cultivated in cemetery
cf. *Adams 11080*

Author with the largest extant
J. bermudiana tree on Bermuda.
at Christine Watlington's house.

J. bermudiana, diseased tree at Church cemetery, cf. *Adams 11082*

Dead *J. bermudiana* trees on Bermuda

Juniperus blancoi

1a. Leaf tips acute to obtuse
   2a. Tree or shrubby tree; female cones not on the underside of planate
      branches; foliage not dense...***Juniperus blancoi*** var. ***blancoi***
   2b. Prostrate shrub forming patches, one to several meters in diameter; female
      cones on the underside of planate branches; foliage dense...
      ....***Juniperus blancoi*** var. ***huehuentensis***
1b. Leaf tips mucronate...***Juniperus blancoi*** var. ***mucronata***

Juniperus blancoi Martinez, Anales Inst. Biol. Univ. Nac. Mexico 17 (1):73 (1946). Tascate. Blanco juniper. Type: Mexico: Durango; Arroyo de Penuelas, El. Salto. C.E. Blanco A-500. (Holotype: MEXU!) named in honor of C. E. Blanco.

Dioecious. Trees Shrub to tree to 15 m, main trunk branching several meters above the base, crown very irregular. **Trunk bark** thin (to 5 mm), brown, of tightly appressed, interconnected, longitudinal strips. **Branches** terminal whip branches spreading to ascending, branch tips straight, bark dark gray-brown, angle of branching of ultimate twig 35-45 degrees. **Leaves,** scale leaves mostly opposite, oval or subelliptic, acuminate with acute to obtuse, appressed tips, 1.5-2.0 mm long, margin entire. **Seed cones** bilobed (reniform), dark bluish black, with a light coat of bloom, 5-7 (-9) mm diam. 3-6 mm long, peduncle straight. **Seeds** 2 (rarely 3, 4, or 5), irregular to subpyramidal, grooved, 2-5 mm long, 3-4 mm wide, brown; hilum about one-half length of seed. **Pollen shed** Jan.- March. **Habitat** along perennial streams and on the flood plains. **Uses** none known. **Dist.**: At the bottoms of arroyos, along stream beds in the pine-oak-juniper forest of Durango, at 2600-2900 m elevation; near El Salvador, state of Mexico. Recently discovered at the summit of Cerro Huehuento, Dgo. at 3227 m. Mexico. **Status**: VU, D2. Due to the rarity of this species it is probably wise to consider it threatened.

This species is very scattered in occurrence in Mexico. See the recent work of Mastretta-Yanes (2009) for detailed analyses of cpDNA. Observations by D. Reyes V. (Compania Maderera de Durango, El Salto, Durango) indicated that it is locally common in the vicinity of El Salto, Durango. Martinez (1946) considered this species to be allied to *J. jaliscana* in his subsection Jaliscanae. Zanoni and Adams (1975, 1976) and Adams (2000a) have shown that this is an unnatural alliance and that *J. blancoi* is related to the entire leaf margined junipers *J. mucronata* and *J. scopulorum*.

J. blancoi, bark

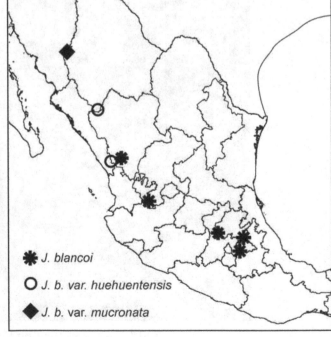

✳ *J. blancoi*

○ *J. b. var. huehuentensis*

◆ *J. b. var. mucronata*

Specimens of *Juniperus blancoi* Martinez var. *blancoi* examined:

Durango: El Salto, *E. C. Blanco A-500* (MEXU!, holotype); 3.2 mi S of town square of El Salto on logging road to Pueblo Nuevo, "tascate", along bottom of arroyo (8 plants only) with pine-oak-Juniper woods with *J. deppeana* var. *patoniana* and *J. deppeana* var. *robusta*, 26.12.1973, *T. Zanoni 2769* (MEXU); Arroyo de Las Adjuntas, 1½-2 mi N of Las Adjuntas (7.2 mi W of El Salto on Rt. 40), scattered in arroyo bottom of pine-*J. deppeana* var. *robusta* forest, 8800-9000 ft, 26.12.1973, *T. Zanoni 2775*; on the banks of a running stream, approx. 7 km from El Salto town square on road to Pueblo Nuevo, 23° 45.241'N, 105° 22.851' W, 2580 m,

9.5.2004, female, past pollination, *Juniperus blancoi*, leaves and reniform (bilobed) seed cones. *R. P. Adams 10257* (BAYLU, CIIDIR); same locality, male, pollen shed in spring, 9.5.2004, *R. P. Adams 10258* (BAYLU, CIIDIR); on the banks of a running stream. approx. 6 km from El Salto town square on road to Pueblo Nuevo, about 23° N, 105° W, 2580 m, 9.5.2004, *R. P. Adams 10259* (BAYLU, CIIDIR).

Michoacán: Ejido San Pedro Tarímbaro, mpio. Tlalpujahua, orilla de canal, 2800 m, 5.02.1993, tree 6-7 m, female, S. Zamudio 9008 (IEB, MEXU); same locality, árbol 6-8 m, male, S. Zamudio 9009 (IEB, MEXU).

Estado de México: El Oro, S of Carmona, R. P. Adams 6849, 6850, 6851, 6903, 6904 (BAYLU); 0.5 km N of El Salvador on road to Carmona; Adams 1486 (CS); Villa Victoria, en la desviación a Agangueo, mpio. San Felipe del Progreso, *Juniperus*, forest, tree 8-9 m, 40-60 cm DAP, 11.06.1992, E. Carranza 4099 (MEXU).

Tlaxcala: 3 km al W de San Felipe de Hidalgo, mpio. Nanacamilpa, *Pinus-Quercus* woods, 3000 m; tree 10-12 m, 16.11.1995, A. Cruz, C. Torres y M. Suarez s.n. (MEXU).

It should be noted that Zanoni and Adams (1979) cite *J. blancoi* from Sonora: Canon de Bavispe, White 3112 (ARIZ, MICH); Rancho de La Nacha, White 4093 (ARIZ, MICH). These specimens appear to be *J. blancoi* with some genes of J. *scopulorum* from past (Pleistocene?) hybridization.

J. blancoi at stream side El Salto, Durango, Mexico with Socorro Gonzales, *Adams 10258*

J. blancoi s of Carmona, Mexico.
cf. *Adams 6849-6904*

Juniperus blancoi Martinez var. **huehuentensis** R. P. Adams, S. González and M. González Elizondo. Biochem. Syst. Ecol.. 34: 207 (2006). Huehue juniper, subalpine blancoi juniper. Type: México. Durango: Cerro Huehuento, cima, al S de Huachichiles, 24°4'30 N, 105°44'22" W, *S. González 6832 with E. Lozano and M. Lozano* (holotype CIIDIR, isotypes BAYLU).

Dioecious shrub, 0.5-1.5 m x 2-4 m wide, brown bark exfoliating in thin plates; foliage yellowish green, branchlets in flattened sprays, the ends lax or pendulous; most female cones bi-lobed, dark blue and glaucous when ripe (1 yr.), on the underside of the branches; pollen shedding mid-March to mid-May.

Other specimens examined: Durango: common on rock at top of Cerro Huehuento, 24° 04.587'N, 105° 44.463' W, 3227 m, 8.5.2004, R. P. Adams 10247 (BAYLU, CIIDIR); same locality, R. P. Adams 10248, 10249, 10250, 10251 (BAYLU, CIIDIR).

Adams et al. (2006) examined morphology, leaf terpenoids and RAPDs for *J. blancoi, J. b.* var. *huehuentensis, J. mucronata* and *J. scopulorum*. The terpenoids of *J. blancoi* and *J. b.* var. *huehuentensis* were very similar. They differed principally in sabinene (25.5%, 39.0%), naphthalene (0.0, 0.5%), germacrene D (0.1%, 0.3%), γ-cadinene (0.0, 0.3%), germacrene B (<0.05%, 0.3%), germacrene D-4-ol (0.2%, 1.0%), γ-eudesmol (0.2%, 0.8%), shyobunol (0.3%, 0.0), manool (12.6%, 0.3%), sempervirol (0.2%, 0.0), and trans-totarol (0.1%, 0.0).

DNA sequencing (Adams et al., 2009b) of nrDNA, 4CL, ABI3 and petN-psbM, revealed that var. *huehuentensis* differed by 8 SNPs from *J. blancoi* (Fig. 5.2.1). Analysis of nrDNA plus 4 cp gene regions showed that var. *huehuentensis* differed by 3 SNPs from *J. blancoi* and 4 SNPs from var. *mucronata* (Fig. 1.16.5, ex. Adams and Schwarzbach, 2013d). These differences of 3 to 8 SNPs are in the range of differences found in many varieties of *Juniperus*.

The prostrate - shrubby habit of the subalpine form of *J. blancoi* might just be an environmentally induced, subalpine ecotype. However, the typical tree forms of *J. deppeana* var. *robusta* grows interspersed with the subalpine *J. blancoi* on the upper slopes of Cerro Huehuento. If the shrubby nature of the subalpine *J. blancoi* was due to frost damage to growing tips, it seems likely that the *J. deppeana* trees would also be shrubby. Furthermore, no individuals of the "typical" tree forms of *J. blancoi* were found on the lower slopes of Cerro Huehuento or nearby. The development of the female cones on the underside of the branches is perhaps an adaptation to protect the cones against the weather. Transplant studies are needed to ascertain if the shrubby type is due to the environment. The differences in DNA data are not due to environmental plasticity, but may be the result of selection to fit this harsh environment.

The DNA data did not support the separation of the subalpine *J. blancoi* taxon as a distinct species.   But the distinct morphology (seemingly not environmentally induced) of the plants from Cerro Huehuento clearly support their recognition as a variety of *J. blancoi*.

The new variety is very similar to *Juniperus blancoi* Martinez var. *blancoi* which is a dioecious tree or shrubby tree, 1.5-15 m tall, with one main stem 15-45(-60) cm diam., sometimes damaged by flood

water and branched close to the base, and then appearing to be a shrub; with brown bark, exfoliating in thin plates, reddish beneath, foliage lax or pendulous, branchlets ~1cm dia., scaly with bronze color beneath scales, ultimate branchlets generally planate; mature fruit with two lobes or globose (if one seeded), dark blue and glaucous upon maturity. *Juniperus blancoi* var. *blancoi* is known from Durango, Michoacán, the state of Mexico and Tlaxcala, forming small, isolated populations.

J. blancoi var. *huehuentensis* as a shrub near the summit of Cerro Huehuento, Durango, Mexico, cf. *Adams 10248*

Habitat of *J. blancoi* var. *huehuentensis* as a shrub near the summit of Cerro Huehuento, Durango, Mexico, cf. *Adams 10247*

Juniperus blancoi* var. *mucronata (R. P. Adams) Farjon, World Checklist & Bibliog. Conifers. ed. 2, 60 (2001)

J. mucronata R. P. Adams, Biochem. Syst. Ecol 28: 158 (2000). Mucronate tipped juniper.
Type: Mexico, Sonora, Maicoba, 1180m, 19 km w of Maicoba on MEX 16, at KM 307
Adams 8704 (holotype: SRCG, Note: SRCG was merged with BAYLU in 2003).

Dioecious. **Trees** with a strong central axis (to 15 m), crown pyramidal. **Trunk bark** brown, exfoliating in strips. **Heartwood** bright purple. **Branches** erect. **Leaves** both scale and decurrent (whip) leaves mucronate tipped, foliage pendulous. **Seed cones** female cones with soft, flesh pulp, globose to reniform (when 2 seeded), dark bluish black, with light white bloom, 5-8 mm wide x 4-6 mm long, mature in one year; peduncle straight. **Seeds** 1-2 (often one aborts, leaving only 1), brown, globose deeply grooved, 3-4 mm long, 2-3 mm wide, hilum scar from 1/3 to 1/2 length of seed. **Pollen shed** early spring? **Habitat** along stream banks. **Uses** none known. **Dist.**: E. Sonora, W. Chihuahua, Mexico. **Status**: This species appears to be rare and is likely subject to cutting for fuel and posts. It is likely to be threatened.

This taxon appeared to be distinct from *J. blancoi* and *J. scopulorum* (Adams, 2000a) in both its leaf terpenoids and DNA fingerprints. Farjon (2001) reduced *J. mucronata* to a variety of *J. blancoi*. A recent study involving sequencing nrDNA, petN-psbM, 4CL and ABI3 (4,182bp) supports the recognition of *J. blancoi* var. *mucronata* (Adams, 2009b). Notice (Fig. 5.1), that *J. scopulorum* is well resolved from *J. blancoi*, but *J. b.* var. *mucronata* is about as resolved (5 SNPs) as var. *huehuentensis* (8 bp). It is thought that *J. scopulorum* - like junipers at Colonia Pacheco and vicinity (north of Maicoba) are intermediate between *J. blancoi* var. *mucronata* and *J. scopulorum* due to past hybridization in the Pleistocene.

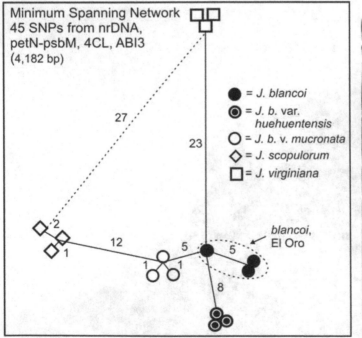

Fig. 5.1. Minimum spanning network (from Adams, 2009a). Notice the diversity among *J. blancoi - mucronata*.

J. blancoi var. *mucronata* bark exfoliating in thin, broad strips or scales.

J. blancoi var. *mucronata*
leaves and seed cones

J. blancoi var. *mucronata*,
along a running stream,
19 km w of Maicoba, Sonora,
Mexico, cf. *Adams 8701-8705*

J. blancoi var. *mucronata*,
along a stream just below
an oak savannah zone, 1180 m
19 km w of Maicoba, Sonora,
Mexico, cf. *Adams 8701-8705*

Juniperus brevifolia (Seub.) Antoine, Cupress. Gatt. 16. tt. 20-22 (1857). Common name short leafed juniper   Type: Azores, Flores, *C. F. F. Hochstetter 123* (holotype TUB?, n. v., isotypes K, P)
  J. oxycedrus L. var. *brevifolia* Seub., Fl. Azor.:26 (1844).
  J. rufescens Link var. *brevifolia* (Seub.) Endl., Syn. Conif.: 11 (1847).

Dioecious. Trees erect shrub or small tree with a broad pyramidal crown. **Trunk bark** brown-purple, exfoliating in strips. **Branches** erect. **Leaves** acicular, 3-10 x 1-2 mm, closely set, strongly curved, with two broad, white stomatiferous bands on adaxial surface. **Seed cones** 7 - 9 mm. ripening the second year, subglobose, green and pruinose while young, dark copper-brown when mature. **Seeds** 3 per seed cone, free, ovoid, triquetrous. **Pollen shed** Spring. **Habitat** wet, volcanic rock. **Uses** none known. **Dist**.: endemic Azores: Santa Maria, Sao Miguel, Terceira, Sao Jorge, Pico, Faial, Corvo and Flores Islands. Mountain slopes 250m to 1500 m. **Status**: EN, B1+2c. Although common in the Azores, it occupies a relatively small range and is subject to a catastrophic event.

        Although *J. brevifolia* has been considered as a variety of *J. oxycedrus* (*J. o.* var. *brevifolia* Seub.), neither terpenoids nor DNA sequencing has placed it with *J. oxycedrus* (Figs. 1.3, 1.11.1) but rather in a clade with *J. navicularis* from coastal Portugal. The Azores are of recent volcanic origin ranging for Santa Maria (8 my) and Sao Miguel (4 my) to very recent (2 my Terceira, Flores) to very recent (0.3 my Pico, 0.8 my Faial, 0.6 my Graciosa, 0.6 my Sao Jorge) origins (Riel et al., 2005). It seems probable that seeds of *J. navicularis*-like plants or their ancestor were brought to the Azores by birds from the Iberian peninsula. Subsequent differentiated in genetic isolation led to the speciation of *J. brevifolia* on the Azores.

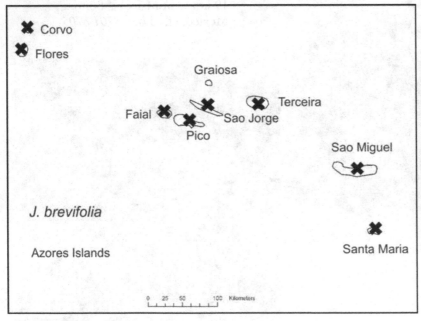

J. brevifolia, bark

J. brevifolia, leaves and seed cones

J. brevifolia, Azores, Pico Verde lookout, San Miguel Island, cf. Adams *8151-8154*

Juniperus californica Carriere, Rev. Hort. Ser. 4, 3: 352 (1854). California juniper, cedro. Type: Illustration in Rev. Hort., ser. 4. 3: 353, f. 21 (1854). Lectotype chosen by Farjon (p. 252, 2005).

    J. pyriformis Lindley A. Murray bis ex Lindl., Gard. Chron. 1855:420 (1855)

    Sabina californica (Carriere) Antoine, Cupress.-Gatt.: 52 (1857)

    J. cedrosiana Kellogg, Hesperian 4:3,4 (1860)

    J. cerrosianus Kellogg, Proc. Calif. Acad. Sci. 2:37 (1863)

    J. californica Carriere f. *lutheyana* J. T. Howell & Twisselm., Four Seasons 2(4): 16 (1968)

    J. occidentalis sensu Parl. *non* W. J. Hooker

Dioecious (rarely monoecious, 1.9%). Shrubs multi- (seldom one) stemmed shrub-tree, 2-8 m, with round crown. **Trunk bark** on twigs (5-10 mm diam.) brown or gray, not exfoliating in scales or flakes. **Branches**, ultimate branchlets approx. as wide as scale-leaf length; scale leaves closely appressed and generally flattened, branchlets terete. **Leaves** both whip and scale. Leaf glands conspicuous. Whip- and scale-leaf margins denticulate (20 X). **Seed cones** bluish-brown, white glaucous, reddish-brown beneath glaucous, (7-) 9-10 (-13) mm. Maturing in 1 yr. **Seeds** 1(2-3) per cone (avg. 1.3), 5-7 mm long. **Pollen shed** Jan. - March. **Habitat** Dry, rocky slopes and flats; 750-1600 m. **Uses** none known, possibly fence posts. **Dist.**: AZ, CA, NV; Baja California, Mexico. **Status**: common and expanding its range.

Two chemical (volatile leaf oils) races were described by Vasek and Scora (1967) and reconfirmed by Adams, von Rudloff and Hogge (1983). These two chemo-types were not found using the volatile wood oils (Adams, 1987). To date, no morphological character appears to be correlated with the chemical races. It is noteworthy that analyses of the leaf volatile oils of all the other 40 taxa of *Juniperus* in the western hemisphere has failed to uncover any other species with chemical races.

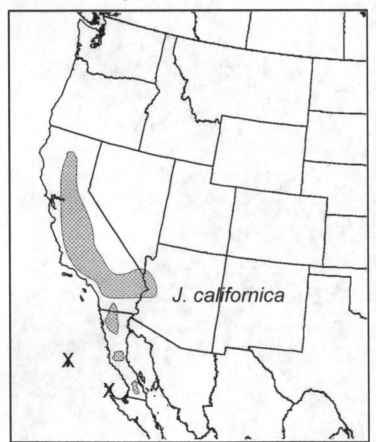

J. californica, bark

J. californica,
leaves and seed cones.

J. californica, southern California. Note that the seed cones are quite apparent due to large amount of waxy bloom that is typical of the species. cf. *Adams 8695-8697*

Juniperus carinata (Y. F. Yu & L.K. Fu) R. P. Adams, Biochem. Syst. Ecol. 28: 541 (2000). Keeled leaf juniper. Type: China, w Sichuan, Yalong River, Yajiang. 4460 m, T.S. Ying *3140*, (holotype PE!).
 Juniperus pingii W.C. Cheng ex Ferre var. *carinata* Y.F. Yu & L.K. Fu, Novon 7:443 (1997, publ. 1998).

Dioecious. **Shrubs** procumbent or erect to 4 m, rarely small trees; branchlets not pendulous, stout, usually not 6-angled. **Trunk bark** tan to purple, exfoliating in thin strips. **Branches** procumbent. **Leaves** both scale and decurrent, scale leaves somewhat decurrent, opposite causing the branchlets to be quadrangular, decurrent leaves often with a keel and mucronate tipped. **Seed cones** axillary, black when ripe, lustrous, ovoid or subglobose, 7-9 mm, 1-seeded, maturing in 2 yrs. **Seeds** ovoid or subglobose, 5-7 mm, with prominent resin pits, base rounded, apex obtuse. **Pollen shed** May – June. **Habitat** rocky areas near timberline. **Uses** none known. **Dist.**: W. Sichuan, China. 4460 m. **Status**: limited distribution, but under no likely threats at present.

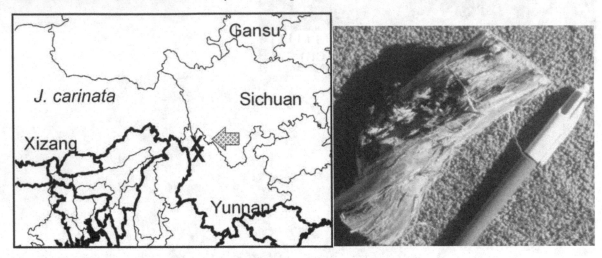

Distribution of J. carinata (presently incompletely unknown).

J. carinata, bark

J. carinata
leaves and seed cones

Juniperus carinata, at timberline, 4380 m, Hengduan Shan (Mts.) near White horse, in Yunnan, with Prof. Yang Fei, now deceased. Note the frequency of shrubby plants. cf. *Adams 8497-8499, 8502-8504*

Juniperus carinata, at timberline, 4380 m, Hengduan Shan (Mts.) near White Horse, Yunnan, cf. *Adams 8497-8499, 8502-8504*

Juniperus cedrus Webb et Berthelot, Hist. Nat. Iles Canaries 3(2), Phytogr. Canar. 3(89): 277 (1847). Canary Islands juniper; Cedro Canario. Type: Canary Islands, *P. B. Webb s. n.* (holotype FI, n. v.; isotype K).

J. webbii Carriere, Traite Gen. Conif.: 13 (1855)

Dioecious. **Trees** to 30 m , or shrub at high altitudes. **Trunk bark** orange-brown, peeling in coarse vertical strings. **Branches** level, upcurved toward tips, with pendulous branchlets. **Leaves** acicular, straight, in whorls of three, 1-2 mm wide, two stomatal bands either side of the midrib on the adaxial surface, adaxial stomatal band wider than midrib. **Seed cones** copper-red, not glaucous, globose, 8-12 mm, green when immature, mature in 2nd year. **Seeds** 3 per cone. **Pollen shed** fall. **Habitat** volcanic rocks and soils. **Uses** very valuable timber extensively harvested in the first 500 years of Spanish occupation of the Canary Islands. **Dist.**: endemic to the Canary Islands.: Gran Canaria, Tenerife, Gomera, and La Palma islands. **Status**: EN, B1+2c, D. This species has a limited distribution on islands that tend to be grazed by goats, so it may be endangered. It is very rare on Gran Canaria Island.

Recent DNA data (Adams et al. 2010b) revealed that *J. cedrus* is not present on Madeira. The taxon previously called *J. cedrus* in Madeira is actually a new species, *J. maderensis* (Figs. 1.3, 1.11.1).

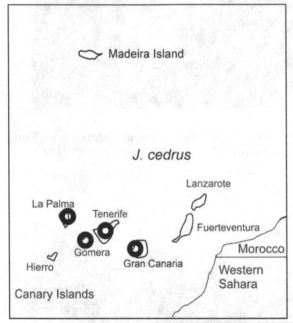

J. cedrus, bark

J. cedrus habitat, La Palma, Canary Islands

J. cedrus, leaves and seed cone
cf. *Adams 8127*

J. cedrus on volcanic rock, La Palma,
Canary Islands, cf. *Adams 11513*
The Canary Islands field trip was
arranged and led by Beatrice R. Ruiz
and her major professor, Dr. Manuel
Nogales Hidalgo

J. cedrus, Fortaleza, Tenerife, Canary Islands, cf. *Adams 11518-11522*.

Juniperus chinensis L.

Key to *J. chinensis* varieties in this treatment:
1a All leaves whip type (decurrent with tips free), no scale-like leaves, procumbent shrubs,
(1-) 2 - 3 seeds per cone, sea shores, Japan and Korea...................*J. chinensis* var. ***procumbens***
1b. Both scale-like and whip (decurrent with free tips) leaves present; whip-leaves
interspersed among sections of scale-leaves in mature branchlets, (1-) 2 seeds per cone
or almost all leaves scale-like, and a decumbent shrub.
2a. Trees; decurrent leaves in whorls of 3, loosely arranged, 8-12 mm .*J. chinensis* var. ***chinensis***
....
2b. Shrubs procumbent; decurrent leaves decussate or in whorls of 3; densely arranged, 3-6 mm.
....*J. chinensis* var. ***sargentii***

Juniperus chinensis L. var. ***chinensis*** Mantl. Pl. 1:127 (1767). Chinese Juniper. Type: Sweden
(cultivated in Uppsala Bot. Garden, Sweden), *leg. ign.* LINN 1198.3 (lectotype LINN).
    J. barbadensis Thunb., Fl. Japon: 264 (1784), *non* L. (1753)
    J. virginica Thunb., Fl. Japon: 264 (1784);   *J. sinensis* J. F. Gmel., Syst. Nat. 2(2): 1005 (1791)
    J. cernua Roxb., Fl. Ind. 3: 839 (1832);   *J. dimorpha* Roxb., Fl. Ind. 3: 389 (1832).
    J. flagelliformis hort. ex Loudon, Encycl. Trees: 1090 (1842).
    J. thunbergii Hook. & Arn., Bot. Beechey Voy. 6: 271 (1838)
    J. sphaerica Lindl., Paxton's Fl. Gard. 1: 58, f. 35 (1851)
    J. fortunei hort. ex Carriere, Traite Gen. Conif.: 11 (1855) [ed. 2: 38 (1867)]
    J. cabiancae Vis., Mem. Reale Ist. Veneto Sc. 6: 246, t. 1 (1856)
    J. chinensis L. var. *pendula* Franch., Nouv. Arch. Mus. Hist. Nat., ser. 2, 7: 101 (1884)
    Sabina chinensis (L.) Antoine, Cupress. Gatt.: 54 (1857)
    S. dimorpha (Roxb.) Antoine, Cupress.-Gatt.: 70 (1857)
    S. cabiancae (Vis.) Antoine, Cupress.-Gatt.: 41 (1857)
    S. sphaerica (Lindl.) Antoine, Cupress.-Gatt.: 52 (1857)

Dioecious, rarely monoecious. Shrubs or trees to 25 m. crown pyramidal to open, broad and
irregular. **Trunk bark** exfoliating in narrow strips, brown to cinnamon. **Branches** spreading or
torrulose (in cv. Kaizuka or Hollywood juniper). **Leaves** both scale and decurrent (whip like) often
interspersed on a single branch. The mosaic of scale and decurrent leaves is characteristic of the
species, decussate or in whorls of 3, decurrent leaves (3-) 6-12 mm, with 2 white stomatal bands
adaxially, scale leaves decussate, closely appressed, 1.5 – 3 mm, abaxial gland elliptical, sunken. **Seed
cones** brown when mature in 1 yr., usually glaucous, subglobose, 4-9 mm in diam. **Seeds** (1-) 2-3 (-4) seeds per cone., ovoid, slightly flattened. **Pollen Shed** spring. **Habitat** rocky seashores (Japan), rocky areas 1400 – 2300 m in China. **Uses** unknown **Dist.:** eastern China, Japan. widely cultivated. **Status**: may be threatened as a natural tree in mainland China, but due to widespread cultivation, it is not endangered.

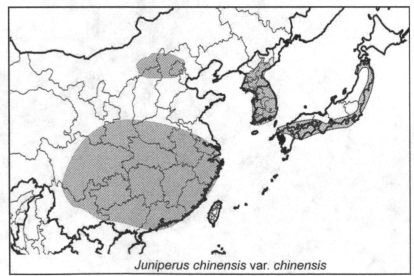

Juniperus chinensis var. chinensis

J. chinensis var. *chinensis* bark, Japan

J. chinensis var. *chinensis*, leaves and seed cones.

J. chinensis var. *chinensis*, Shizuoka Pref., Japan, cf. *Adams 8535-8537*

J. chinensis var. *chinensis* growing along a rocky beach in Japan. cf. *Adams 8535-8537*

Juniperus chinensis var. **sargentii** A. Henry in Elwes & Henry, Trees of Great Brit. Ireland 6: 1432 (1912). Sargent's juniper, yan bai (Chin.), Miyama-byakushin (Jap.). Type: Japan, Hokkaido, Kitami Prov., Rebun-jima. *M. Furuse 9385* (neotype K, designated by Farjon (p. 259, 2005), orig. mat. not found by Farjon).

    J. procumbens Sarg. For. Fl. Japan, 78 (1894)
    J. sargentii (A, Henry) Takeda ex Nakai, Bot. Mag. (Tokyo) 44: 511 (1930)
    Sabina sargentii (A. Henry) Miyabe & Tatewaki, Trans. Sapporo Nat. Hist. Soc. 15: 128 (1938)
    S. pacifica Nakai, Chosen Sanrin Kaiho 163: 31 (1938)
    S. chinensis (L.) Antoine var. *sargentii* (A. Henry) W. C. Cheng & L.K. Fu, Fl. Reipubl. Pop. Sin.
       7: 363 (1978)

Dioecious, rarely monoecious. **Shrubs** trailing with a long stem and a dense mass of lateral branches. **Trunk bark** brown, exfoliating in thin plates. **Branches** ascending. **Leaves** scale and decurrent, light green or glaucous, decurrent leaves lanceolate, convex on the lower side, stiffly pointing toward the tips of branchlets; scale-like leaves closely approximate, obtusish, glandular; fruit globose, dark blue or black, 5-7 mm long. **Seed cones** brown, glaucous, subglobose. **Seeds** 2 or 3, rarely 4 or 5 oblong, subtrigonous, glossy, brown. **Pollen shed** spring? **Habitat** seaside rocks all along the coast in the S part of Sakhalin; also collected on the Sichuan River on the rocks of the Chandalaz Ridge. **Uses** recommended for cultivation as a soil cover, commonly cultivated for landscaping. **Dist.**: China (N.E.); Japan; Korea; USSR: Sakhalin. **Status**: Little is known about the abundance of this taxon by the author.

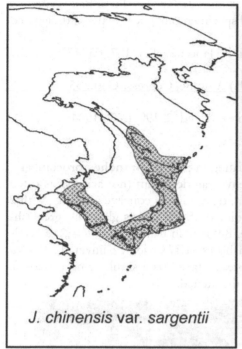

J. chinensis var. sargentii

J. chinensis var. *sargentii*, with scale and whip leaves on the same branchlets.

J. chinensis var. *sargentii* as a cultivated shrub

Juniperus chinensis var. ***procumbens*** Siebold ex Endl., Syn. Conif. 21 (1847). Pu di Bai. Type: Japan, *P. F. von Siebold s. n.* [ex Herb. Bentham 1854, right-hand specimen only] (lectotype K, designated by Farjon (p. 352, 2005).

> *J. procumbens* (Sieb. ex Endl.) Miquel in Siebold & Zuccarini, Fl. Japon 2: 59, t. 127, f.3 (1870).
>
> *J. japonica* hort. ex Carriere, Traite Gen. Conif.: 33 (1855)
>
> *Sabina chinensis* (L.) Antoine var. *procumbens* (Siebold ex Endl.) Antoine, Cupress.-Gatt.: 55 (1857)
>
> *S. procumbens* (Siebold ex Endl.) Iwata & Kusaka, Conif. Japon. Ill., ed. 2: 199, t. 79 (1954)
>
> *J. recurva* D. Don var. *squamata* Mast., non Parlat

Dioecious. Shrubs procumbent. **Trunk bark** brown, exfoliating in plates. **Branches** procumbent, much elongated, leading shoots and branchlets ascending. **Leaves** all decurrent (not acicular as often incorrectly reported), in whorls of 3, unequal in length, 6-8 mm, rigid, slightly concave adaxially, with 2 white stomatal bands adaxially, base decurrent apex sharply pointed. **Seed cones** glaucous, greenish-tan when immature, glaucous, black when mature, subglobose, 8-9 mm in diam. **Seeds** (1-)2 -3 seeds per cone, ca. 4 mm, ridged. **Pollen shed** spring? **Habitat** seashores. **Uses** widely cultivated, cv. *nana* is very common. **Dist.**: s. coast Japan, s. and w. coasts of Korea. **Status**: as a shrub of no economic importance and with some seashore habitat protected, it is not threatened.

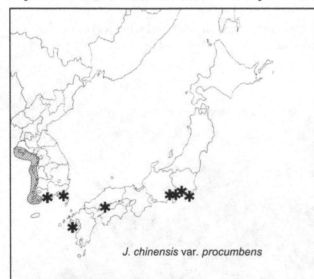

J. chinensis var. procumbens

J. chinensis var. *procumbens* cv. *nana*, Royal Bot. Gardens, Kew, UK

J. chinensis var. *procumbens* leaves

1 cm

Juniperus coahuilensis (Martinez) Gaussen ex R. P. Adams, Phytologia 74: 413 (1993). Rose fruited juniper. Type: Mexico, Coahuila, Sierra de los Hechiceros, Canon de la Madera, *I. M. Johnston (with C. H. Muller) 1290* (holotype MEXU!, isotypes GH, NA, TENN, TEX)

    J. erythrocarpa var. *coahuilensis* Martinez
    J. pinchotii var. *erythrocarpa* (Cory) J. Silba
    J. erythrocarpa Cory. Rhodora 38:186

Dioecious. **Trees** large shrub to small tree, 3-8 m, often with a single stem to 1 m, when shrubs, branched at the base, with flattened-globular or irregular crowns. **Trunk bark** brown, thin, exfoliating in long ragged strips. **Branches** ascending to erect in shrubs, spreading in trees. Branch bark scaly, ashy-gray. Stumps sprouting after burning or cutting. **Leaves** both whip and scale. Whip- and scale-leaf margins denticulate (20 X), white glaucous on adaxial leaf surface. At least ¼ or more of the whip leaf glands with a white crystalline exudate. **Seed cones** rose to pinkish but yellow-orange, orange or dark red beneath the white-blue glaucous, soft and juicy, globose to ovate, 6-7 mm, 1(-2) seeded, the hilum scar pale brown, approx. ½ as long as seed. **Seeds** 4-5 mm long. **Pollen shed** late fall - early winter. **Habitat** *Bouteloua* grasslands and adjacent rocky slopes. **Uses** fence posts. Sprouts from cut stumps and is thus a pest in grasslands. **Dist.**: 980-1600 (-2200) m, trans-Pecos Texas, common in northern Mexico around the margins of the Chihuahuan Desert. **Status**: abundant and increasing.

    Hybridization between *J. coahuilensis (erythrocarpa)* and *J. monosperma* appears likely in Arizona (see *J. monosperma* above). Hybridization between *J. coahuilensis (erythrocarpa)* and *J. pinchotii* occurs in the Big Bend Natl. Pk., Tex. (Kistler, 1976) and possibly near Saltillo, Mexico. Previous reports of hybridization with *J. ashei* have been negated (see *J. ashei* above).

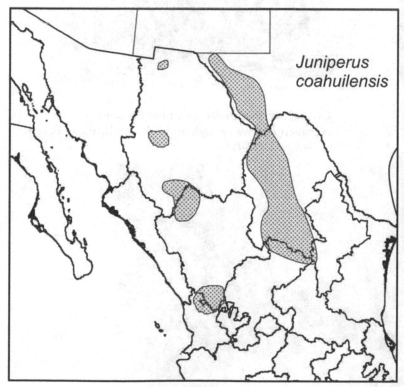

Juniperus coahuilensis

J. coahuilensis, bark

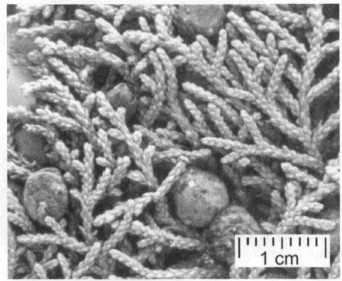

J. coahuilensis
leaves and seed cones

1 cm

J. coahuilensis var. *coahuilensis*, north of Durango, Mexico in *Bouteloua* grassland, cf. *Adams 6829-6831*

J. coahuilensis var. *coahuilensis*, near Presidio, TX in degraded *Bouteloua* grassland in the Chihuahuan desert. cf. *Adams 4988-4994*

J. coahuilensis habitat at Hueco Tanks, Texas in the Chihuahuan desert.

Juniperus comitana Martinez. Anal. Inst. Biol. Mexico 15:12, 13. (1944). Cedro; cipres; sicop (Baja Verapaz, Guatemala); bac'iul nuhkupat (Tzeltal at Tenejapa, Chiapas); Comitan Juniper. Type: Mexico: Chiapas: 12 km S. of Comitan. *M. Martinez 6700* (holotype: MEXU!).

Dioecious. **Trees** (to 10 m), single-stemmed, branching several meters above the base, crown rounded or broadly conic. **Trunk bark** about 5 mm thick, light ashy-brown, of ragged longitudinal strips. **Branches** terminal whip branches ascending to erect, branch tips straight, reddish-brown with scaly bark; angle of branching of ultimate twig 35-45 degrees. **Leaves** decurrent and scale, scale leaves mostly opposite, ovate, with acuminate, appressed tips, 1.5-2 mm long, margin finely toothed, green; foliage very slender (ultimate twig diam. less than 1 mm), with very weak odor from unusually low content of aromatic (terpenoid) compounds. **Seed cones** globose, glaucous, bluish, mature in 1 yr. **Seeds** 1(2) per cone. **Pollen shed** Nov. – Dec. **Habitat** on limestone hills, in pine-oak forests, and in *Ficus-Acacia* forests in the mountains of Chiapas, Mexico. On dry rocky hills in the pine-oak forest of Depto. Baja Verapaz, in the Sierra de los Cuchumatanes of Depto. Huehuetenango and on dry dolomitic slopes and bluffs of Depto. Zacapa, Guatemala. **Uses** none known. **Dist**.: Mexico/Guatemala border, 1300-2300 m elevation. **Status**: VU, B1+2c. This taxon occurs in a limited region that is likely to be cleared for agriculture, so it is vulnerable.

J. comitana, 14 km s of Comitan, Chiapas, Mexico. cf. *Adams 6858-6862*

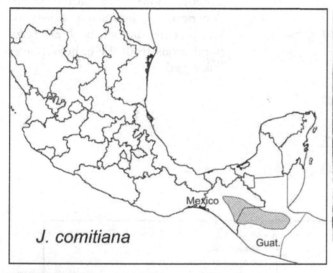

1 cm

J. comitana with bark in narrow strips.

Juniperus communis L.

A study of Arctic populations of *J. communis* (Adams et al., 2003), revealed (Figure 5.2) that these Arctic populations clustered by continent with the populations in Greenland and Iceland showing the highest affinities to populations from Europe, not those from North American. The North American populations were all *J. c.* var. *depressa*, whereas the eastern hemisphere populations included *J. c.* var. *communis* (CC), *J. c.* var. *saxatilis* (GR, IC, SW, UR, KA). Adams et al. (2003) concluded that the post-Pleistocene populations on Greenland and Iceland came from Europe and not North America.

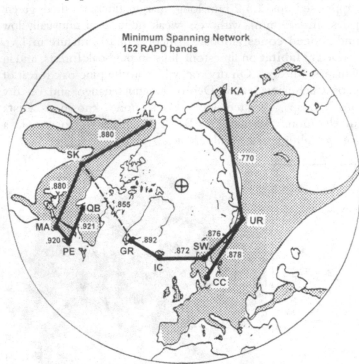

Figure 5.2. Minimum spanning network showing that all the North American *J. communis* populations link together and all the *J. communis* populations from the e. hemisphere link together.

Analysis of most *Juniperus communis* varieties (Adams and Pandey, 2003), resolved these taxa (Fig. 5.3) into six major groups: *J. communis* from Europe and central Asia (*J. communis* L. var. *communis*, *J. c.* var. *depressa* Pursh, N. America; *J. c.* var. *saxatilis* Pall.); *J. c.* var. *megistocarpa* Fern. & St. John, Quebec; *J. c.* var. *nipponica* (Maxim.) E. H. Wilson, Japan; and putative *J. c.* var. *saxatilis* Kamchatka, Russia. However, Adams and Pandey (2003) did not include *J. c.* var. *jackii*, nor putative *J. c.* var. *saxatilis* from the Pacific the Pacific northwest, USA/Canada in their analysis.

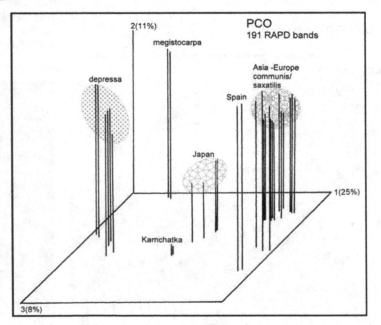

Figure 5.3 PCO based on 191 RAPD bands.
Adapted from Adams and Pandey (2003).

Ashworth, et al. (1999, 2001) used DNA fingerprinting to examine *J. communis* plants identified as *J. c.* var. *depressa*, *J. c.* var. *jackii* Rehder, *J. c.* var. *montana* Aiton (= *J .c.* var. *saxatilis* Pall. see Adams, 2004) collected from California, Oregon, Nevada or Utah in the southwest and west coast of the United States. They did not get a clear pattern separating these taxa, and concluded that their samples represent a single varietal taxon. However, it not clear if they utilized population samples to remove spurious variation in RAPD bands.

The major trend (figure 5.4) among the taxa is the separation of the eastern hemisphere plants (*J. communis* var. *communis*, *J. c.* var. *saxatilis*, and putative *J. c.* var. *saxatilis*, Kamchatka) from the western hemisphere plants (*J. c.* var. *depressa*, *J. c.* var. *jackii*, *J. c.* var. *megistocarpa*, and putative var. *saxatilis*). The resolution (figure 5.4) of *J. c.* var. *jackii* (and plants from nearby Mt. Hood) is in contrast to the report by Ashworth, et al. (1999, 2001). The Banff, Alberta individuals (putative hybrids) are intermediate between the coastal, short, curved leaved plants (Queen Charlotte Islands plants, var. *jackii*) and *J. c.* var. *depressa* (figure 4). *Juniperus c.* var. *megistocarpa* is distinct from *J. c.* var. *depressa*.

The most interesting facet of this PCO is that Queen Charlotte Islands (*J. c.* var. *charlottensis*) and *J. c.* var. *jackii* plants do not cluster with *J. c.* var. *saxatilis* (Norway, mountain, figure 5.4). It appears that the short, curved leafed taxon from the Pacific northwest thence into Alaska is part of several taxa (*J. communis* var. *charlottensis*, *J. c.* var. *jackii* and *J. c.* var. 'saxatilis' (now var. *kelleyi* in N. American).

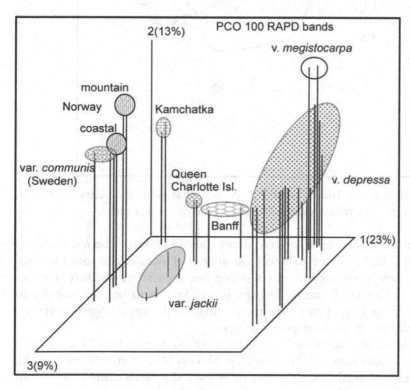

Figure 5.4. PCO based on 100 RAPD bands. See text for discussion.
Adapted from Adams and Nguyen (2007).

It is possible that the Alaska (A) population was not glaciated during the Wisconsin (figure 5.5), but all the other northern populations were glaciated. Only the Alaska (D), Mt. Charleston

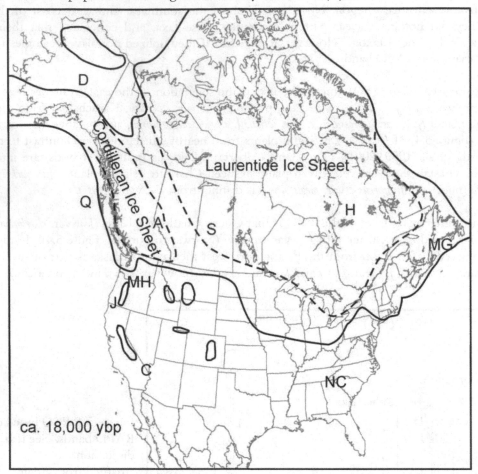

Figure 5.5. Maximal ice cover during the late Wisconsin. Notice that only the Alaska (D), Mt. Charleston (C) and North Carolina (NC) populations may have survived *in situ* or nearby.

(C) and North Carolina (NC) populations may have survived during the Wisconsin. The *J. c.* var. *jackii* (now *J. jackii*) populations (J, MH) likely moved to lower elevations. However, the northwestern California population of *J. c.* var. *jackii* presently occurs on serpentine, so it seems unlikely that this edaphic type grew on serpentine at a lower elevation. *Juniperus* is well known to be very adaptive to edaphic conditions, so Wisconsin era genotypes may have merely invaded the largely open habitat on the serpentine of northwestern California and southwestern Oregon.

There seem to be four possible refugia during the Wisconsin: southern Appalachian Mts. (cf. NC); southern Rocky Mountains (cf. Mt. Charleston and Arizona/New Mexico Mts.; central Sierra Nevada; and possibly an ice free corridor in central Alaska. It is easy to imagine that birds carried seeds from plants from the southern Appalachians northward into northern US and Canada. It appears more likely that the southern Appalachians were the source of germplasm in re-colonization of Canada than the southern Rocky Mountains.

Recently the North American *J. communis* taxa have been examined by use of Single Nucleotide Polymorphisms (SNPs) of nr DNA (Adams, 2008). Analyses of 1119 bp of nrDNA (ITS) sequences revealed 23 SNPs among the taxa including 2 and 4 bp deletions in *J. c.* var. *communis* and var. *saxatilis* from Norway as well as *J. c.* var. *jackii* (NW CA and Mt. Hood, OR). The 2 bp and 4 bp deletions were each coded as single deletion events in making comparisons.

Factoring the association matrix yielded five eigenroots before they began to asymptote, implying that six groups were present. These eigenroots accounted for 48.9, 15.8, 12.0, 6.9, and 5.7% of the variation among the 58 individuals analyzed. PCO of the eigenvectors (Fig. 5.6), clearly shows that the differentiation of *J. jackii* from NW CA and Mt. Hood, OR accounts for 49% of the variance among the 58 individuals. *Juniperus c.* var. *jackii* had 5 bp differences, plus a 4 bp deletion and a 2 bp deletion. Interestingly, the 4 bp and 2 bp deletions were shared with *J. c.* vars. *communis* and *saxatilis* from Norway. The Queen Charlotte Island junipers are separated by 2 bp (Fig. 5.6). These junipers grow in muskeg bogs that are very atypical of *J. communis*. An individual plant of *J. communis* found growing on a sand dune on Whidbey island has 3 mutations that separate it from all other individuals (Fig. 5.6). The leaves and habit of the plant are similar to *J. c.* var. *depressa*. All of the other samples (27 individuals) of *J. c.* var. *depressa* (N. A.) and v. *saxatilis* (Asia) form a fifth group.

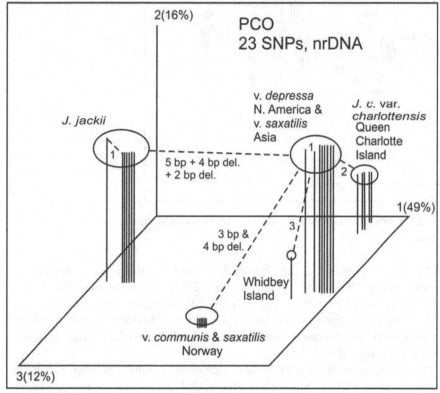

Figure 5.6. PCO of 58 individuals based on 23 SNPs. Identical bars, closely spaced indicate no variation among these individuals. The bars in the largest group are symbolic, as that group contains 27 individuals! From Adams (2008).

Because six groups were indicated by the presence of five significant eigenroots, additional variation is hidden in the PCO in figure 5.6. Therefore, the *J. jackii* individuals (NW CA and Mt. Hood, OR) were removed and the PCO re-run. This resulted in four eigenroots accounting for 35.5, 23.7, 13.4 and 11.0% of the variance before asymptoting. This implies the presence of five groups as seen in figure 5.7. Each of the groups are separated by 2 - 3 bp (plus 4 bp and 2 bp deletion in the Norway group). The large group (27 individuals) of *J. c.* var. *depressa* and *J. c.* var. *saxatilis* is relative uniform, with only 1 bp difference between members of that group. All of the individuals that morphologically appear to be *J. c.* var. *saxatilis* (short, curved leaves, with a stomatal band that is twice as wide the green leaf margins) are within the large group. Therefore, the addition of *'saxatilis'* (J. c. var. *kelleyi*) plants from NW North America to this analysis did not change the PCO in comparison with the RAPDs data.

However, it is interesting that *J. c.* var. *saxatilis* from Japan is within this group, in contrast to a previous RAPDs study (Adams and Nguyen, 2007). Differentiation in the nrDNA sequence was not as great as found in RAPDs. However, the pattern of considerable differentiation in the RAPDs of the *J. jackii* and Queen Charlotte Islands (*J. c.* var. *charlottensis*) plants (Fig. 5.4) is clearly concurrent with the nrDNA (Figs. 5.6, 5.7).

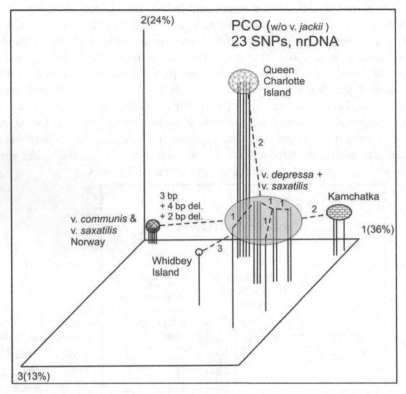

Figure 5.7. PCO of 23 SNPs of nrDNA without *J. c.* var. *jackii* individuals.
From Adams (2008).

Adams and Nguyen (2007) recognized *J. c.* var. *depressa*, *J. c.* var. *jackii* and *J. c.* var. *megistocarpa* based on leaves, female cones, and RAPDs data. Later, Adams (2008) showned that the Queen Charlotte Islands junipers are distinct in their nrDNA (Fig. 5.7) from the junipers on the mainland of British Columbia and recognized the plants growing on muskeg on Queen Charlotte Islands (Graham Island) (and elsewhere) as a new variety: ***Juniperus communis*** var. ***charlottensis*** R. P. Adams. These junipers grow in a muskeg bog that is atypical of *J. communis*.

A large study using DNA sequences from nrDNA and 5 cp gene regions (6862 bp) involving all the taxa of *Juniperus* section *Juniperus* (Adams and Schwarzbach, 2012a) as well as 14 *J. communis* infraspecific taxa. *Juniperus jackii*, which is morphologically quite similar to *J. communis* var. *kelleyi*, comes out in a clade with *J. mairei* from China. The Kamchatka juniper (*J. c.* var. *kamchatkensis*) is in an outlying position. *J. c.* var. *kelleyi* (formerly treated as *J. c.* var. *saxatilis*) is clearly in the clade with *J. communis* from North America, not with typical *J. c.* var. *saxatilis* from Europe (Figs. 5.8, 5.9).

Juniperus jackii was found to be in a clade with *J. mairei* from China, suggesting an ancient migration pathway via the BLB (Bering Land Bridge).

The distributions of *J. communis* and *J. jackii* in North America are shown in Fig. 5.10. Recently, I have analyzed plants from timberline on Hurricane Ridge, Olympic Park, WA and found these to be *J. jackii* rather than *J. c.* var. *kelleyi* (old var. *saxatilis*). Thus, the range of *J. jackii* has been extended from Oregon to Washington. Due to the only slight differences in morphology, DNA sequencing of petN-psbM is an efficient method to identify *J. jackii* as its petN-psbM region is 836 bp vs. 718 bp in *J. c.* var. *kelleyi* (and *J. c.* var. *charlottensis*, and *J. c.* var. *depressa*). It seems likely that *J. jackii* will prove to be more widespread that originally thought (Adams, 2011).

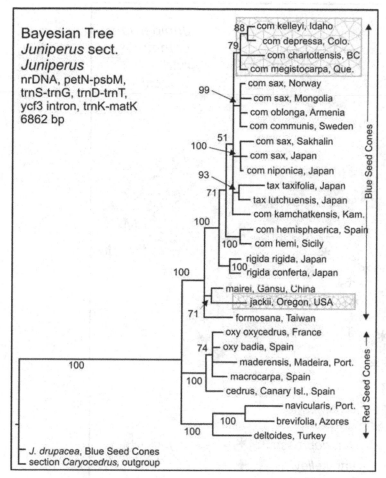

Bayesian Tree
Juniperus sect.
Juniperus
nrDNA, petN-psbM,
trnS-trnG, trnD-trnT,
ycf3 intron, trnK-matK
6862 bp

Figure 5.8. Bayesian tree of *Juniperus* sect. *Juniperus* taxa. Notice that *communis* taxa of North America are in a clade. The Kamchatka juniper (*J. c.* var. *kamchatkensis*) is in an outlying position. Note that *J. jackii* is in a clade with *J. mairei*, Gansu, China.

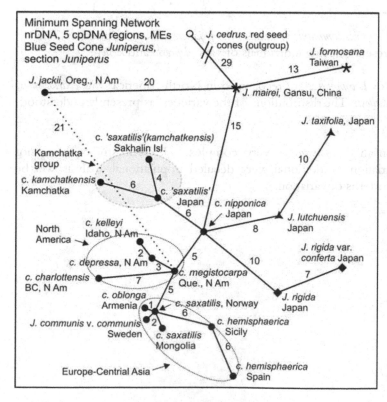

Figure 5.9. Minimum spanning network of the blue seed cone *Juniperus* sect. *Juniperus*. Note the positions of *J. jackii*, *J. c.* var. *kamchatkensis*, and *J. c.* var. *kelleyi*.

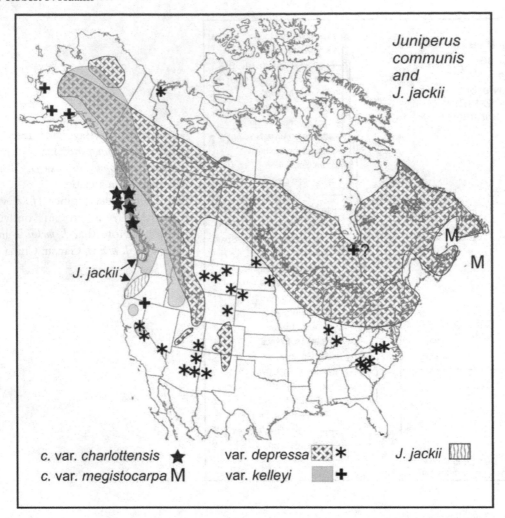

Figure 5.10. Distribution of *Juniperus communis* varieties plus *J. jackii* in North America. The asterisk and plus symbols represent outlying individuals of vars. *depressa* and *kelleyi*.

At present, four (4) varieties of *J. communis* are recognized in North America: var. *charlottensis*, var. *depressa*, var. *megistocarpa*, and var. *kelleyi*. The distributions of the varieties, as presently understood, are shown in figure 5.10.

Clearly infraspecific variation in *J. commuis* is very complex. Present studies have not completely resolved the complex variation. Additional, very detailed populational analysis will be needed to more fully understand the patterns of variation.

Key to *J. communis* varieties and *J. jackii*.

Eastern Hemisphere:

1a. tree or upright shrub, leaves 15 – 20 (25) mm, straight (not curved), acuminate – subulate, restricted to the eastern hemisphere ....……….....................….........…..........…..var. **communis**

1b. prostrate or small shrub, leaves 8 -15 mm, curved (upturned), closely set, linear to linear-oblong, acuminate to subulate,

  2a. stomatal band approx. 2 or more times as wide as each green leaf margin, leaves not deeply concave, curved to straight

    3a. bushy shrub to 2.5 m, leaves mostly straight, 4-12 (20) x 1.3 - 2 mm.......var. **hemispherica**

    3b. low spreading mat-like shrub, leaves upturned or upcurved, sometimes almost imbricate, 15 x 2 mm...var. **saxatilis**

  2b. stomatal band approx. 0.25 to 1.5 times as wide as each green leaf margin,

    4a. stomatal band approx. 0.25 to 0.5 times as wide as each green leaf margin, leaves very concave, nearly folded, restricted to the far east: Korea, Japan......................var. **nipponica**

    4b. stomatal band approx. 1 to 1.5 times as wide as each green leaf margin, leaves flat to somewhat concave, restricted to the far east: Kamchatka peninsula, Sakhalin Island ....var. **kamchatkensis**

Western Hemisphere:

  4a. seed cones 10 – 13 mm diam., much larger than leaf length, known only from southeastern Canada……..…….............var. **megistocarpa**

  4b. seed cones 6 – 9 mm diam., smaller or about equal to leaf length, other than se Canada

    5a. glaucous stomatal band about as wide to 1.5 x as wide as each green leaf margin, prostrate or low shrub with ascending branchlet tips (or occasionally a spreading shrub), leaves upturned, rarely spreading, linear to curved, rarely a small, strict tree (to 2-4 m) in the New England...……....................…......var. **depressa**

    5b. glaucous stomatal band twice or more as wide as each green leaf margin, upright shrubs or spreading, mat-like shrubs

      6a. glaucous stomatal band twice or more as wide as each green leaf margin, boat-shaped, curved leaves, mature seed cones length greater than leaf length, spreading, mat-like shrub, grows in muskeg bogs, Calvert Island to Queen Charlotte Islands..var. **charlottensis**

      6b. glaucous stomatal band 3 to 4 times as wide as each green leaf margin,

        7a. immature seed cones elongated-subglobose, leaves curved, boat-shaped, appressed to stem or leaf above on branchlet, gland on brown sheath long, narrow, raised, usually prostrate or mat-like on serpentine or ultramafic rock (some volcanic rocks, rarely on granite..***J. jackii***

        7b. immature seed cones globose, leaves most straight to slightly curved, not usually boat-shaped, free (not appressed to stem or leaf above on branchlet), gland on brown sheath elongated oval to long narrow gland, with rounded bottom end, usually shrubs to 0.5 m tall with upturned to elevated branchlets..var. **kelleyi**

Table 3. Comparison of the leaf morphology of *J. communis* var. *kelleyi*, *J. c.* var. *depressa* and *J. jackii*.

	J. c. var. *kelleyi*	*J. c.* var. *depressa*	*J. jackii*
Stomatal band width vs. green leaf margin (GM)	1.5 - 2x GM	1-1.5x GM	2-4x GM
Leaf cross-section	concave	very concave	concave,
Leaf shape	curved	straight	curved, boat shaped
Leaf blades	free, 30° to 80°	free, 45° to 20°	mostly appressed to stem
Mature seed cones vs. leaf length	cones about as long as leaves	cones much shorter than leaves	cones as long as or longer than leaves
Seed cone shape	ovoid	ovoid	elongated oval (ellisoid) esp. in immature cones

Juniperus communis L. var. ***communis*** Sp. Pl. 2: 1040 (1753). Common juniper. Type: (Europe, Alps?), leg. ign., (lectotype BM-HSC, see Jarvis et al., 1993).

    J. suecica Mill., Gard. Dict., ed. 8: *Juniperus* No. 2 (1768)

    J. communis L. var. *vulgaris* Aiton, Hort. Kew. 3:414 (1789)

    J. communis L. var. *suecica* (Mill.) Aiton, Hort. Kew. 3:414 (1789)

    J. communis L. var. *montana* Neilr., Fl. Nieder-Osterreich: 227 (1859), *non* Aiton (1789)

    J. difformis Gilib., Exerc. Phytol. 2:216 (1792)

    J. borealis Salisb., Prodr. Chap. Allerton: 397 (1796)

    J. communis L. var. *typica* Fomin in Jaczewski, Mel. Biol Borodine Jubile: 363 (1927), *nom inval.*, Art. 24:3

    J. communis L. var. *erecta* Pursh, Fl. Amer. Sept. 2:646 (1814)

    J. communis L. var. *aborescens* Gaudin, Fl. Helvet. 6:301 (1830)

    J. communis L. var. *oblonga* hort. ex Loudon, Arbor. Frut. Brit. 4:2490 (1838)

    J. occidentalis hort. ex Carriere, Traite Gen. Conif.: 54 (1855), *non* Hook. (1840)

    J. communis L. var. *hispanica* Endl., Syn. Conif.: 15 (1847)

    J. communis L. var. *stricta* Carriere, Traite Gen. Conif.: 22 (1855)

    J. microphylla Antoine, Cupress.-Gatt.: s.n., t. 31-32 (1857)

    J. communis L. subsp. *eucommunis* Syme in Sowerby, Engl. Bot., ed. 3, 8:273, t. 1382 (1868)

    J. communis L. var. *fastigiata* Parl. in Candolle, Prodr. 16(2):479 (1868)

    J. cracovia hort. ex K. Koch, Dendrol. 2(2): 115 (1873)

    J. x kanitzii Csato, Magyar Novenyt. Lapok 10:145 (1886)

    J. communis L. var. *genuina* Formanek, Kvet. Morav. Slezska 1:66 (1887)

    J. vulgaris Tragus ex Bubani, Fl. Pyren. 1:45 (1897)

    J. communis L. var. *brevifolia* Sanio in Hegii, Ill. Fl. Mitteleuropa 1:91 (1907)

    J. niemannii E. L. Wolff, Bot. Mater. Gerb. Glavn. Bot. Sada RSFSR 3:37 (1922)

    J. communis L. subsp. *brevifolia* (Sanio) Penzes, Bot. Kozlem. 57 (1):49 (1970)

    J. albanica Penzes, Bot. Kozlem. 57 (1): 49 (1970)

    J. communis L. subsp. *cupressiformis* Vict. & Sennen ex Penzes, Bot. Kozlem. 57(1): 49 (1970)

    J. communis L. subsp. *pannonica* Penzes, Bot. Kozlem. 57 (1): 48 (1970)

    J. communis L. var. *pannonica* (Penzes) Soo, Acta Bot. Acad. Sci. Hungarica 16 (3-4): 365 (1971)

    J. communis L. f. *crispa* Browicz & Zielinski, Arbor. Kornickie 19:42 (1974)

Dioecious. **Trees** to 4-5 m or shrubs. **Trunk bark** brown, exfoliating in thin strips. **Branches** stout, erect. **Leaves** acicular, 15-20 (25) mm, with a single white band above, which is rarely divided by a faint midrib in the basal half. **Seed cones** 6-9 mm, ripening in the second or third year, ovoid to globose, green the first year, then pruinose, and black when fully ripe. **Seeds** free, usually 3 per cone. 2n = 22. **Pollen shed** spring – early summer. **Habitat** various. very adaptable from rocks, to sandy areas to old fields. **Uses** the berries are used for flavoring gin, common in cultivation for landscaping. **Dist.**: Europe to Armenia and to central Russia. **Status**: common and increasing in some areas where goats are not present. Not threatened.

J. communis var. *communis* leaves and seed cones.

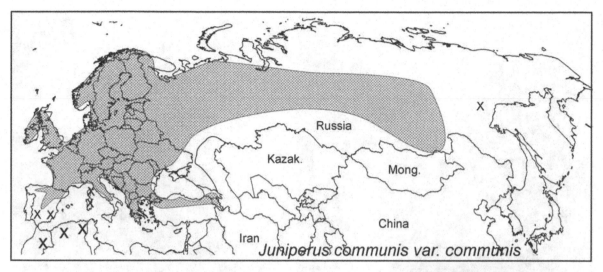

J. communis var. *communis* bark

J. communis var. *communis*, a small tree in Sweden, cf. *Adams 7846-7848*

J. communis var. *communis*, trees and shrubs, Hungary, cf. *Adams 9027-9029*

J. communis, Norway, natural tree growing at the home of Jan Karlsen.

J. communis Spain, cf *Adams 10384*

Juniperus communis var. *charlottensis* R. P. Adams, Phytologia 90(2): 187 (2008), Queen Charlotte Island juniper. TYPE: Canada, Queen Charlotte Islands, Graham Island, 9 km s of Masset, on hwy 16, in muskeg bog, 53° 55.511'N, 132° 06.471'W, 61m, 8-July-2007, *R. P. Adams 10306* (holotype: BAYLU, paratypes: *R. P. ADAMS 10304, 10305, 10307, 10308* (BAYLU).

Dioecious. Shrub, procumbent to 50 cm. **Trunk bark** thin, brown, exfoliating in strips. **Branches** upturned. **Leaves** acicular, in whorls of 3, ascending, curved, 5 - 7 x 1-2 mm, boat shaped with a single white stomatal band 2 times as wide as the green marginal bands, keeled abaxially, base jointed, not decurrent. **Seed cones** dark blue when ripe, glaucous, globose, 7-9 mm diam. **Seeds** 1-2 per cone, 3-4 mm. 2n=22? **Pollen shed** spring. **Habitat** muskeg bogs. **Uses** none known. **Dist.**: coastal islands and mainland coast, BC, Canada and Alaska, USA. **Status**: this taxon grows very specific areas and thus could become threatened by habitat degradation.

Additional specimens examined: **UNITED STATES, ALASKA, Duke Island**, Lat. 54.9409°N; Long. 131.376° W, *Adams 11550 (Mary Stensvold ns)*, ID by trnSG.
CANADA, BRITISH COLUMBIA, Khtada swamp, 60 miles sw of Terrace, 18 June 1966, *G. Mendel 196* (V), **Calvert Island, Kwakshua**, on muskeg, 14 July 1939, *I. McT. Cowan ns* (V), **Queen Charlotte Islands, Langara Island**, sphagnum bog, 21 May 1952, *F. L. Beebe ns* (V), **Dewdney Island**, w of Pemberton Bay, on rock outcrop in bog, 10 July 1984, *R. T. Ogilvie & Hans Roemer 8471067* (V), **Vancouver Island, Green Mountain**, 35 degree slope, rocky, 1431 m, 16 July 2002, *R. Hebda, N. Hebda, L. Kennedy 02-57* (V).

The distribution in Alaska is uncertain as DNA analyses of collections from Petersburg (Mitoff Isl.), and Biorka Island proved to be *J. communis* var. *kelleyi*, whereas two samples from Duke Island were *J. communis* var. *kelleyi* and a third sample was *J. communis* var. *charlottensis*. It appears that var. *kelleyi* can also grow on muskeg.

Most the specimens examined grew on muskeg or sphagnum bogs, except the unusual specimen from Green Mountain, Vancouver Island that grew on rocks at 1431 m. Whether this specimen is truly var. *charlottensis* is not known.

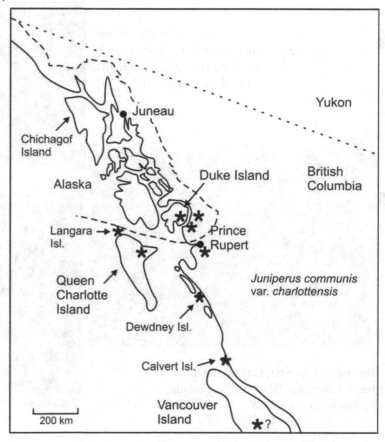

Distribution of *Juniperus communis* var. *charlottensis*. The star with a ? on Vancouver Island is a location that putative var. *charlottensis* does not grow in a muskeg bog, but in a rocky area.

The present range of var. *charlottensis* was glaciated during the Wisconsin (Flint, 1971). Because the taxon seems mostly restricted to muskeg bogs, it is difficult to determine a refugium for this taxon during the late Pleistocene Wisconsin glacial maximum. However, suitable habitat may have become available as new land emerged off the present west coast of Washington and Oregon.

J. c. var. *charlottensis*,
leaves and seed cones.

J. communis var. *charlottensis*
growing in muskeg bog, Graham Island,
Queen Charlotte Islands, BC, Canada
cf. *Adams 10306* (holotype)

Muskeg bog habitat, Graham Island,
Queen Charlotte Islands, BC, Canada.
cf. *Adams 10304-10308*

Juniperus communis var. ***depressa*** Pursh, Fl. Amer. Sept. 2:646 (1814). Depressed juniper. Type: Type: not located, (Coll. F. T. Pursh?), said to be from New York, and particularly in the province of Maine.

> *J. canadensis* Lodd. ex. Burgsd., Anleit. Sich. Erfzieh. Holzart. 2:124 (1787), *nom. nud.*
>
> *J. depressa* Raf. ex M'Murtrie, Florula Louisvill, 219 (1819)
>
> *J. depressa* (Pursh) Raf., Med. Fl. 2:13 (1830)
>
> *J. communis* L. var. *canadensis* (Lodd. ex Burgsd.) Loudon, Arbor. Frut. Brit. 4:2490 (1838), fig. 2347.
>
> *J. intermedia* Schur. Verh. Mitth. Siebenburg. Vereins Naturwiss. Hermannstadt 2:169 (1851)
>
> *Sabina multiova* Goodwyn, Amer. Botanist 37(4): 152 (1931)
>
> *J. communis* L. subsp. *depressa* (Pursh) Franco in Bol. Soc. Broteriana Ser. 2, 36:117 (1962)
>
> *J. communis* subsp. *depressa* (Pursh) E. Murray

Dioecious. Prostrate or low shrubs with ascending branchlet tips (or occasionally a spreading shrub to 3 m). **Trunk bark** brown, exfoliating in wide strips or plates. **Branches** erect to ascending. **Leaves** acicular, upturned, rarely spreading, linear, acuminate, tips acute to mucronate, to 15 x 1.6 mm. Glaucous stomatal band approx. as wide as each green leaf margin. **Seed cones** 6-9 mm, smaller than leaf length. $2\underline{n}= 22$ (Hall, Mukherjee and Crowley, 1979). **Seeds** 3 per cone. 2n = 22? **Pollen shed** spring. **Habitat** Rocky soil, rocky slopes and summits, sea level to 2800 m due to latitudinal range. **Uses** none known. **Dist.**: Canada, all provinces; Alaska, Ariz., Calif., Colo., Conn., Ga., Ill., Ind., Idaho, Md., Mich., Minn., Mont., Maine, Mass., N.C., Nev., N. Dak., N.H., N. Mex., N.Y., Ohio, Oreg., Pa., R.I., S.C., S. Dak., Utah, Va., Vt., Wash., Wisc., Wyo. **Status**: common and expanding into disturbed areas. Not threatened.

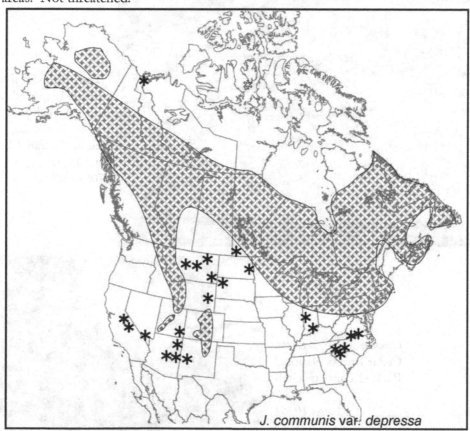

Distribution of *J. communis* var. *depressa*. Asterisks (*) represent outlying populations.

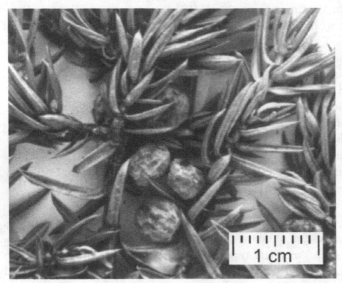

J. communis var. *depressa*, bark

J. communis var. *depressa*, leaves and seed cones.

J. communis var. *depressa*, on granite (exposed from the Glacial shield), Nemeiben Lake, Saskatchewan, Canada cf. *Adams* *7094-7095*

J. communis var. *depressa*, Cedar Breaks National Park, Cedar City, Utah USA cf. *Adams 10361*

J. communis var. *hemisphaerica* **(J. & C. Presl) Parl.** in Candolle, Prodr. 16(2): 479 (1868). hemisperical juniper. Type: Mt. Etna, Sicily, Italy, *Todaro 458* (K).

J. hemisphaerica J. & C. Presl, Delic. Prag.: 142 (1822)
J. communis L. subsp. *hemisphaerica* (J. & C. Presl) Nyman, Consp. Fl. Europ. 3:676 (1881)
J. oxycedrus L. subsp. *hemisphaerica* (J. & C. Presl) E. Schmid, Virteljahresschr. Naturf. Ges. Zurich 78:237 (1933)
J. communis L. var. *depressa* (Steven) Boiss., Fl. Orient. 5:707 (1884), *non* Pursh (1814)
J. depressa Steven, Bull. Soc. Imp. Naturalistes Moscou 30:398 (1857), *non* Raf. (1830)

Dioecious. Shrubs bushy shrub to 2.5 m. **Trunk bark** thin, cinnamon to brown, exfoliating in wide strips. **Branches** upright, densely arranged, 3-angled, thick, ca. 2 mm in diam. **Leaves** acicular, in whorls of 3, , leaves mostly straight, 4-12 (20) x 1.3 - 2 mm, with a single white stomatal band broader than green marginal bands, keeled abaxially, base jointed, not decurrent. **Seed cones** brownish black when ripe, glaucous, globose or subglobose, 4-7 mm in diam. **Seeds** 1-3 per cone, 3-4 mm. **Pollen shed** late spring. **Habitat** rocky areas. **Uses** female cones ('berries') used to flavor gin. **Dist.:** Circum-Mediterranean mountains, up to 3000 m. **Status**: this taxon is widespread so it seems robust.

Franco (1962) recognized this taxon as subsp. *hemisphaerica* (J. & C. Presl.) Nym. and examined specimens from Turkey, Bulgaria, Greece, Sicily, France, Spain, and Algeria. It seems likely that some of these specimens may prove to be var. *saxatilis* or var. *communis* when DNA analyses are

done. The author collected from the type locality at 1,750 m on Mt. Etna in 2000 (*Adams 9045, 9047*). The type locality was destroyed in a lava flow when Mt. Etna erupted in 2011-2012.

Juniperus communis var. *hemispherica*, at type locality, Mt. Etna, Sicily cf. *Adams 9045, 9046, 9047* , (29 May 2000, 37° 47.81' N, 15° 2.83' E, 1750 m).

Leaves and seed cones
of *J. communis* var. *hemispherica*

1 mm

Juniperus communis var. ***kamchatkensis*** **R. P. Adams,** Phytologia 94(2): 294 (2012). Type: Russia, near Esso, Kamchatka peninsula, 56° N, 159° E, 550-700 m., *J. W. Leverenz 5* (= *Adams 9164*) (HOLOTYPE: BAYLU).

Shrubs, similar to *J. communis* var. *saxatilis,* but differing in having straight to slightly curved leaves, with cross section concave with keel and stomatal band 1 - 1.5 x width of green leaf margins. Seed cones purple-blue when mature.

Other specimens studied: TOPOTYPES: *J. W. Leverenz 6* (*Adams 9181*), *J. W. Leverenz 7* (*Adams 9182*), *J. W. Leverenz 8* (*Adams 9183*).

Juniperus communis var. *kamchatkensis* is known only from the type locality and vicinity in Kamchatka; usually beneath *Betula platyphylla, Populus tremula* and *Salix* sp. in ravines with large rocks.

In spite of the fact that DNA sequence data (Fig. 1.11.2) shows var.'*saxatilis*', Japan is much more closely related to var. *kamchatkensis* than to var. *saxatilis* of Europe and central Asia (Fig. 1.11.2), no reliable morphological character was found to separate the Europe - central Asia plants from those on Japan and Sakhalin Island.

Holotype of *Juniperus communis* var. *kamchatkensis. Adams 9164.*

Juniperus communis var. *kelleyi* R. P. Adams, Phytologia 95(3): 215 (2013). Type: USA, Idaho, Blaine Co., on shore of Little Redfish Lake, 44° 09.588' N, 114° 54.372' W, 1997m, *Adams 10892* (HOLOTYPE: BAYLU).

Dioecious. **Shrubs** to 50 cm. **Trunk bark** thin, brown, exfoliating in wide strips. **Branches** upturned, densely arranged, 3-angled, thick, ca. 2 mm in diam. **Leaves** curved to slightly curved, acicular, in whorls of 3, ascending, lanceolate or linear, usually subfalcate, 4-10 x 1-2 mm., slightly concave adaxially with a single white stomatal band 1.5-2X as wide as green marginal bands, keeled abaxially, base jointed, not decurrent, leaf blades free, 30° to 80° to the stem, **Seed cones** purple-blue when ripe, glaucous, globose or subglobose, 4-7 mm in diam. about as long as leaves. **Seeds** 1-3 per cone, 3-4 mm. **Pollen shed** late spring. **Habitat** rocky areas. **Uses** none known. **Dist.**: common in Rocky Mtns., and Coastal ranges of western North America (Fig. 2). **Status**: this taxon is widespread so it seems robust.

Previously called *J. c.* var. *saxatilis* Pall., DNA work has shown var. *kelleyi* is a part of the clade of North America *J. communis* and not the European clade containing *J. c.* var. *saxatilis* (*sensu stricto*) (see Figs. 5.8, 5.9). In British Columbia and Alaska, var. *kelleyi* and var. *depressa* appear to intergrade.

The new variety is named in honor of my former student, Walter A. Kelley, Ph. D. 1976, Colorado State University. Walt passed away, unexpectedly with a heart attack in Costa Rica, Dec. 31, 2010, while on one of his many trips to the rainforest with his wife, Jan. Walt worked on isozymes of *Juniperus*. Photo (right) shows Walt collecting samples of *J. saltillensis* (Nuevo Leon, Mexico) on a trip with Tom A. Zanoni and RPA. Walt's keen interest in plants and sense of humor will be missed.

Holotype of
Juniperus communis
var. *kelleyi*,
Adams 10892

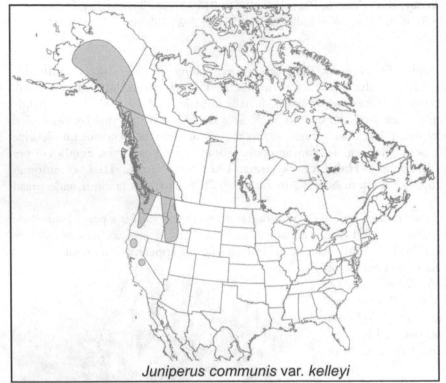

Distribution of *Juniperus communis* var. *kelleyi*.

Juniperus communis var. *kelleyi*

Leaves and seed cones of *J. communis* var. *kelleyi* to scale from holotype, *Adams 10892*.

J. *communis* var. *kelleyi*,
on granite boulders,
Tioga Pass, CA.

J. *communis* var. *kelleyi*, Redfish Lake,
ID, type locality.
cf. *Adams 10890 - 10894.*

Comparison of the leaf morphology of *J. communis* var. *kelleyi*, *J. c.* var. *depressa* and *J. jackii*.

	J. c. var. *kelleyi*	*J. c.* var. *depressa*	*J. jackii*
Stomatal band width vs. green leaf margin (GM)	1.5 - 2x GM	1-1.5x GM	2-4x GM
Leaf cross-section	concave	very concave	concave,
Leaf shape	curved	straight	curved, boat shaped
Leaf blades	free, 30° to 80°	free, 45° to 20°	mostly appressed to stem
Mature seed cones vs. leaf length	cones about as long as leaves	cones much shorter than leaves	cones as long as or longer than leaves
Seed cone shape	ovoid	ovoid	elongated oval (ellipsoid) esp. in immature cones

Juniperus communis L. var. *megistocarpa* Fernald & H. St. John, Proc. Bos. Soc. Nat. Hist. 36:58 (1921). Large fruited common juniper. Type: Canada, Quebec, Magdalen (Madeleine, Fr.) Islands, Alright Island, Narrows, *M. L. Fernald (with B. H. Long) 6729*, (holotype GH!)

Dioecious. Prostrate shrubs. Trunk bark cinnamon, exfoliating in wide strips or plates. **Branches** prostrate on the ground. **Leaves** acicular, boat shaped, curved, 7 – 10 mm, stomatal band 1.5 x as wide as green leaf margins. **Seed cones** very glaucous, purple-blue, mature in 2 yrs., 9-13 mm, larger than leaf length. **Seeds** 1 – 3 per cone. 2n=22? **Pollen shed** spring? **Habitat** sand dunes, serpentine and limestone barrens; 0-500 m. **Uses** none known. **Dist.:** Newfoundland, N.S.: Sable Isl., Que.: Magdalen Isl. (type locality). **Status:** this is a very restricted taxon and can easily become threatened.

This is the most distinct variety of *J. communis*, in its seed cones, habitat, and DNA fingerprints, yet it appears to be of only recent (Pleistocene) origin (Adams, 2003).

J. communis var. *megistocarpa* bark

J. c. var. *megistocarpa* habitat (center, on sand dune) on Isle de la Magdalen. cf. *Adams 8576-8577*.

J. communis var. *megistocarpa*, leaves and very large seed cones (largest cones in *J. communis* complex).

J. communis var. *megistocarpa*, on sand dunes at Magdalen Island, Quebec, Canada (type locality). cf. *Adams 8575-8577*

Juniperus communis var. ***nipponica*** (Maxim.) E. H. Wilson, Conifer and Taxads of Japan, Publ. Arnold Arbor. 8: 81 (1916). Miama-nezu, Type: Japan, Honshu (Prov. Nambu in alpibus, *Tschonoski [ex herb. C. J. Maximowicz] s. n.* (holotype: LE; isotypes BM; NY).

    J. nipponica Maxim., Bull. Acad. Imp. Sci. St. Petersbourg 12 :230 (1868)

    J. rigida Siebold & Zucc. subsp. *nipponica* (Maxim.) Franco, Bol. Soc. Brot. ser. 2, 36: 119 (1962)

Dioecious. **Procumbent shrubs**. **Trunk bark** exfoliating strips. **Branches** yellowish brown, glabrous. **Leaves** acicular, crowded, spreading, short, linear oblong, curved, 1-1.2 mm wide, deeply grooved on adaxial (upper) surface, stomatal band (1) very narrow, 1/2 to 1/4 width of green margin, keeled below. **Seed cones** globose, rounded at apex, 7-9 mm, bluish-black. **Seeds** 1-3 per cone. 2n = 22? **Pollen shed** spring. **Habitat** alpine slopes. **Uses** none known. **Dist.:** Hokkaido, Honshu, Japan and Korea, Sakhalin, Kamchatka peninsula. **Status**: the exact distribution is not known, but due to the protected areas that it occurs in and the non-economic use, it is probably not threatened.

    DNA sequencing shows *J. c.* var. *nipponica* (Fig. 1.11.2) to hold a central position in the Japanese/eastern Asia *J. communis* group. Additional work is needed on the Japanese junipers.

Leaves and male cones of *Juniperus communis* var. *nipponica*.

Distribution of *Juniperus communis* var. *nipponica*.

Juniperus communis L. var. *saxatilis* Pall., Fl. Ross. 1 (2): 12 (1789). Rocky juniper Type: none, (lectotype: Illustration in Pallas, Fl. Rossica 1(2): 12, t. 54 (1789), designated by Farjon (p.271, 2005). Farjon notes (p. 274, 2005) "The validation of Pallas' name remains somewhat problematic because strictly speaking the plate lacks diagnostic features (details) and the description is unclear".

J. sibirica Burgsd., Anleit. Sich. Erzieh. Holzart. 2:124 (1787)

J. nana Willd., Berl. Baumz.: 159, *nom. nud.* (1796)

J. communis L. var. *montana* Aiton, Hort. Kew 3:414 (1789)

J. communis L. var. *alpina* Suter, Fl. Helvet. 2:292 (1802)

J. oblonga M.-BIeb., Fl. Taur.-Cauc. 2:426 (1808)

J. communis L. var. *nana* (Willd.) Baumg. Enum. Stirp. Transsilv. 2:308 (1816)

J. alpina S. F. Gray, Nat. Air. Brit. Pl. 2: 226 (1821)

J. communis L. var. *oblonga* (M.- Bieb.) Parl. in Candolle, Prodr. 16 (2): 479 (1868), *non* Loudon (1838)

J. communis L. var. *caucasica* Endl., Syn. Conif.: 16 (1847)

J. nana Willd. var. *alpina* (Aiton) Endl., Syn. Conif.: 14 (1847)

J. pygmaea K. Koch, Linnaea 22:302 (1849)

J. montana (Aiton) Lindl. & Gordon, J. Hort. Soc. London 5:200 (1850)

J. caesia Regel. Gartenflora 6:346 (1857), *non* Carriere (1855)

J. communis L. subsp. *alpina* (Suter) Celak., Prodr. Fl. Bohmen: 17 (1867)

J. communis L. subsp. *nana* (Willd.) Syme in Sowerby, Engl. Bot., ed. 3, 8:275, t. 1383 (1868)

J. communis subsp. *alpina* (Smith) Celakovsky (1869)

J. communis var. *sibirica* Rydbg., Contr. U. S. Natl. Herb. 3(8): 33 (1896).

J. sibirica Burgsd. var. *montana* (Aiton) Beck, Blatt. Verein. Landesk. Niederosterreichs 1890: 78 (1890)

J. rebunensis Kudo & Suzaki, Med. Pl. Hokaido, No. 6, t.6 (1920)

J. communis L. subsp. *oblonga* (M.-Bieb.) Galushko, Mat. Izuchenya Stavrop. Kraya 2-3:165 (1950)

J. communis L. subsp. *saxatilis* (Pall.) E. Murray, Kalmia 12:21 (1982)

J. communis L. subsp. *pygmaea* (K. Koch) Imkhan., Novosti SIst. Vyssh. Rast. 27: 10 (1990)

Dioecious. Shrubs procumbent, to 70 cm. **Trunk bark** thin, cinnamon to brown, exfoliating in wide strips. **Branches** procumbent, densely arranged, 3-angled, thick, ca. 2 mm in diam. **Leaves** acicular, in whorls of 3, ascending, lanceolate or linear, usually subfalcate, 4-10 x 1-2 mm., slightly concave adaxially with a single white stomatal band 2X as wide as green marginal bands, keeled abaxially, base jointed, not decurrent. **Seed cones** brownish black when ripe, glaucous, globose or subglobose, 4-7 mm in diam. **Seeds** 1-3 per cone, 3-4 mm. **2n = 22. Pollen shed** late spring. **Habitat** rocky areas. **Uses** female cones ('berries') used to flavor gin. **Dist.**: Mountain areas: 600-2400 m. Europe, Russia, Mediterranean, C and W Asia to W. Himalaya, Mongolia, Russia. Korea, Japan is not typical (see diss. below).

Status: this taxon is widespread so it seems robust.

J. communis var. *saxatilis*, leaves and seed cones, ex Mongolia

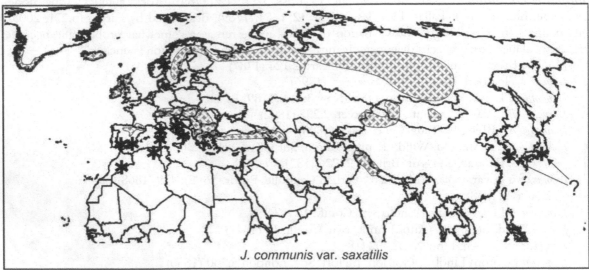

J. communis var. *saxatilis*

Note that recent DNA sequencing shows that *J. c.* var. *saxatilis* is confined to the eastern hemisphere (Figs. 1.3, 1.11.2). ? on distribution map in Korea, and Japan indicates that typical *J. c.* var. *saxatilis* does not seem to be present (see discussion p. 10 and Fig. 1.11.2).

J. communis var. *saxatilis*, bark

J. communis var. *saxatilis*,
Norway
cf. *Adams 11196-11200*

J. communis var. *saxatilis*,
Somosierra, Spain, N. of Madrid.
cf. *Adams 10373-10377*

Juniperus convallium

Key to two varieties:

1a. Trees, rarely shrubs, branches and foliage erect.........................*J. convallium* var. *convallium*

1b. Trees, branches and foliage drooping or flaccid............................*J. convallium* var. *pendula*

Juniperus convallium Rehder & Wils. in Sargent, Pl. Wilson 2:62 (1914). Mi zhi ynan bai Type: China, Sichuan, NW Sichuan, *E. H. Wilson 3010* (holotype A!, isotypes BM, K).

    J. mekongensis Kom., Bot. Mater. Gerb. Glavn. Bot. Sada RSFSR 5: 28 (1924)

    J. ramulosa Florin, Acta Hort. Gothoburg. 3: 5 (1927)

    Sabina convallium (Rehder & Wils.) W.C. Cheng & L.K. Fu, Fl. Reipubl. Pop. Sin. 7: 372 (1978)

Dioecious or monoecious. Trees rarely shrubs, branchlets densely arranged, straight or curved, terete, rarely slightly 4-angled, thin, ultimate cones usually about 1 mm in diam. **Trunk bark** cinnamon, exfoliating in thin plates. **Branches** erect. **Leaves** grayish green, both scale and decurrent (whip); decurrent leaves present only on young plants, decussate or in whorls of 3, ascending, 3-8 mm, concave adaxially; scale leaves decussate, rarely in whorls of 3, closely appressed, 1.5-2 x 0.8-1 mm. abaxial gland near center, convex or concave. **Seed cones** reddish brown to purplish black when ripe, glaucous or not, terminal on short, curved or erect branchlets, ovoid, conical-ovoid to slightly turbinate or globose, 5-8(-10) x 5-6 mm. **Seeds** 1 per cone, conical-globose or flattened ovoid, 3-5 mm in diam., with or without resin pits. **Pollen shed** spring. **Habitat** high mountains. **Uses** incense. **Dist.**: central western China, see map. **Status**: LRnt. The distribution is rather limited and the populations generally small. This taxon may become threatened.

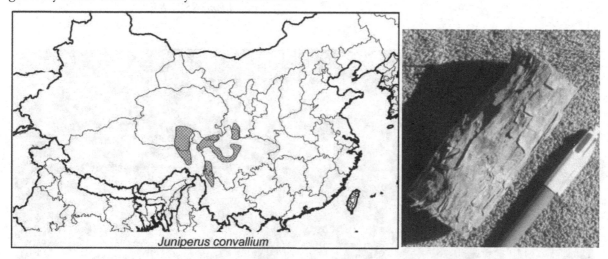

J. convallium bark

Juniperus convallium forma *pendula* (Cheng & L-K. Fu) R. P. Adams, pendulous convallium.
    Type: PE. *Juniperus przewalskii* f. *pendula* R. P. Adams and G-L. Chu, J. Essential Oil Res. 6:17. 1994.
    Sabina przewalskii (Kom.) W-C., Cheng & L-K. Fu f. *pendula* Cheng & L-K. Fu, Acta Phytotax. Sin. 13(4): 86. 1975.

This form has pendulant foliage and occurs occasionally in southern Gansu and Qinghai. DNA sequencing clearly assigns (Fig. 1.15.1) this taxon to *J. convallium* rather than *J. przewalskii* as previously treated (Adams and Chu, 1994).

J. convallium, leaves and seed cones.

J. convallium, Sichuan, China, cf. *Adams 8525-8527*

J. convallium on hillside, Sichuan, China, cf. *Adams 8525-8527*

Juniperus coxii A. B. Jacks, New Fl. & Silva 5:33 (1932). Coffin juniper, xiao guo chui zhi bai Type: England, cultivated in Exbury Gardens, Hampshire, from R. Farrer & E. H. M. Cox, 1407 collected in Myanmar (Burma), Chimili Valley (26.3N 98.6E) in 1920, *R. Farrer 1407* (holotype K!)

    J. recurva Buch.-Ham. ex. D. Don var. *coxii* (A. B. Jacks.) Melville, Kew Bull. 1958: 533 (1959)

    Sabina recurva (Buch.-Ham. ex D. Don) Antoine var. *coxii* (A.B. Jacks.) W. C. Cheng & L.K. Fu, Fl. Reipubl. Pop. Sin. 7:352 (1978)

Dioecious. Trees to 10-12 m, or large shrubs. **Trunk bark** brown to cinnamon, exfoliating in wide strips or plates. **Branches** branchlets long, pendulous. **Leaves** decurrent and scale (often becoming somewhat decurrent), 6-10 mm. adaxial surface of leaves with 2 greenish white stomatal bands and a prominent, green midvein. **Seed cones** turbinate, greenish brown, turning nearly black upon maturity in 2 yrs., 6-8 x 5-6 mm. **Seeds** 1 per cone, conical-ovoid, 5-6 x 3-4 mm, 3-ridged. **Pollen shed** spring. **Habitat** forests in high mountains. **Uses** coffins, furniture. **Dist.:** N. Burma to Yunnan, China to 3000 m. **Status**: VU, A1c. This species is vulnerable.

Juniperus coxii is often treated as *J. recurva* var. *coxii*, but Adams and Schwarzbach (2013c) showed *J. coxii* much included with *J. uncinata*, not with *J. recurva* (Fig. 5.11, Fig. 1.15.1).

J. coxii bark

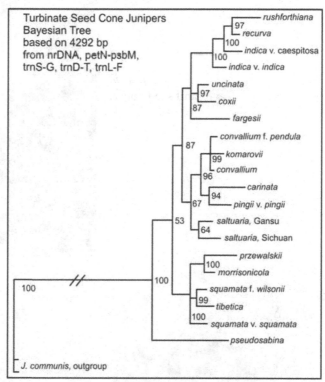

Figure 5.11. Bayesian tree (Adams and Schwarzbach, 2013c) shows *J. coxii* as a distinct clade from *J. uncinata* and J. *fargesii*. *J. recurva*, previously thought to be closely related to *J. coxii*, is far removed in a clade with *J. indica*.

Large (~1 m DBH) *J. coxii* tree, Yunnan, China, cf. *Adams 8508.*

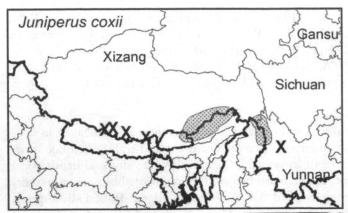

The distribution of *J. coxii*. The outlying specimens (X) are suggestive of a larger range. However, they might be mis-dentified. Addition field work is needed to adequately map the distribution of *J. coxii*.

J. coxii, leaves and seed cones.

1 cm

J. coxii, Yunnan, China near the Vietnam border cf. *Adams 8509*.
This field trip in 1998 was led by Prof. Wang Fei, Kunming Botanical Garden, who passed away suddenly a few years later. He did training in DNA work in my lab and was a jovial colleague. He will be missed.

Juniperus davurica

Farjon's (2001) transferred *J. davurica* into *J. sabina*, but recent DNA sequencing supports the recognition of *J. davurica* (Fig. 5.11.1). *Juniperus davurica* is separated from: *J. sabina* by 25 MEs, *J. virginiana* group by 22 MEs, *J. erectopatens* by 34 MEs and *J. tsukusiensis* by 24 MEs (Fig. 5.11.1). It occupies an unusual position being almost equally related to Europe/central Asia (*J. sabina*), western China (*J. erectopatens*), Japan (*J. tsukusiensis*) and North America (*J. virginiana* and allies). It ranges from Mongolia - northern China, which places it (or an ancestor) in a strategic position (Fig. 5.11.2) to act as an ancestor in the migration of a cool-temperate juniper to North America via the Bering Land Bridge (ca. 17.6 - 5.5 Mya) (see Chapter 1 for discussion).

Unfortunately, little is known about the biology of *J. davurica*, as it occupies a region that has not been well explored and has been inaccessible to botanists for many decades.

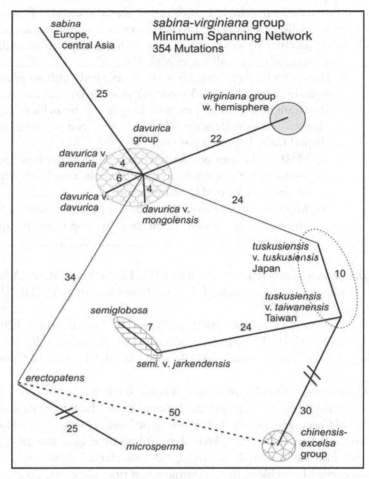

Figure 5.11.1. Minimum spanning network from Adams and Schwarzbach (2013d). The numbers next to lines are the number of MEs for that link.

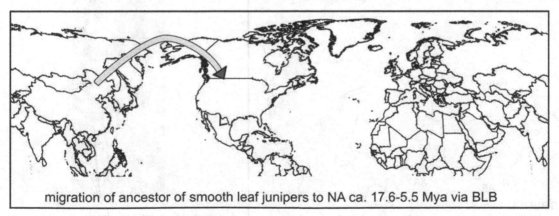

Figure 5.11.2. Putative Bering Land Bridge (BLB) for the migration of smooth leaf junipers to North America (from Adams and Schwarzbach, 2013d).

Key to **Juniperus davurica** varieties: (**J. sabina** is included in this key, because it is very similar)

1a. Both scale-like and whip- (decurrent with free tips) leaves found on mature plants, interspersed on a single branchlet, not just at the ends of rapidly growing branchlets, occasionally, one leaf type predominates..**J. davurica** var. **davurica**

1b. Whip (decurrent) leaves found only on juvenile plants or at the ends of rapidly growing branchlets, otherwise all leaves scale like

    2a. Shrubs to 1 m; growing on rocky areas, seeds with an acute tip that tapers into a globose shaped base, 1-3 seeds per globose cone..**J. sabina**

    2b. Prostrate to shrubs to 1 m; with long lateral branchlets, growing on sand, seeds with an obtuse (occasionally acute when only one seed per cone) tip on an ellipsoid to globose shaped base, 1-4 seeds per cone

        3a. Most seed cones bi-lobed, seeds (1) 2 (3-4) per cone, seeds an elongated ellipsoid with an acute tip that resemble a duck bill, ultimate branchlets radially distributed around lateral branches...**J. davurica** var. **arenaria**

        3b. Most seed cones globose, seeds 2-4 per cone, seeds a flattened globose with an obtuse tip, ultimate branchlets growing from the top (upper) side of long lateral branches..**J. davurica** var. **mongolensis**

Juniperus davurica var. **arenaria (E. H. Wilson)** R. P. Adams, sand loving savin. **Type**: China, Qinghai, Qinghai Lake, *J. F. Rock 13346* (holotype HUH! (A!); isotypes E, K).

J. chinensis L. var. *arenaria* E. H. Wilson, J. Arnold Arbor. 9:20 (1928)
J. arenaria (E. H. Wilson) Florin, Acta Horti Berg. 14 (8): 353 (1948)
J. sabina var. *arenaria* (E. H. Wilson) Farjon, Checklist of Conifers, ed. 2, 73 (2001).

Dioecious. **Shrubs to 1 m**. **Trunk bark** purplish-brown, exfoliating in thin plates. **Branches** branchlets densely arranged, ca. 1 mm in diam. **Leaves** scale and decurrent, decussate; decurrent leaves loosely arranged, narrowly lanceolate or linear-lanceolate, (3-)4-6(-9) mm. concave, with broad, white stomatal bands adaxially, arched and obtusely ridged abaxially, apex acuminate or occasionally acute; scale-like leaves densely arranged, 1-3 mm, abaxial gland central, elliptic or oblong. **Seed cones** borne on curved branchlets, bluish purple when ripe, glaucous, irregularly globose, 4-6 x 6-8 mm. **Seeds** (1-) 2 (3-4) seeds per cone, ovoid, 3-5 mm, elongated ellipsoid with an acute tip that resembles a duck bill. **Pollen shed** spring. **Habitat** sand. **Uses** none known. **Dist**.: N. Gansu, Inner Mongolia?, Shaanxi?, China.

J. davurica var. *arenaria* bark exfoliating in smooth, thin plates

J. davurica var. *arenaria* as a shrub
on sand dune at Lake Qinghai,
Qinghai, China.
cf. *Adams 10347-10352*

J. davurica var. *arenaria*
leaves and seed cones.
cf. *Adams 10347*

1 cm

J. davurica var. *arenaria* habitat
on top or near the top of sand dunes at Lake Qinghai,
Qinghai, China.
cf. *Adams 10347-10352*

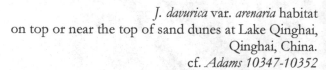

J. davurica var. *arenaria* on sand dunes at Lake
Qinghai, Qinghai, China.
cf. *Adams 10347-10352*

Juniperus davurica **Pallas**, Fl. Ross. 1(2):13, t. 55 (1789). Note: originally publ. as *J. dauurica* but uu was subsequently changed to vu (*davurica*), also spelled as *J. dahurica*. Dahurian juniper, Type: Russia, Russian Far East, Amur River, *P. S. Pallas [ex herb. Pallas] s. n.* (lectotype, BM) designated by Farjon (p. 367, 2005) as 'left-hand specimen on sheet mounted with Gmelin 35'.

Sabina davurica (Pall.) Antoine, Cupress.-Gatt. 56 (1857)
J. davurica Pall. subsp. *maritima* V. M. Urussov, Bjull. Glavn. Bot. Sada (Moscow) 122: 55 (1981)
J. sabina L. var. *davurica* (Pall.) Farjon, World Checklist of Conifers, ed. 2: 73 (2001), Kew Gardens, UK

Dioecious. **Procumbent shrubs**. **Trunk bark** purplish brown, exfoliating in thin strips. **Branches** branchlets densely arranged, ca. 1 mm in diam. **Leaves** scale and decurrent, decussate; decurrent leaves loosely arranged, narrowly lanceolate or linear-lanceolate, (3-)4-6(-9) mm. concave, with broad, white stomatal bands adaxially, arched and obtusely ridged abaxially, apex acuminate or occasionally acute; scale-like leaves densely arranged, 1-3 mm, abaxial gland central, elliptic or oblong. **Seed cones** borne on curved branchlets, dark brown or bluish purple when ripe, glaucous, irregularly globose, 4-6 x 6-8 mm. **Seeds** (1-)2 - 4 seeds per cone, ovoid, 3-5 mm, slightly flattened, apex acute. **Pollen shed** spring. **Habitat** rock outcrops. **Uses** none known. **Dist.**: far eastern Russia, and n. Mongolia. **Status**: unknown.

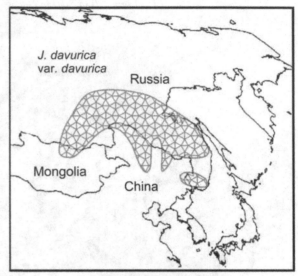

J. davurica var. davurica, bark

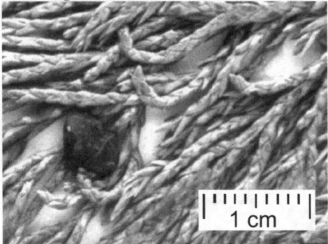

J. davurica var. *davurica*, leaves and seed cones. cf. *Adams 7252*.

J. davurica var. *davurica*, 15 km se of Ulan Batar, Mongolia, cf. *Adams 7252-7253*. 1994 field trip arranged by Prof. S. Shatar, Ulan Batar, Mongolia.

Juniperus davurica var. ***mongolensis*** (R. P. Adams) R. P. Adams, Phytologia 94(3): 360 (2012) Mongolian juniper. Type: Mongolia, 80 km sw of Ulan Batar, N 47° 36.691', E 105° 09.332', 1230 m, on sand dunes, 16 Jun 1994, *Adams 7255* (holotype BAYLU, topotypes *Adams 7254, 7256,* BAYLU).

   Juniperus sabina var. *mongolensis* R. P Adams,
      Phytologia 88(2): 182 (2006).

Dioecious. Procumbent shrubs. Trunk bark brown, exfoliating in thin strips. **Branches** branchlets densely arranged, ca. 1 mm in diam. **Leaves** scale and decurrent, decussate; decurrent leaves loosely arranged, narrowly lanceolate or linear-lanceolate, (3-)4-6(-9) mm. concave, with broad, white stomatal bands adaxially, arched and obtusely ridged abaxially, apex acuminate or occasionally acute; scale-like leaves densely arranged, 1-3 mm, abaxial gland central, elliptic or

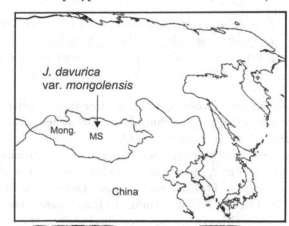

oblong **Seed cones** borne on curved branchlets, bluish purple when ripe, glaucous, irregularly globose, 4-6 x 6-8 mm. **Seeds** 2 - 4 seeds per cone, 3-5 mm, flattened globose, apex obtuse. **Pollen shed** spring. **Habitat** sand dunes. **Uses** none known. **Dist.:** Mongolia. **Status**: unknown.

   This variety is supported by the DNA sequencing (Fig. 5.11.1).

J. davurica var. mongolensis,
leaves and seed cones, cf. Adams 7255

J. davurica var. *mongolensis* a prostate plant, Mongolia cf. *Adams 7254-7256* 1994 field trip arranged by Prof. S. Shatar, Ulan Batar, Mongolia.

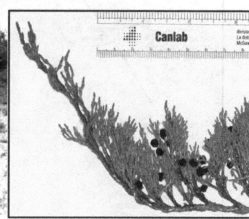

J. davurica var. *mongolensis* showing the prostate nature of the stems.

Juniperus deltoides R. P. Adams, Phytologia 86: 49 (2004). Type: Greece. 14 km e. of Arachova (Arahova), 420 m, *R. P. Adams 9436* (holotype: BAYLU!)

   Representative Specimens Examined. Bulgaria, s. Varna, 7Dec. 1923, B. Gilliat-Smith 373 (K).

   J. oxycedrus f. *parvifolia* Novak, Preslia 4: 53 (1926)

   J. oxycedrus var. *microcarpa* Neilr., Veg. Croat. 52 (1968)

   J. oxycedrus var. *parvifolia* (Novak) Jovan. in Saric (ed.), Fl. Serbia 1: 218 (1992)

   J. oxycedrus var. *fastigiata* Jovan. in Saric (ed.), Fl. Serbia 1: 413 (1992)

Dioecious. Trees to 12 m or large shrubs, often with pyramidal crowns. **Trunk bark** cinnamon to brown, exfoliating in thin plates. **Branches** erect. **Leaves** acicular, 9-17 mm long, 1.5-2.4 mm wide, base of the leaf nearly as wide as the blade. Leaves with two glaucous bands on the adaxial surface, generally not sunken. **Seed cones** with protruding leaf scales, ripening in second year, globose, dark red when ripe, green turning to brownish yellow during ripening. **Seeds** 3-5 per cone. **Pollen shed** late spring. **Habitat** rocky areas. **Uses** wood destructively distilled to produce Oil of Cade. **Dist.**: Italy, Greece, Balkans, Turkey to Iran. **Status**: this taxon is abundant and reproducing, not threatened.

   This is a species recently described (Adams, 2004) as differing from *J. oxycedrus* in its terpenes, RAPDs, leaf morphology and range. DNA sequencing supports the specific status of *J. deltoides* (Adams et al., 2005) as shown in figure 5.12 (and Figs. 1.3, 1.11.1 in chapter 1).

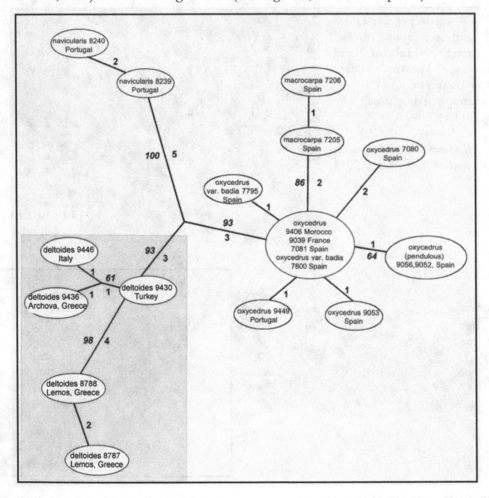

Figure 5.12. Haplotype tree based on ITS sequences. From Adams et al. (2005).

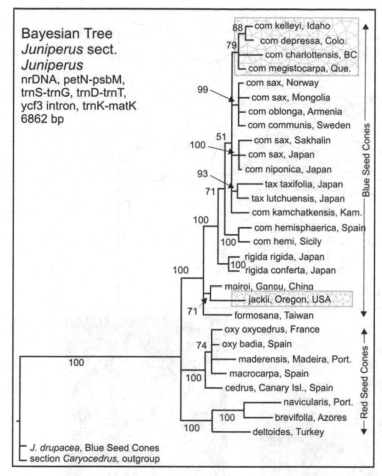

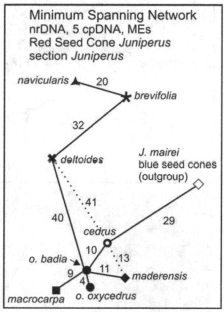

Figure 5.13. Bayesian tree based on combined nrDNA and and 5 cp gene sequences from Adams and Schwarzbach (2013a) for *Juniperus* section *Juniperus*.

Fig. 5.13.1. Min. spanning network. for the red seed cone junipers.

Juniperus deltoides is in a clade with *J. brevifolia* and *J. navicularis*, with 100% support for being distinct from *J. oxycedrus* and allies (Fig. 5.13). *Juniperus deltoides* is 40 MEs removed from *J. oxycedrus* var. *badia* and 41 MEs from *J. cedrus* as well as 32 MEs from *J. brevifolia* (Fig. 5.13.1).

Recently, Yaltırık et al. (2007) described a new shrubby variety, *J. oxycedrus* L. var. *spilinanus* Yalt., Eliçin & Terzioğlu from Spildağı National Park of western Turkey. The occurrence of *J. oxycedrus* in Turkey seems problematical as recent studies (Adams, 2004; Adams, et al., 2005) utilizing nrDNA sequencing, RAPDs, leaf terpenoids and morphology, clearly indicate that *J. oxycedrus* (*sensu stricto*) is restricted to the western Mediterranean; another, sibling species, *J. deltoides* R. P. Adams occupies the eastern Mediterranean region, including Turkey.

Adams et al. (2010c) compared SNPs of nrDNA and petN-psbM of *J. deltoides*, *J. oxycedrus*, and *J. o.* var. *spilinanus* revealed that *J. o.* var. spilinanus is allied with *J. deltoides* (Turkey), not *J. oxycedrus* (France and Spain). Leaf terpenoids showed the same pattern, supporting the recognition of *J. o.* var. *spilinanus* as **J. deltoides** R. P. Adams **var. spilinanus** (Yalt. et al.) Terzioğlu, **comb. nov.** *Phytologia 92(2): 156-166 (August 2, 2010).*

However, the principal character to distinguish var. *spilinanus*, shape of detached leaf base (sagittate), does not appear to consistently separate var. *deltoides* and var. *spilinanus*. It seems that var. *spilinanus* being a shrub, is the only character that distinguishes it from the trees of var. *deltoides*.

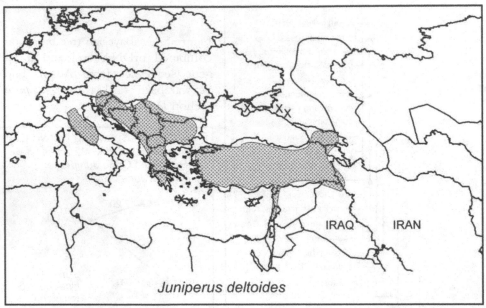

Juniperus deltoides

J. deltoides, bark

J. deltoides, leaves and seed cones.

J. deltoides, 30 km n of Eskisehir, Turkey
cf. *Adams 9430-9432*

J. deltoides, 14 km e of Arachova, Greece (type locality) cf. *Adams 9436-9438*

J. deltoides, shrubs in Italy. It is not known if the shrubby form is natural or caused by goats browsing. cf. *Adams 9445.*

Juniperus deppeana Steudel

A recent re-examination of *J. deppeana* resulted in several taxonomic changes based on sequence data (Adams and Schwarzbach, 2013a,b). As seen in Fig. 1.14, Chpt. 1, a Bayesian tree showed *J. deppeana* as a well support clade. However, within *J. deppeana*, the situation is poorly resolved. A minimum spanning network of the DNA data reveals the taxa differ by only a few MEs (Fig. 5.14). Thus, infraspecific taxa are recognized, primarily, on morphology, ecology and distribution ranges.

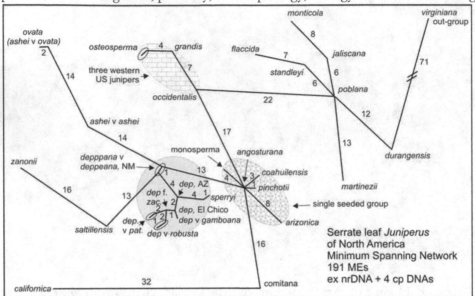

Figure 5.14. Minimum spanning network: serrate junipers, North America (Adams and Schwarzbach, 2013a,b).

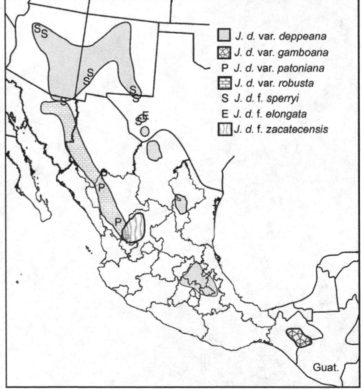

Adams, Zanoni and Hogge (1984b), using leaf terpenoids examined the varieties of *J. deppeana*. They found that samples from Arizona (BA, SA) as a group rather distinct from the other *J. deppeana* varieties.

Adams et al. (2007) expanded this research on J. deppeana using RAPDs. Figure 5.16 shows PCO based on 97 RAPD bands and indicates a similar pattern to that found in DNA sequencing (cf. Fig. 5.14).

Notice (Fig. 5.16) that *J. saltillensis* is well resolved and *J. deppeana* is resolved into three groups: *J. deppeana* var. *robusta*, var. *gamboana*, and all other *J. deppeana* varieties.

Fig. 5.15. The distribution of *J. deppeana*, its forms and varieties.

Removing *J. saltillensis* and reanalysis of the PCO resulted in the separation of var. *gamboana*, var. *robusta*, but var. *patoniana* and f. *zacatecensis* were interspersed with var. *deppeana* (Fig. 5.17). There is considerable variation in *J. deppeana* and infraspecific variation in *J. deppeana* is still not completely understood (Adams, et al., 2007).

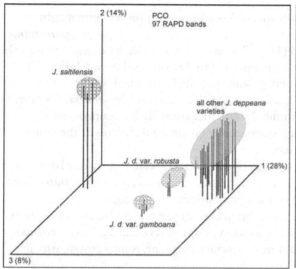

Figure 5.16. PCO, 97 RAPDs of *J. saltillensis* and *J. deppeana* varieties.

Figure 5.17. PCO, 97 RAPDs of *J. deppeana* varieties. H and L are from El Chico, Natl. Park, Hidalgo and Los Liros, Coah.

J. d. var. deppeana

J. d. var. patoniana

J. d. var. robusta

J. d. var. gamboana

Figure 2. Comparison of bark exfoliation patterns. Note the checked bark of *J. d.* var. *gamboana* and the phylogenetically closely related *J. d.* var. *robusta*. (The photos of *J. d.* var. patoniana and *J. d.* var. *robusta* are from T. A. Zanoni).

Key to *J. deppeana* varieties and forms:

1a. Terminal whips long (15 - 30 cm) and pendulous, all (or nearly all) leaves on adult plants juvenile (decurrent, or whip type).. var. **deppeana** f. **elongata**

1b. Terminal whips short (5 - 10 cm) and not pendulous, all leaves on adult plants scale-like (except on new growth where whip leaves occur)

  2a. Seed cones small (5-8 mm diam.), with soft pulp and 1(2) seeds, reddish brown with a light bloom, Chiapas, Mexico and adjacent Guatemala..var. **gamboana**

  2b. Seed cones large (8-20 mm diam.), woody and (1) 2 - 7 seeds, brown, reddish brown, or purplish with little to copious bloom, from central Mexico northward to Arizona and New Mexico

    3a. Stem bark longitudinally furrowed into long, interconnected strips, terminal whip branches often flaccid and pendulous.. var. **deppeana** f. **sperryi**

    3b. Stem bark in quadrangular plates or in longitudinal strips (occasionally interconnected, if exfoliating in strips, then foliage not weeping), occasionally quadrangular plates at the trunk base, terminal whip branches ascending to erect

      4a. Stem bark exfoliating in longitudinal strips (occasionally interconnected) or with plates near the trunk base...var. **patoniana**

      4b. Stem bark exfoliating in square or oblong quadrangular plates, not in strips

        5a. Trees with a strong central axis, no major side branches, crown pyramidal, and open as in *Cupressus*, often with 2 (3-4) trunks rising at (or below) ground level..................var. **robusta**

        5b. Trees with round crown, branching at 1-4 m to produce irregular, round crown, usually with a single trunk

          6a. Mature female cones larger, 10-20 mm. diam., heavy bloom (glaucous waxy coating) on cone surface causes cone to appear white; shrub/small round topped tree (to 8m) ....var. **deppeana** f. **zacatecensis**

          6b. Mature female cones smaller, 8-15 mm diam., glaucous or not, if glaucous not appearing as white, small to large trees...var. **deppeana**

J. deppeana var. *deppeana* with quadrangular bark McKittrick Canyon, Texas

J. deppeana var. *deppeana*, Davis Mtns., Texas, cf. *Adams 4974*

Juniperus deppeana Steudel var. **deppeana**, Nom. Bot. ed. 2, 1:835 (1840). Alligator bark juniper, Cedro, cedro chins (Puebla), sabino, Tascate (Chihuahua and Durango), Tlascal or Tlaxcal (Hidalgo), Huata, Agoziza (Sonora). Type: Mexico, Veracruz, Llanos de Perote, *C. J. W. Schiede in 1828* (holotype location unknown or destroyed). lectotype: MO, designated by Zanoni and Adams, Bot. Soc. Mex. 38: 83 (1979).

> *J. thurifera* Spach, Ann. Sci. Nat. Bot., ser. 2, 16: 298 (1841), *non* L. (1753)
>
> *J. mexicana* Schiede ex Schltdl. & Cham., Linnaea 5: 77 (1830), *non* Spreng. (1826)
>
> *J. foetida* Spach, Hist. Nat. Veg. Phan. 11: 314 (1841)
>
> *Sabina mexicana* (Schltdl. & Cham.) Antonine, Cupress.-Gatt.: 38 (1857)
>
> *J. gigantea* Roezl, Cat. Grain. Conif. Mexic. 8 (1857)
>
> *Sabina gigantea* (Roezl) Antoine, Cupress.-Gatt.: 36 (1857)
>
> *J. pachyphlaea* Torr., US Rep. Expl. Survey Miss. Pacific 4(5): 142 (1857)Thick bark juniper, alligator bark juniper, oak bark juniper, cedros. Type: Zuni Mts., NM, USA, Bigelow in 1853, NY!.
>
> *Juniperus deppeana* Steud. var. *pachyphlaea* (Torr.) Martinez. Anales Inst. Biol. Univ. Nac. Mex. 17(1): 53 (1946).
>
> *Sabina pachyphlaea* (Torr.) Antoine, Cupress.-Gatt.: 39 (1857)
>
> *S. plochyderma* Antoine, Cupress.-Gatt.: 40 (1857).[*bason = nom nud.* (*J. plochyderma*)]

Dioecious. **Trees** 10-15 (-30) m, with rounded crown. **Trunk bark** exfoliating in rectangular plates. **Branches** erect, often gray green or light green, branchlets (1 cm) exfoliating to reveal copper color. **Leaves** both decurrent (whip) and scale. Decurrent and scale leaf margins denticulate (20 X), whip and scale leaves usually with ruptured glands (clear, yellow or white exudate). **Seed cones** globose, 8-15 mm, fibrous to obscurely woody, maturing in the second year, reddish-tan to dark reddish-brown with glaucous bloom. **Seeds** 2-4 per cone, 6-9 mm long. **Pollen shed** late winter - early spring. **Habitat** rocky soils, slopes and mountains; 2000-2900 m. **Uses** fence posts. Wood et al. (2012) reported 47% sprouted from cut stumps and is thus a pest in grasslands (). **Dist.**: AZ, NM, TX, northern Mexico, see Fig. 5.15. **Status**: common, not threatened.

J. deppeana var. *deppeana* leaves and seed cones.

United States National Big Tree for *Juniperus deppeana* var. *deppeana* in the Prescott National Forest, AZ.
Craig Walton is on the left and David Emerson is on the right. Photo courtesy of Craig Walton.

Fence posts made from *J. deppeana* in
Mexico.

Habitat of *J. deppeana*, 2580 m in the White Mtns.,
AZ, with Ponderosa pine. cf. *Adams 10926, 10927*.

Juniperus deppeana f. *elongata* R. P. Adams. Phytologia 87(2): 100 (2005). Elongated branch deppeana. Type: USA, Texas, Jeff Davis Co., 4.2 km w of entrance to Lawrence E. Wood Madera Ck. park, 1845 m, N 30° 43.437', W104° 08.255', 11 Mar 2005. *R. P. Adams 10627* (holotype BAYLU, isotype SRSC) (note: SRSC has been merged with BAYLU).

This form differs from typical *J. deppeana* in having terminal whips long (15 - 30 cm) and pendulous, and all (or nearly all) leaves on adult plants are juvenile (decurrent, or whip type). Another plant was found in the Davis Mtns. at Brown Mtn., 2190 (summit), *R. P. Adams 10629*. It seems this form involves only a few genes. Dist.: only 2 trees known. Davis Mtns., TX, see map above.

J. deppeana f. elongata, Davis Mtns. cf. *Adams 10627*, holotype.

J. deppeana f. elongata, Davis Mtns., Brown Mtn. cf. *Adams 10629*, isotype.

J. deppeana f. elongata, Davis Mtns., bark exfoliating in rectangular plates as is typical of *J. deppeana*. cf. *Adams 10627*.

J. deppeana f. elongata, elongated branchlets and sparse leaves on holotype, *Adams 10627*.

Juniperus deppeana Steudel f. ***sperryi*** (Correll) R. P. Adams. Brittonia 25 (3): 289 (1973). Sperry's juniper. Type: United States: Jeff Davis County: Dry Canyon of Davis Mountains, about 8 mi. (13 km) from Sproul Ranch Headquarters. *Sperry T870* (holotype GH, isotype: US!).
Named after O. E. Sperry.

> *J. deppeana* Steud. var. *sperryi* Correll, Wrightia 3:188 (1966)
> *J. deppeana* Steud. subsp. *sperryi* (Correll) E. Murray, Kalmia 13:8 (1983)

Dioecious. **Trees** 10-15 m, with rounded crown. **Trunk bark** stem bark longitudinally furrowed into interconnected strips. **Branches** terminal whip branches and larger branches flaccid and pendulous. **Leaves** both decurrent (whip) and scale. Decurrent and scale leaf margins denticulate (20 X). **Seed cones** globose, 8-15 mm, fibrous to obscurely woody, maturing in the second year, reddish-tan when immature, then reddish-blue with very light bloom (glaucous) when mature. **Seeds** 5-6 per cone or 1(2) in Sonora, 6-9 mm long. **Pollen shed** spring. **Habitat** rocky soils, slopes and mountains. **Uses** none known. **Dist.**: Davis Mts., Gila Natl. Forest, SW NM, Coconino NF, AZ, n. Sonora, MEX. **Status**: rare, threatened. This form as been treated as a variety, but the only character separating it from *J. deppeana* is the bark. As the form always consists only a few isolated individuals, forma is more fitting that variety status.

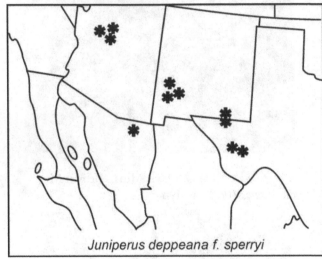

J. deppeana f. *sperryi* with bark exfoliating in strips.

1 cm

J. deppeana f. *sperryi* leaves and seed cones.

Andy Allgood with largest known tree of *J. d.* f. *sperryi*, Coconin NF, AZ.

Juniperus deppeana f. sperryi

The author by the type tree of *J. deppeana* f. *sperryi* (1968) on the H. E. Sproul Ranch, Ft. Davis, Texas, cf. *Adams 352*

Juniperus deppeana f. *sperryi* tree with furrowed bark (insert photo) in Gila National Forest, NM. Photo by Lew Stockman.

Juniperus deppeana* var. *gamboana (Mart.) R. P. Adams Phytologia 88: 229 (2006) Type: Mexico: Chiapas: near Teopisca, *M. Martinez 6701* (holotype: MEXU!)

    J. gamboana Martinez., Anales Inst. Biol. Univ. Nac. Mexico 15(1): 7 (1944).

Cedro, cipres, cipres comun, bac'il nuhkupat (Tzeltal at Tenejapa, Chiapas), K'uk",ton, nukul pat (Tzotzil at Zinacantan, Chiapas), gamboa juniper.

Dioecious. Trees (to 12 m), stem branching 1 or 2 m above base, crown rounded or very broadly conic. **Trunk bark** reddish-brown to dark gray brown, exfoliating in quadrangular plates. **Branches** ascending to erect, terminal whip branches ascending to erect, with straight tips, angle of branching of ultimate twig 35-45 degrees. **Leaves** decurrent and scale like, scale leaves mostly opposite, ovate to elliptic, 1.5-2 mm long, with acute or obtuse, appressed tips, margin finely-toothed, yellow-green to green. **Seed cones** with soft pulp, globose, reddish-brown, with a light coat of bloom (glaucous), 5-8 mm diameter. **Seeds** 1(or 2) per cone, ovoid, 4-6 mm long, 3-4 mm wide, several large grooves, brown; hilum about two/thirds length of seed. **Pollen shed** spring. **Habitat** on limestone soils in pine-oak pine-oak-juniper forests in the sierras at 1670-2200 m elevation in Chiapas, Mexico; limestone hillsides near San Miguel Acatan at 1920-2134 m elevation in the Sierra de los Cuchumatanes of Depto. Huehuetenango, Guatemala. **Uses** none known. **Dist.**: Mexico-Guatemala border. **Status**: VU, B1+2c. This taxon is narrowly distributed in areas that may be converted to grazing. It is vulnerable.

    Recent DNA sequencing has indicated that *J. deppeana* var. *gamboana* (*J. gamboana*) nested well within the *J. deppeana* clade (Fig. 1.14), so it appears that the checkered (quadrangular) bark is a conserved character in this group.

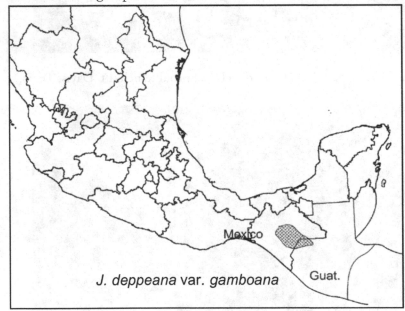

J. deppeana var. *gamboana* bark exfoliating in rectangular plates.

J. deppeana var. *gamboana* leaves and seed cones.   Habitat of *J. deppeana* var. *gamboana*, Chiapas.

J. deppeana var. *gamboana*,
Chiapas, Mexico. cf. *Adams
6863-6867*

164 Robert P. Adams

Juniperus deppeana Steudel var. ***patoniana*** (Martinez) T. A. Zanoni. Biochem. Syst. & Ecology 4: 152 (1976). Cedros, tascate, sabino. Type: Mexico, Durango, El Salto. *M. Martinez and C.E. Blanco 6710* (holotype: MEXU!).

J. patoniana Martinez, Anal. Inst. Biol. Mexico 17:62, 63 (1946)
J. patoniana f. *obscura* Martinez. Anal. Inst. Biol. Mexico 17:68 (1946)
J. deppeana var. *obscura* (Martinez) Gaussen, Trav. Lab. Forest. Toulouse Tome II, Sect. 1.
    Vol. 1 partie II. Fase. 10.150.152 (1968)

Dioecious. Trees to 15 m, very sparsely branched, with very irregular, round crown, crown as wide as tall. **Trunk bark** exfoliating in longitudinal strips (occasionally interconnected). **Branches** erect. **Leaves** both decurrent (whip and scale). Decurrent and scale leaf margins denticulate (20 X), gland on whip leaves usually not ruptured (with white exudate). **Seed cones** globose, 8-15 mm, fibrous to obscurely woody, maturing in the second year, reddish-tan (when immature) turning dark reddish-brown with glaucous bloom. **Seeds** (1-)2 – 3 (-6) per cone. **Pollen shed** early spring. **Habitat** rocky soils. **Uses** fence posts. Sprouts from cut stumps and is thus a pest in grasslands. **Dist.**: Occurs as a single or a few scattered trees in the pine-oak-juniper forests, usually with *Juniperus deppeana* var. *robusta*, at 2400-2800 m elevation in the Sierra Madre Occidental of Durango, Mexico. **Status**: VU, B1+2ce. Uncommon, probably vulnerable.

Juniperus deppeana var. patoniana

J. deppeana var. *patoniana*, with bark exfoliating in interlaced strips, Mexico. (photo by T. A. Zanoni)

1 cm

Author with *J. deppeana* var. *patoniana*, (1974), Mexico. cf. *Zanoni 2752* (photo by T. A. Zanoni)

Juniperus deppeana Steudel var. ***robusta*** Martinez, Anales Inst. Biol. Univ. Nac. Mexico 17(1):47 (1946). Robust juniper, cedros. Type: Mexico, Durango, Pueblo Nuevo, Pino Gordo, *Estevez 4502* (holotype: MEXU!)

Dioecious. **Trees** to 25 m, with strong central axis, often with 2 trees arising from a single base. **Trunk bark** exfoliating in square to rectangular plates. **Branches** erect. **Leaves** both decurrent (whip) and scale. Decurrent and scale leaf margins denticulate (20 X), gland on whip leaves usually not ruptured. **Seed cones** globose, 8-15 mm, fibrous to obscurely woody, maturing in the second year, tan when immature, to dark reddish-brown with glaucous bloom when mature. **Seeds** (1-)2 - 3 (-6) per cone, 6-9 mm long. **Pollen shed** early spring. **Habitat** rocky soils, slopes and mountains. **Uses** fence posts. Sprouts from cut stumps and is thus a pest in grasslands. **Dist.**: In the pine, pine-oak, oak, pine-oak-juniper-*Arctostaphylos* forests and occasionally in grassland of the Sierra Madre Occidental at 1500-3200 m elevation from Chihuahua s. to Jalisco, Mexico. **Status**: VU, B1+2ce. Common, but probably threatened due to extensive logging operations.

Juniperus deppeana var. robusta

J. deppeana var. *robusta*, bark
(photo T. A. Zanoni)

J. deppeana var. *robusta*,
Cerro Huehuento, Durango, Mexico,
cf. *Adams 10252*

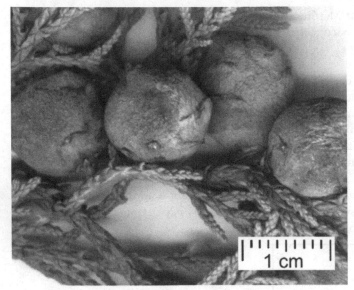

J. deppeana var. *robusta* leaves and seed cones.

1 cm

Large (12-15 m) dead tree, Durango, Mexico showing the 'twin' trunks commonly seen in *J. deppeana* var. *robusta*.

J. deppeana var. *robusta*, north of El Salto, Durango, Mexico cf. *Adams 10255-10256*

Juniperus deppeana f. *zacatecensis* (Mart.) R. P. Adams, Phytologia 88(3): 229 (2006).
 Zacatecas juniper, cedros. Type: Mexico, Zacatecas, 10 km. W. of Sombrerte. *Martinez A503*
(holotype: MEXU!).

> *Juniperus deppeana* Steudel var. *zacatecensis* Martinez, Anales Inst. Biol. Univ. Nac. Mexico
> 17(1): 57, 58 (1946).
> *J. zacatecensis* (Martinez) Gaussen. Trav. Lab. Forest. Toulouse Tome II. Sec. I Vol. 1. partie
> II 2. fasc. 10. 151. 1968

Dioecious. Trees to 8 m with a round crown or large shrubs. **Trunk bark** exfoliating in square to oblong plates. **Branches** erect. **Leaves** both decurrent (whip) and scale. Decurrent and scale leaf margins denticulate (20 X), gland on whip leaves usually not ruptured (with white exudate). **Seed cones** globose, 13-20 mm diam. with heavy bloom on cones, causing cone to appear white, reddish-brown beneath bloom. **Seeds** 1-4(-7) per cone. **Pollen shed** fall. **Habitat** rocky areas and grasslands. **Uses** fence posts. Sprouts from cut stumps and is thus a pest in grasslands. **Dist.:** In oak-pine-juniper and pinyon-juniper woodlands, and on grasslands on hills at 1980-2470 m elevation, in w. Zacatecas and adjacent Durango, Mexico. **Status:** VU, B1+2ce. This taxon is known from a limited area and is likely vulnerable due to land clearing.

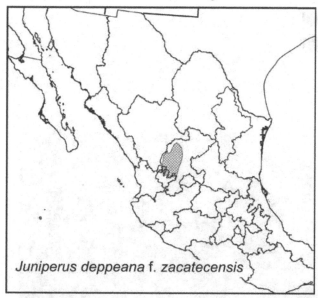

Juniperus deppeana f. zacatecensis

Figure 5.18. Distribution of *J. d.* forma *zacatecensis*.

J. deppeana f. *zacatacensis* bark

Juniperus deppeana
f. *zacatecensis*
leaves and seed cones.
Notice the copious amounts
of waxy bloom on seed cones.

1 cm

J. deppeana f. *zacatecensis*, notice white bloom on seed cones.

J. deppeana f. *zacatacensis*, 18 km W. of Sombrerete, Mexico, cf. *Adams 6840-6844*

J. deppeana f. *zacatacensis*, habitat, Mexico, cf. *Adams 6840-6844*

Juniperus drupacea Labill., Icon. Pl. Syriae 2: 4, t.8 (1791). Syrian juniper, drupe-like fruited juniper,
Type: Turkey, Hatay, Jebel Akra, *J. J. H. de la Labillardiere s. n.* (holotype G-DEL?, n.v.; isotypes FI, K)
    Arceuthos drupacea (Labill.)) Antoine & Kotschy, Oesterr. Bot. Wochenbl. 4(31): 249 (1854).

Dioecious. **Trees** up to 15 m, with a broad pyramidal crown (but assuming a columnar habit under
cultivation). **Trunk bark** brown, cinnamon beneath, exfoliating in narrow strips. **Branches** erect.
Leaves acicular, 10-25 x -4 mm, linear-lanceolate, acuminate, patent, with two white bands above.
Seed cones 20-25 mm, ripening in the second year, sub-globose, brownish-purple or bluish-black,
pruinose, glaucous. **Seeds** 3 per cone, united to form a single large stone. **Pollen shed** fall. **Habitat**
rocky areas. **Uses** none known. **Dist.**: S. Greece (N. end of Parnon Oros, Arkadhia), Asia Minor and
Syria. **Status**: common in areas, but it may become threatened.

    This species in the monotypic section, *Caryocedrus*, is thought to be the most primitive of the
junipers (see Fig. 1.5, Chpt. 1).

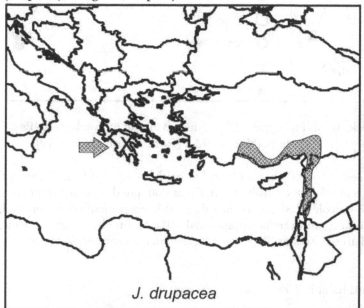

J. drupacea

J. drupacea bark

1 cm

J. drupacea, 15 m tall tree, e. of Tripolis
Greece, cf. *Adams 8795-8796*

Seed cone and leaves of *J. drupacea.*

Juniperus durangensis

Adams, Elizondo and Zanoni (2012) examined variation in the leaf essential oils, DNA sequences and morphology in *Juniperus durangensis*. For the DNA sequencing data, they found the Topia shrubs to be intermingled with typical *J. durangensis* samples (Fig. 5.19.1). However, petN-psbM SNPs revealed they differed by 3 SNPs from all other *J. durangensis* samples (Fig. 5.19.2). In addition, they found the plants

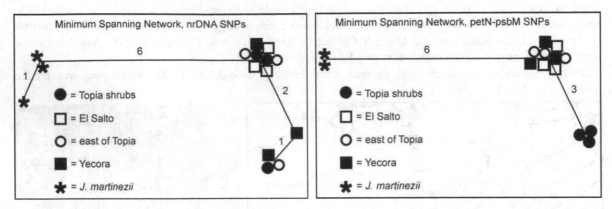

Figure 5.19.1. Minimum spanning network, 10 SNPs in nrDNA. (from Adams et al., 2012).

Figure 5.19.2. Min. spanning network, 9 SNPs in petN (cpDNA). (from Adams et al., 2012).

Topia plants differed in the number of seeds/ cone (Table 5.17) and in habit, being shrubs in Topia, vs. trees/ shrubs for typical *durangensis*. The leaf oil of the shrubs from Topia contained several compounds not found in other *J. durangensis* oils: 1,8-cineole (2.8%, trace in other oils), cis-p-menth-2,8-dien-1-ol, germacrene B, patchouli alcohol, hexadecanol and sandaracopimarinal. Based on these data, Adams, Elizondo and Zanoni (2012) recognized the shrubs at Topia as a new variety: *Juniperus durangensis var. topiensis* R. P. Adams & S. Gonzalez.

Table 5.17. Comparison of seeds/cone and habit in *J. durangensis*.

	seeds/cone	habit
Typical *durangensis*	1-3 (4)	small tree/shrub
var. *topiensis*	(3) 5-9, 6.5 avg.	shrub
Santiago Papasquiraro(SP)	1.7 - 5 very variable	small tree/shrub

Key to *J. durangensis* varieties:
1a. Trees or shrubby trees, 1-3 (4) seeds per cone..........................***J. durangensis* var. *durangensis***
1b. Shrubs, (3) 5-9, avg. 6.5 seeds per cone...***J. durangensis* var. *topiensis***

Juniperus durangensis Martinez, Anales Inst. Biol. Univ. Nac. Mexico 17(1): 94-95 (1946). Cedro, tascate, Durango juniper. Type: Mexico, Durango, Puerto de Santo Domingo, 30 km. from El Salto, *M. Martinez 7015* (holotype: MEXU!).

Dioecious. Trees shrub to small tree (to 5 m), usually branching near base, trees have irregular crowns. **Trunk bark** think ashy-brown, long fibrous strips or scales. **Branches** terminal whips recurved so the final 2-4 cm appears as hooks, ascending-erect to strict, spreading in trees, angle of branching of ultimate twig about 60 degrees. **Leaves** scale leaves mostly opposite on ultimate twigs, on ultimate twig appear like a chain of beads, 1 - 2 mm long, margins finely toothed, dark gray-green. **Seed cones** very glaucous, bluish-reddish, with soft pulp, gibbous, 6-7 mm long, 4-6 mm wide. **Seeds** 1-3 (-4) per cone, subconic to oval, acuminate or obtuse, 3-4 mm long, 2-3 mm wide, dark reddish-brown with shallow grooves; hilum to one-half length of seed. **Pollen shed** Jan. – March. **Habitat** Rhyolite rocks and in openings in pine-oak or pine-oak-juniper-Arctostaphylos forests at 1600-2900 m elevation. **Uses** none known. **Dist.**: Sonora, Chihuahua, Durango, extreme w. and s Zacatecas, n. Jalisco and Aguascalientes, Mexico. **Status**: VU,B1+2c. This taxon is found on unusual Rhyolite rocks. It is not common, but probably not threatened at present.

Juniperus durangensis

J. durangensis, bark

Juniperus durangensis showing the curved or 'hooked' ends of branchlets that are characteristic of the species.

J. durangensis
leaves and seed cones.
Notice the scale leaves appear
as a 'chain of beads'.

J. durangensis on Rhyolite rocks, La
Ciudad, Durango, Mexico, cf.
Adams 10253, 10254.

J. durangensis, habitat on Rhyolite with
Pinus in background.
cf. *Adams 10253, 10254*

Juniperus durangensis var. ***topiensis*** R. P. Adams & S. Gonzaléz, Phytologia 94(1): 46 (2012). Topia juniper. TYPE: Mexico, Durango, Topia, 25° 12.894' N, 106° 33.891' W, 1840 m, *Adams 11923*. (HOLOTYPE: BAYLU). Fig. 5.17.3.

Similar to *Juniperus durangensis*, but differing in having 5-9 seeds per cone and being a small, multi-branched shrub with less crowded branchlets.

Juniperus durangensis var. *topiensis* is currently known only from the type locality where, it is common on hillsides around Topia at about 1800-1900 m.

Other specimens studied: TOPOTYPES: *Adams 11924, 11925*, BAYLU, *S. Gonzaléz 7268a, 7268b*, BAYLU, CIIDIR, others to be distributed).

Holotype of *J. durangensis* var. *topiensis* (*Adams 11923*).

Juniperus durangensis var. *topiensis* as a shrub, at the type locality, Topia, Durango, Mexico. cf. *Adams 11923, 11924, 11925*.

Juniperus erectopatens (Cheng and L. K. Fu) R. P. Adams, Biochem. Syst. Ecol. 27: 723 (1999). Song pan yuan bai (Songpan juniper). Type: China, Sichuan, Songpan, 2700m, *T. T. Yu 2702* (holotype PE!

Sabina vulgaris Antoine var. *erectopatens* Cheng and L.K. Fu. Acta Phytotax. Sin. 3(4):86(1975).
Juniperus sabina L. var. *erectopatens* (Cheng and L.K. Fu) Y.F. Yu and L. K. Fu, Novon 7: 444 (1997).
Juniperus chinensis L. var. *chinensis* in Farjon (2005).

Dioecious rarely monoecious. Small **Trees** to 4 m. **Trunk bark** grayish brown exfoliating strips. **Branches** branchlets densely arranged, ascending, slender, 0.8-11 mm in diam. **Leaves** both scale-like and decurrent; decurrent leaves usually present on young plants, rarely present on adult plants, decussate or in whorls of 3, closely appressed, 3-7 mm, concave adaxially, convex abaxially, apex sharply pointed; scale like leaves decussate, rhombic or rhombic-ovate, 1-2.5 mm, abaxial gland central, prominent, elliptic. **Seed cones** light brownish green when immature, purplish blue to black when mature, often glaucous, usually irregularly globose, (3-) 4-5 mm, (1-) 2-seeded. **Seeds** ovoid, slightly flattened, 4-5 mm, ridge, with resin pits, apex blunt or slightly pointed. **Pollen shed** summer. **Habitat** on rocky slopes around Songpan, Sichuan. **Uses** none known. **Dist.**: around Songpan, Sichuan, China. **Status**: distribution and abundance is not known, so it should be considered threatened at present. I collected it at the edge of a field south of Songpan, Sichuan, China. Its distribution is not fully known at present.

DNA sequencing (Adams and Schwarzbach, 20133) has shown (Fig. 1.3) that *J. erectopatens* is one of the most distinct species of *Juniperus*. Sequence data is in concordance with oil and RAPDs data showing that *J. erectopatens* is not conspecific with *J. chinensis* as previously thought (Farjon, 2001, 2005), but in a well supported clade with *J. microsperma* (Fig. 1.3). A minimum spanning network shows *J. erectopatens* differs by 25 MEs from *J. microsperma* and 34 MEs from *J. davurica* (Fig. 5.20).

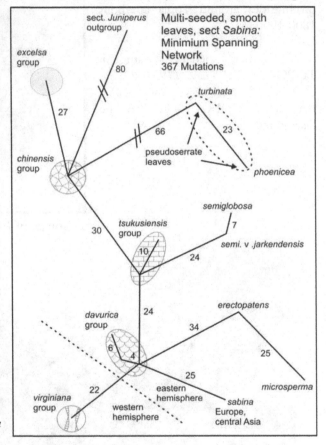

Fig. 5.20. Minimum spanning network. (Adams and Schwarzbach, 2013d).

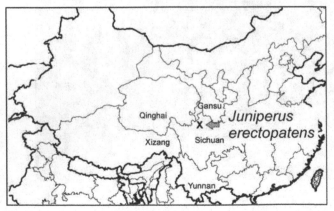

Distribution of *J. erectopatens*. Known only from the type locality.

J. erectopatens leaves and seed cones.

J. erectopatens, bark

J. erectopatens, Songpan, Sichuan, China, cf. *Adams 8532-8534*

Juniperus excelsa M.-Bieb.

Juniperus excelsa has had a variable taxonomic history as Farjon (2005, 2010) treated it as a species with two subspecies: *J. excelsa* subsp. *excelsa* and *J. excelsa* subsp. *polycarpos*. Whereas, Adams (2008, 2011) treated it as *J. excelsa* and *J. polycarpos* K. Koch. Indeed these taxa are quite similar in their morphology. However in the recent phylogeny of *Juniperus* based on DNA sequence data, Adams and Schwarzbach (2013d) found 91% support for *J. excelsa* as a species, distinct from *J. polycarpos* (Fig. 1.16.1, Chpt. 1).

A minimum spanning network shows (Fig. 5.21) *J. excelsa*, *J. procera* and *J. polycarpos* var. *turcomanica* separated by 17 MEs. *Juniperus polycarpos* is separated by 8 MEs (Fig. 5.21).

It is interesting to note that *J. seravschanica*, which as been treated as *J. excelsa* subsp. *polycarpos* (Farjon 2005, 2010) differs by 27 MEs from *J. polycarpos* and 20 MEs from *J. thurifera* (Fig. 5.21).

Because *J. procera* is consistently recognized at the species level, it is reasonable to recognize *J. excelsa* subsp. *polycarpos* at the specific level (*J. polycarpos*). An alternative would be to treat *J. procera* and *J. polycarpos* as varieties or subspecies of *J. excelsa*. In this treatment, *J. excelsa* and *J. polycarpos* are maintained as separate species.

Douaihy et al. (2011) recent used 3 nuclear microsatellites to study *J. excelsa*. They reported 3 subgroups: Turkey, Greece, Ukraine, Cyprus; 4 populations from Lebanon; and 2 divergent populations from Lebanon. Adams et al. (2013) examined leaf essential oils from Greece, Bulgaria, Turkey and Cyprus and found the Greece - Bulgaria very similar and the Turkey - Cyprus oils similar.

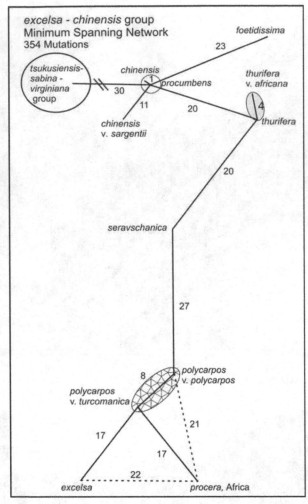

Fig. 5.21. Min. spanning network *excelsa - chinensis* group (from Adams and Schwarzbach, 2013d).

Juniperus excelsa M.-Bieb., Beschr. Land. Terek & Kur, Bot. Anhang: 204 (1800). Grecian juniper.
Type: Ukraine, Crimea, Krymskiye Gory, *P. S. Pallas [ex herb. Pallas] s. n.* (lectotype LE! see Imkhanitskaya, 1990).

J. lycia Pall, Fl. Ross. 1(2): 14, t. 56 (1789), *non* L. (1753)

J. sabina L. var. *taurica* Pall., Fl. Ross. 2:15 (1789)

J. sabina L. var. *excelsa* (M.-Bied.) Georgi, Beschreib. Russ. Reiches 3(5):1358 (1800)

J. foetida Spach var. *excelsa* (M.-Bieb) Spach (proparte, sine syn. = *J. occidentalis* (Hook.) Ann. Sci. Nat. Bot. ser.2, 16:297 (1841)

J. isophyllos K. Koch, Linnnaea 22: 304 (1849)

J. olivieri Carriere, Traite Gen. Conif.: 57 (1855)

Sabina excelsa (M.-Bieb.) Ant., Cupress.-Gatt.:45, t. 60 (1857)

Sabina religiosa Antoine, Cupress.-Gatt.:47, t.61 (1857)

Sabina isophyllos (K.Koch)Ant., Cupress.-Gatt.:48, tt. 64-66 (1857)

Sabina oliviera (Carriere) Antoine, Cupress.-Gatt.70 (1857)

J. aegaea Griseb., Veg. Erde: 378, 572 (1871)

J. taurica (Pall.) Lipsky (no. Lindl.,1850) in O. E. Knorring & Z. A. Minkvich, Rastit. Aulie-
Atinsk. u. s. Dar'inskoi obl./Travaux d'expedition pur exploration des regions de
Colonization Russe d'Asie; 2. Explor. bot. 1909, 6: 185-186. pl. (1912)

J. excelsa M. –Bieb. subsp. *excelsa* var. *depressa* O. Schwarz, Fedde's Repert. Spec. Nov. Regni
Veg. 36(1): 66 (1934)

Monoecious or Dioecious. **Trees** up to 20 m, with a conical crown when young, later broad and
open. **Trunk bark** light brown, exfoliating in narrow strips. **Branches** twigs 0.6-0.8 mm in diameter,
terete, scaly. **Leaves** scale and decurrent (juvenile), decurrent leaves 5-6 mm, only on juvenile growth;
scale-like leaves 1-1.5 mm, ovate-rhombic, closely appressed, acute, with a central, ovate or linear gland
on the back, entire, the border not scarious. **Seed cones** 8 mm, ripening in the second year, globose,
slightly glaucous, dark purplish-brown when ripe. **Seeds** 4 - 6 per seed cone, free. **Pollen shed** spring.
Habitat rocky areas. **Uses** none known. **Dist.**: Albania, Arabian Peninsula, Bulgaria (S). Greece (N),
Lebanon, Iran, Turkey, Caucasus Mtns., Yugoslavia, 300 m to 1000 m in Crimea, up to 3400 m in Iran.
Status: abundant in parts of its range so it is not threatened.

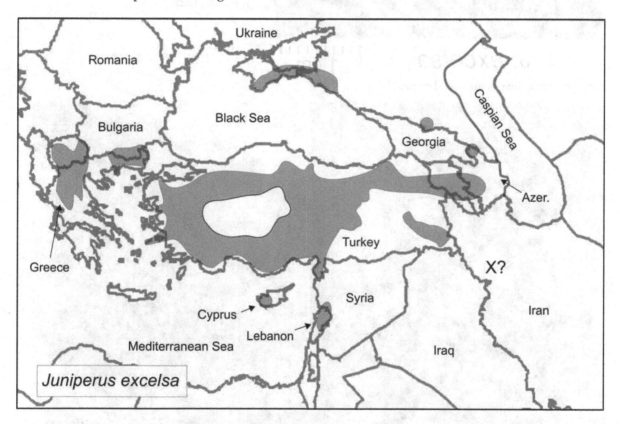

Distribution of *J. excelsa*. Based on Douaihy et al. (2011), who used data from Browicz (1982,
Boratynski et al. 1992 and Farjon (2005). The X? in Iran appears to be the Lushan, Iran population
that Adams and Hojjati (2012) showed by DNA sequencing to be *J. polycarpos*, not *J. excelsa*.

J. excelsa, bark

J. excelsa |''''|''|''''| 1 cm |

Juniperus excelsa leaves and seed cones.

J. *excelsa*, 8 km w of Lemos, Greece,
cf. *Adams 8787, 8788*

J. *excelsa*, 30 km n. of Eskisehir, Turkey
cf. *Adams 9430-9432*

Juniperus fargesii (Rehder & E. H. Wilson) Kom

Previously, I treated *J. fargesii* as *J. squamata* var. *fargesii* (Adams, 2011). However, the recent phylogeny of *Juniperus* shows that *J. fargesii* is actually in a separate clade from *J. squamata* (Fig. 1.15, Chpt. 1). Closer examination of the turbinate seed cone junipers (Fig. 5.21.1) confirms that *J. fargesii* is quite removed from *J. squamata* that is in the *J. przewalskii - tibetica* group. Note that *J. fargesii* is 19 MEs from *J. convallium* and 21 MEs from *J. coxii* (Fig. 5.21.1). *Juniperus squamata* is 9 MEs from *J. s.* var. *wilsonii* and 19 MEs from *J. przewalskii*.

My samples of *J. fargesii* used for DNA sequencing (*Adams 6769-6771*) came from Jone, Gansu, China. Several other specimens have been examined from central and eastern China (see distribution), but it may be that the material from Gansu is not typical of all the specimens in its DNA.

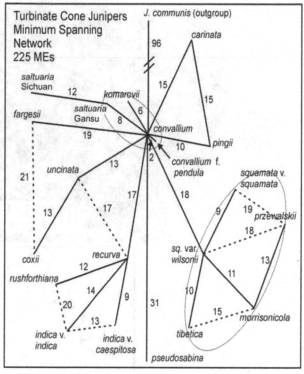

Fig. 5.21.1. Min. spanning network of the turbinate junipers. (Adams and Schwarzbach, 2013c)

To aid in identification of *J. fargesii* and *J. squamata*, a key is presented:
1a. Leaf blades 6-7 mm long, seed cones 8-9 mm long..............***J. squamata*** var. ***squamata***
1b. Leaf blades 3-5 mm long, seed cones 4-5 mm long
   2a. Leaf blades 3-4 mm long, seed cones 5 mm long..***J. fargesii***
   2b. Leaf blades 4-5 mm long, seed cones 4-5 mm long...................***J. squamata*** f. ***wilsonii***

Juniperus fargesii (Rehder & E. H. Wilson) Kom., Bot. Mater. Gerb. Glavn. Bot. Sada RSFSR 5:30 (1924). Chang y gao shan bai. Type: China, Sichuan, Chengkou, Tchen keou tin, *Rev. Pere P. G. Farges 153*, (holotype A!, isotypes E, K, LE, NY)
   J. squamata Buch.-Ham. ex D. Don var. *fargesii* Rehder & Wils. in Sarg., Pl. Wils. 2: 59 (1914)
   J. lemeeana Lev. & Blin in Leveille, Fl. Kouy-Tcheou: 111 (1914-15)
   J. kansuensis Kom., Bot. Mater. Gerb. Glavn. Bot. Sada RSFSR 5: 31 (1924)
   Sabina squamata (Buch.-Ham. ex D. Don) Antoine var. *fargesii* (Rehder & E. H. Wilson) L.K.
     Fu & Y. F. Yu, Higher Plants of China 3: 86 (2000)
   J. changii Hu & Cheng

Dioecious. Shrubs to small trees to 4 m. **Trunk bark** light brown, exfoliating in thin plates or strips. **Branches** erect. **Leaves** all decurrent (not acicular), spreading, straight or slightly curved, leaf blades 3 mm long. **Seed cones** green when immature, black when mature in 2 yrs., 5 mm long, elongated subglobose. **Seeds** 1 per seed cone, 4 mm long, subglobose. **Pollen shed** fall. **Habitat** Forests, valleys, roadsides, 1600-4500 m. **Uses** none known. **Dist.**: S. Anhui (Huang Shan,), W Fujian (Liancheng Xian), S. Gansu, E. Guizhou (Jiangkou), SW Shaanxi (Taibai Shan), Sichuan, Yunnan, China. **Status**: appears to be rare, it may be threatened.

Comparison of the type specimens for *J. fargesii* and *J. lemeeana* revealed that they are the same taxon.

J. fargesii, leaves and seed cones.

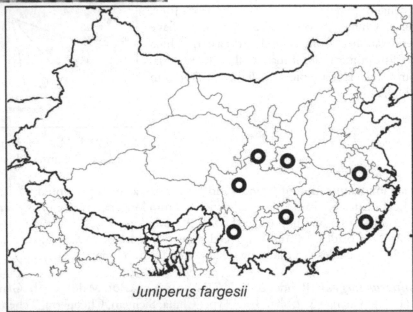

Juniperus fargesii

J. fargesii Zhongdian Co., Yunnan, China. cf. *Adams 8491-8493*, led by Prof. Yang Fei.

J. fargesii Xian Botanic Garden, China, with Prof. Ge-Lin Chu, who trained in my lab. A fine colleague.

Juniperus flaccida Schltdl., Linnaea 12: 495 (1838). Weeping juniper. Type: Mexico, Hidalgo, Mineral del Monte, Regla, *C. A. Ehrenberg 8* (lectotype MO 2085919, B lost, see Zanoni and Adams, 1979).

> *J. gracilis* Endl., Syn. Conif.: 31 (1847)
> *J. gigantea* Roezl in part
> *J. flaccida* var. *gigantea* (Roezl) Gaussen
> *J. foetida flaccida* (Schlecht.) Spach
> *Sabina flaccida* (Schlecht.) Antoine, Cupress.-Gatt.: 37 (1857)
> *S. flaccida* (Schlecht.) A. A. Heller
> *S. flaccida* (Schlecht.) I. M. Lewis

Dioecious. **Trees** to 12 m, trunk branching at 1-2 m. **Trunk bark** cinnamon reddish brown or gray reddish brown, exfoliating in broad interlaced fibrous strips. **Branches** spreading and forming a globular crown. Ultimate branchlets drooping, flaccid. **Leaves** both decurrent (whip) and scale. Scale leaves often appearing somewhat decurrent, 1.5-2 mm, opposite, narrowly ovate, acuminate. Whip- and scale-leaf margins appearing entire at 20 X but with irregular teeth at 40 X. **Seed cones** spherical (4-) 6-10 (-13) seeded, tan-brown to brownish-purple with white glaucous, 9-12 mm in diam., maturing in 1 yr. **Seeds** 5-6 mm long. **Pollen shed** late winter-early spring. **Habitat** rocky soils and slopes. **Uses** none known. **Dist.**: Mexico, Big Bend Natl. Park., Tex., Texas, USA. **Status**: widespread range in Mexico and reproducing, not threatened.

Recent DNA sequencing (Adams Schwarzbach, 2013a,b) has shown that *J. martinezii* and. *J. poblana* are distinct at the species level (see Fig. 5.22). For *J. flaccida* var. *martinezii*, see **J. martinezii** and for *J. flaccida* var. *poblana*, see **J. poblana**.

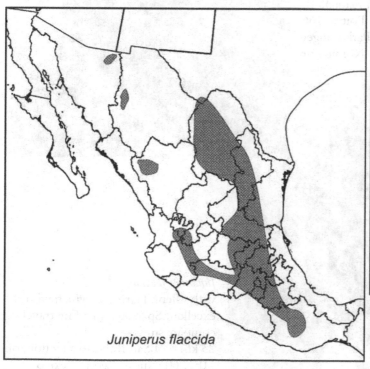

Juniperus flaccida

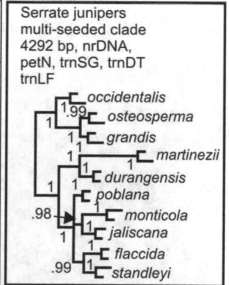

Figure 5.22. Partial Bayesian tree (cf. Fig. 1.3) for *flaccida/poblana/martinezii* clades

J. flaccida, bark (above)
leaves and seed cones (left)

1 cm

US National Register of Big Trees
Champion *Juniperus flaccida*
Chisos Mtns.,
Big Bend National Park.
Circumference: 102 inches
Height: 55 ft.
Spread: 35 ft.
Points: 166
James E. Liles, Chief Park Ranger
Nominator

Juniperus flaccida,
with Glenn Harris, a layman with
excellent Spanish and a fine traveling
companion.
23 km e of San Roberto Junction, on
MEX 60, Nuevo Leon, Mexico

Juniperus foetidissima Willd., Sp. Pl. 4(2):853 (1806). Fetid smelling juniper, stinking juniper, Type: Turkey, *J. P. de Tournefort B-W 18547* (holotype B-W, isotypes BM, M, P-TRF).

   J. phoenicea Pall., Fl. Ross. 1(2): 16, t. 57 (1789), *non* L. (1753)
   J. sabina Sibth. & Sm., Fl. Graeca Prodr. 2:264 (1816), *non* L. (1753)
   J. foetida Spach var. *squarrulosa* Spach, Hist. Nat. Veg. Phan. 11:321 (1841)
   J. sabinoides Griseb. Spicil. FL. Rumel. 2:352 (1846)
   Sabina foetidissima (Willd.) Antoine, Cupress.-Gatt.: 49 (1857)
   S. grisebachii Antoine, Cupress.-Gatt.:61 (1857)
   J. foetidissima Willd. var. *pindicola* Formanek, Beitr. Fl. Serbien 2, Brunn Verh. 33:20 (1895)
   J. foetidissima Willd. var. *squarrosa* Medw., Trudy Bot. Sada Imp. Jur'evsk Univ. 3:229 (1903)

Monoecious or dioecious. Trees up to 20 m, with a straight stem, and a narrow conical crown. **Trunk bark** light brown to tan, exfoliating in narrow strips. **Branches** erect, twigs 1 mm wide, distinctly quadrangular, scaly, irregularly arranged. **Leaves** decurrent (only on juvenile foliage) and scale like, fetid when crushed, all decussate; the adult scale leaves 1.5 mm, ovate-rhombic, free at their acuminate apices, mostly eglandular, the border entire, not scarious. **Seed cones** 7-12 mm, globose, pruinose when young, dark reddish-brown to nearly black when mature in 2 yrs. **Seeds** 1 or 2 (rarely 3) free. **Pollen shed** March – April. **Habitat** rocky areas. **Uses** none know, possible future source of cedarwood oil. **Dist.:** Mountains of Balkan peninsula from Albania and Macedonia southwards, Turkey, Caucasus Mts. **Status**: common and reproducing, not threatened.

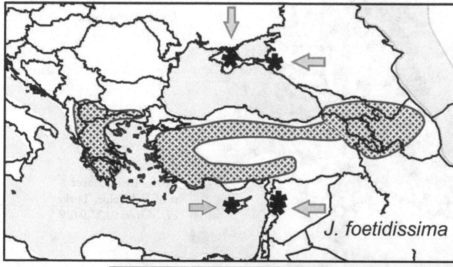

J. foetidissima

J. foetidissima near Lemos, Greece

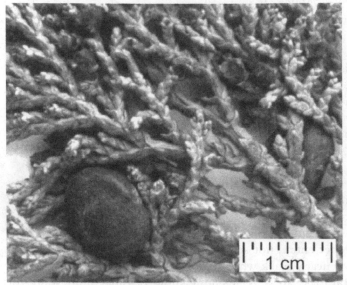

Juniperus foetidissima leaves and seed cone

J. foetidissima, bark

J. foetidissima with
Prof. Husnu Baser
n of Eskisehir, Turkey
cf. *Adams 9427-9429*

J. foetidissima, Mt. Parnassus, Greece,
cf. *Adams 5646*

Juniperus formosana

Recent DNA sequencing (Adams and Schwarzbach, 2012a) has shown that *J. formosana* and *J. f.* var. *mairei* are quite distinct (Figs. 1.3, 1.10). For *J. formosana* in inland China, see *J. mariei*.
Notes: Farjon (p.304, 2005) recognizes *J. formosana f. tenella* Hand.-Mazz. I have not examined this taxon. *J. formosana* **var.** *concolor* Hayata; Icon. Pl. Form. 7: 39, f.25 (1918). Full ref.: B. Hayata, Icones Plantarum formosanarum, Vol. 7, p.39. Type: BM . The type locality was searched on a field trip in 2000 and no junipers were found. It is presumed to be extinct.

Juniperus formosana Hayata var. *formosana*, J. Coll. Sci. Imp. Univ. Tokyo 25(19): 209 (1908). Formosan juniper. Type: Taiwan, Nantou, Chia-i Pref., Yu-Shan (Mt. Morrison), *T. Kawakami (with U. Mori) 2039* (lectotype TI!, designated by Farjon (p. 303, 2005).

Dioecious. **Shrubs** or trees to 15 m tall. Crown usually pyramidal or cylindrical. **Trunk bark** brown. **Branches** spreading or ascending; branchlets pendulous, 3-angled. **Leaves** acicular, usually with raised elongate to linear gland on the persistent branch (brown) leaf sheath, in whorls of 3, linear-lanceolate or linear-needlelike, 12-20 x 1.2-2 mm, slightly concave adaxially, with 2 white, broad stomatal bands separated by a narrow, green midvein, green and obtusely keeled abaxially, base jointed, not decurrent, apex sharply pointed. **Seed cones** axillary, glaucous, light reddish brown when immature, turning black when ripe, subglobose or broadly ovoid, 6-9 x 6-8 mm, with 6 fused scales in 2 alternating whorls, with a single seed on each scale of apical whorl, mature in 2 yrs. **Seeds** 3 seeds per cone. Ovoid-triangular, 4-5 x 3-3.5 mm, 3- or 4-ridged, base with 3 or 4 resin pits, apex pointed. **Pollen shed** spring. **Habitat** rocky area in high mountains of Taiwan and 400m in Kushan, China. **Uses** none known. **Dist.**: Taiwan and e. mainland China (Kushan). **Status**: in protected area of Taiwan, so it is not threatened at present.

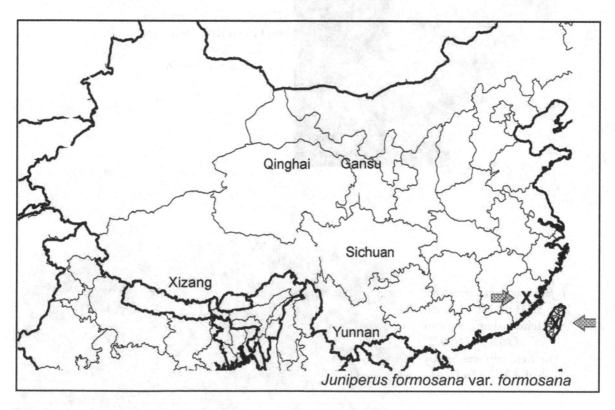

Juniperus formosana var. *formosana*

Distribution of *J. formosana* var. *formosana*

J. formosana var. *formosana* leaves and seed cones

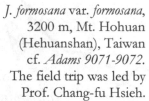

J. formosana var. *formosana* bark exfoliating in plates

J. formosana var. *formosana*, on Green River trail near Lushui, Taiwan cf. *Adams 9065-9066*

J. formosana var. *formosana*, 3200 m, Mt. Hohuan (Hehuanshan), Taiwan cf. *Adams 9071-9072*. The field trip was led by Prof. Chang-fu Hsieh.

Juniperus gracilior

Recent DNA sequencing has shown the *J. gracilior* varieties are in a well supported clade with *J. saxicola* (Figs. 1.3, 1.16.1, Chpt. 1). The *J. gracilior* varieties are very uniform in morphology and in their DNA sequences, such that they differ by only 3 to 7 MEs (substitutions + indels) and *J. saxicola* only differs by 3 MEs from *J. gracilior* var. *urbaniana* (Fig. 1.22.1). It appears that too much emphasis has been placed on the presence of decurrent leaves and *J. saxicola* was recently recognized as *J. gracilior* var. *saxicola*. (Adams and Schwarzbach, 2012).

It might be noted that although *J. barbadensis* and *J. b.* var. *lucayana* form a well supported clade (Figs. 1.3, 1.16.1, Chpt. 1), they differ by only 7 MEs from *J. gracilior* (Fig. 5.22.2). It is possible that additional DNA sequencing may support the recognition of only one species in the Caribbean, *J. barbadensis* L. (1753).

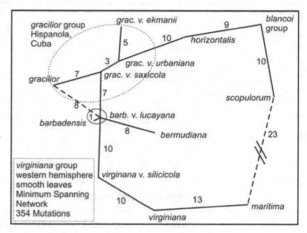

Fig. 5.22.2. Minimum spanning network. *virginiana* group. Note the few mutations (3-7) separating *J. gracilior* varieties.

Key to **Juniperus gracilior** varieties:

1a. All leaves decurrent (whip like), no scale-like leaves on adult trees, endemic to Cuba...
....*J. gracilior* var. **saxicola**

1b. Leaves scale-like with decurrent leaves found only on young (juvenile) trees or at the ends of rapidly growing shoots.....2a.

2a. Branchlets pendulous, ultimate leafy branching angle 20-30 degrees; glands on old, brown, persistent whip leaves not conspicuous...*J. gracilior* var. **gracilior**

2b. Branchlets erect or procumbent, ultimate leafy branching angle 35-40 degrees; glands on old, brown, persistent whip leaves conspicuous

3a. Erect trees; whip leaf glands oval (to twice as long as wide)..............*J. gracilior* var. **ekmanii**

3b. Prostrate shrubs; whip leaf glands elongated (3-4 times as long as wide)...............
....*J. gracilior* var. **urbaniana**

Juniperus gracilior Pilger var. **ekmanii** (Florin) R. P. Adams, Phytologia 78: 144 (1995). Ekman's juniper. Type: Haiti, Ouest Dept., Massif de la Selle, Badeau, above Croix-des-Bouquets, 2100m, *Ekman H3140* (holotype: S, isotypes: GH, IJ!, K!, US, NY)
J. ekmanii Florin, Ark. Bot. 25-A (5): 14 (1934)

Dioecious. Trees to 15m, subglobose crown. **Trunk bark** brown, exfoliating in strips. **Branches** erect, grayish-green, ultimate leafy branching angle large (37 - 42 °), on main branches exfoliates in irregular plates. **Leaves** decurrent and scale like, Glands on old brown persistent whip leaves conspicuous, dark brown; scale leaves mostly opposite but ternate on some branchlets, leaf tips acuminate to mucronate, Glands on scale leaves ovate but not very conspicuous. **Seed cones** reddish-brown underneath the bluish bloom. **Seeds** 1 - 2 seeds per cone (Seed cones have not been found on the type or any other specimens examined). **Pollen shed** in winter? **Habitat** forest in Mts. of Haiti. **Uses** none known. **Dist.**: Haiti: Massif de la Selle and Pedernales region of the Dominican Republic (in the Massif de la Selle mountain chain). **Status**: CR, D. Only a few trees known in the Massif de la Selle, Haiti. The area is extremely disturbed and cutting of trees for lumber and firewood continues. This taxon is very endangered.

Distribution of *J. gracilior* var. *ekmanii*, known only from the mountains on the Haiti - Dominican Republic border.

J. gracilior var. *ekmanii* bark

J. gracilior var. *ekmanii* leaves, no seed cones have ever been collected, even on the type specimen.

1 cm

Stump of *J. g.* var. *ekmanii* near Mare Rouge, Massif de la Selle, Haiti.

Large (ca. 2 m in diam.) standing stump of *J. gracilior* var. *ekmanii*. Note the four men standing at the base of the tree stump. near Mare Rouge, Massif de la Selle, Haiti.

J. gracilior var. *ekmanii*, 2013 m, Massif de la Selle, Haiti. cf. *Adams 7653-7654*. Notice the depauperate foliage on the limbs. In an expedition in Feb. 1996, only 2 trees were found. Most have been cut for their red wood for use in carving.

Juniperus gracilior Pilger in Urban var. **gracilior**, Symbol. Antill. 7: 481 (1913).   Graceful juniper.
Type: Dominican Republic, Azua Prov. near Las Canitas, 1300 m, *M. Fuertes (with D. Goodwin) 1939*
(lectotype NY; see Adams (1995), isolectotype US; syntypes: *Tuerckheim 2981, Eggers 2320*).

Dioecious. **Trees** to 10 m. **Trunk bark** cinnamon color, exfoliating in narrow strips. **Branches** lax
and branchlets thin, branching angle 25°. Leaves decurrent (whip) and scale like, light green.   Scale
leaves 1.0 - 1.5 mm with acuminate to mucronate tips, opposite, with some branchlets appearing
quadraform. Glands on scale leaves not conspicuous, if visible then oval and often sunken. **Seed
cones** globose (if with one seed) or reniform (if with 2 seeds), with bloom, reddish-blue, mature in 1
year. **Seeds** 1-2 seeds/cone, borne on short, straight peduncle (2 mm). **Pollen shed** spring. **Habitat**
rocky area of Mts. **Uses** none known. **Dist.**: *Juniperus gracilior* is endemic to Hispaniola occurring at
elev. of 1000 - 1700 m. It appears to be cultivated in n. Haiti as an ornamental tree. **Status**: EN,
B1+2c. Common in western parts of the Dominican Republic, but these areas are being cleared for
pastures. It should be considered endangered. Pilger also cited W. Buch p. 313. from near St. Michel,
Haiti as *J. gracilior* but this should be referred to *J. lucayana.*, now thought to be extinct in natural
populations in Haiti.

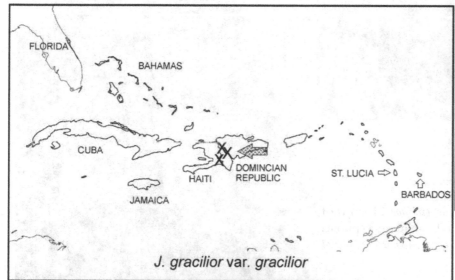

J. gracilior var. *gracilior*

J. gracilior var. *gracilior*
bark exfoliating in
thin, narrow strips.

J. gracilior var. *gracilor*
leaves and seed cones

J. gracilior var. *gracilior*, 14 km w of
Constanza,
Dominican Republic.
cf. *Adams 2785-2794*

J. gracilior var. *gracilior*, 14 km w of
Constanza, Dominican Republic.
cf. *Adams 2785-2794*

J. gracilior var. *gracilior*
in its natural habitat,
14 km w of Constanza,
Dominican Republic

Juniperus gracilior var. *saxicola* (Britton & P. Wilson) R. P. Adams. Phytologia 94(3): 361 (2012).
Type: Cuba, Granma Prov., Sierra Maestra, Oriente, *J. S. S. (Frere) Leon 10798* (holotype NY!)

   J. saxicola Britton & P. Wilson, Bull. Torrey Bot. Club 50: 35 (1923).
   J. barbadensis L. subsp. *saxicola* (Britton & P. Wilson) Borhidi, Acta Bot. Acad. Sci. Hungarica
      37 (1-4):90 (1992)
   J. barbadensis L. var. *saxicola* (Britton & P. Wilson) Silba, J. Int. Conifer Preserv. Soc. 7(1): 25 (2000)

Dioecious. Trees to 3 - 8 m. **Trunk bark** brown, exfoliating in strips. **Branches** erect. **Leaves** all decurrent, spreading, 5 - 7 mm long and ca. 1 mm wide, scale leaves never present on adult trees, gland not obvious as it appears as ridge on the leaves. **Seed cones** dark blue with bloom, subglobose to reniform, ca. 5 mm long, 3-4 mm diam., **Seeds** 2 seeds/cone. **Pollen shed** winter (?). **Habitat** rocky areas on Pico Turquino. **Uses** none known. **Dist.**: Endemic to Cuba. crest of Sierra Maestra and Pico Turquino, Granma Prov. (previously Oriente Prov.), Cuba. **Status**: VU, D2. known only from Pico Turquino, so it could become threatened.

   The species has apparently been fixed by neoteny in the juvenile-leafed (decurrent) growth stage, because even the adult, reproductive individuals have only juvenile (decurrent) leaves (an awn-shaped blade that diverges from the stem at 45 to 60 degrees and a sheath portion that clasps the stem). Generally, in *Juniperus* section *Sabina*, only young (up to 4 or 5 years old) junipers have the awn-shaped (decurrent) leaves. Later, the scale (adult) leaves are produced throughout their life span, except that juvenile (decurrent, whip) leaves are produced at the tips of branches during a rapid growth period or on damaged branches. Having only juvenile leaves on adult trees is very rare in section *Sabina*, in the western hemisphere, where field examination of thousands of junipers has resulted in finding only 8 to 10 plants that have only juvenile foliage on otherwise mature trees (personal observation). However, several species in the eastern hemisphere (*J. coxii, J. recurva, J. morrisonicola,* and *J. squamata*) are also fixed in the juvenile state and some cultivars of *J. chinensis* have almost all juvenile (decurrent) leaves.

Juniperus gracilior var. saxicola

J. gracilior var. *saxicola* bark exfoliating in thin, narrow strips.

J. gracilior var. *saxicola* leaves and seed cones notice that there are no scale (adult) leaves for this species.

J. gracilior var. *saxicola* on a very steep slope with Agave in the cloud forest, Pico Turquino, Cuba (photo by Burkhard Witt, 2008)

Juniperus gracilior var. *urbaniana* (Pilger & Ekman) R. P. Adams, Phytologia 78: 144 (1995). Haitian creeping juniper. Lectotype: Haiti, Ouest Dept.; Morne de la Selle, limestone area, 2500-2600 m, *Ekman H3157* (holotype: B, destroyed; lectotype designated by Adams, Phytologia 78: 144 (1995)). Farjon (2005) notes a holotype at S, with isotypes at F, K, MICH, NY, US). I have not confirmed his notes.

  J. urbaniana Pilg. & Ekman, Ark. Bot. 20-A (15): 9 (1926)
  J. barbadensis L. var. *urbaniana* (Pilg. & Ekman) Silba, Phytologia 56: 341 (1984)
  J. barbadensis L. subsp. *urbaniana* (Pilg. & Ekman) Borhidi, Acta Bot. Acad. Sci. Hungarica 37 (1-4): 90 (1992)

Dioecious. Prostrate shrub. Trunk bark brown, exfoliating in small plates. **Branches** procumbent. **Leaves** decurrent (whip) and scale-like, scale leaves opposite with leaf tips acuminate to mucronate. Branchlets mostly quadraform. Glands on old, brown, persistent whip leaves elongated (3 - 4 times as long as wide), conspicuous, dark brown, glands on scale leaves ovate. **Seed cones** with bloom, reddish-brown, 5-7 mm diam., mature in 1 year. **Seeds** one and sometimes 2 seeds/cone. **Pollen shed** spring. **Habitat** on peculiar limestone. **Uses** none known. **Dist.**: Pic la Selle, Massif de la Selle, Haiti on peculiar limestone region in the forest at 2510 m and on adjacent mountains in the Dominican Republic. **Status**: EN, B1+2c. This taxon is now known from two populations (Adams et al., 2010e). The Haiti population is estimated to be only less than 100m in diameter with about 30 plants rediscovered by Adams in 1996. Because it is a prostrate shrub, it is of no economic importance, so it is not likely to be cut. However, the area is subject to both natural and man caused fires and the population could be destroyed by fires. The population in Dominican Republic (Adams et al., 2010e) appears to grow in a small opening in the forest on limestone soil. Both populations in Haiti and Dominican Republic are endangered.

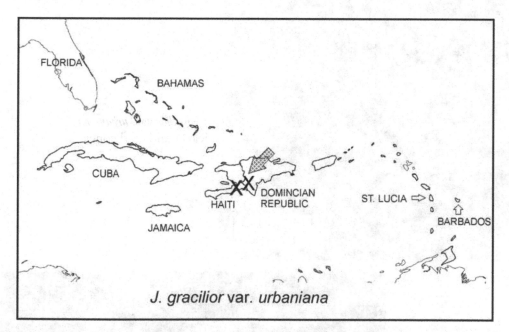

Distribution of *J. gracilior* var. *urbaniana*, known only from the type locality (Pic La Selle), Haiti and in the mountains across the border in the Dominican Republic. The arrow shows the location of the new population found in 2010 (see Adams et al., 2010e).

Adams et al. (2010e) reported on the discovery of a new population in the Dominican Republic. These plants are more shrub-like, than the plants in Haiti (see photo). Analyses of the leaf oils and nrDNA and trnC-trnD sequences indicated that the Dominican Republic shrubs are essentially the same as those from Haiti.

Juniperus gracilior var. *urbaniana* as a prostate shrub in the Dominican Republic. Photo by Francisco Jimenez Rodriquez, Dept. Botany, Jardin Botanico Nacional, Santo Domingo, Dominican Republic (see Adams et al., [2010e] for analyses).

cf. *Adams 12005.*

J. gracilior var. *urbaniana* leaves and seed cones

1 cm

J. gracilior var. *urbaniana* bark, cf. *Adams 7656*

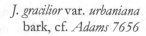

J. gracilior var. *urbaniana* small plant, from Pic La Selle, 2510 m, Haiti cf. *Adams 7656-7658*

J. gracilior var. *urbaniana*, several prostrate shrubs growing on a peculiar limestone at 2510 m, in an open forest below the summit of Pic la Selle, Haiti, cf. *Adams 7656-7658*

Juniperus grandis R. P. Adams, Phytologia 88(3): 306 (2006). Grand juniper, big western juniper, Type: United States, California, San Bernardino Mtns., along Poligue Canyon road to Holcomb Valley, *Vasek 610929 -38* (holotype: RSA).

 J. occidentalis W. J. Hooker subsp. *australis* Vasek, Brittonia 18: 352 (1966)
 J. occidentalis var. *australis* (Vasek) A. & N. Holmgren., Intermountain Fl. 1: 239 (1972).

Dioecious, approx. 5% trees **Monoecious** (Vasek, 1966). **Trees** to 30 m. **Trunk bark** reddish-brown to tan to cinnamon color, exfoliating in shaggy strips. **Branches** erect to pendulous. **Leaves** decurrent (whip) and scale-like, scale and whip leaves with visible glands but not usually ruptured, if ruptured, then with clear, pale yellow or whitish exudate on surface. **Seed cones** blue to blue-black, with resinous pulp, maturing in 2 yrs., 5 – 9 mm long (avg. 7.6). **Seeds** 1-2(3) per cone (avg. 1.5). **Pollen shed** spring. **Habitat** Sierra Nevada on dry rocky slopes; 1000-3000 m. Calif. **Uses** fence posts. **Dist.**: Sierra Nevada of California. **Status**: occurs in protected areas, it is not threatened.

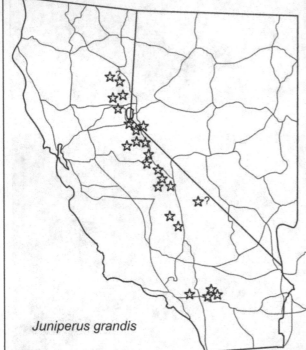

Juniperus grandis

J. grandis, with reddish-brown to cinnamon-tan bark exfoliating in thin, broad strips or plates. cf. *Adams 11134*

  Adams and Kauffmann(2010b) recently examined variation in *J. grandis* using SNPs of nrDNA and cpDNA and found some geographical differences. See Adams and Kauffmann (2010a) for a recent comparison of leaf oils of *J. grandis*, *J. occidentalis* and *J. osteosperma*.

1 cm

J. grandis, leaves and seed cones. Notice glands on leaves.

J. *grandis*,
Onyx Summit, on CA Hwy. 38,
2588m, San Bernardino Mtns. California.
cf. *Adams 12319-12322*.

The author with a large *Juniperus grandis* tree at Monitor Pass, CA.

198 Robert P. Adams

Juniperus horizontalis Moench, Methodus Plantas 699 (1794).   Creeping juniper, prostrate juniper.
Sabina horizontalis (Moench) Rydb., Bull. Torrey Bot. Club 39:100 (1912). Type: Canada, Nova Scotia,
Halifax, *M. Hultgren s. n.* (neotype BM, designated by Farjon (p. 308, 2005).

    J. sabina Michx., Fl. Bor. Amer. 2:246 (1803), *non* L. (1753)
    J. prostrata Pers., Syn. Pl 2(2): 632 (1807)
    J. sabina Michx. var. *procumbens* Pursh, Fl. Amer. Sept. 2: 647 (1814)
    J. repens Nuttall, Gen. N. Amer. Pl. 2: 245 (1818)
    J. sabina Michx. var. *humilis* Hook., Fl. Bor. Amer. 2(10):166 (1838)
    J. hudsonica Forbes, Pinetum Woburn.: 208 (1839)
    J. virginiana L. var. *prostrata* (Pers.) Torr., Fl. New York 2: 235 (1843)
    Sabina prostrata (Pers.) Antoine, Cupress.-Gatt.: 57 (1857)
    J. foetida multicaulis Spach in part
    J. horizontalis forma *alpina* (Loud.) Rehder
    J. horizontalis forma *lobata* O.W. Knight
    J. horizontalis var. *douglasii* Hort.
    J. horizontalis var. *variegata* Beissn.
    J. sabina B. *humilis* Carr. in part
    J. sabina β *procumbens* Pursh
    S. vulgaris Antoine in part

Dioecious. **Prostrate to decumbent shrub**. **Trunk bark** brown, exfoliating in plates. **Branches** procumbent, forming large mats often several meters across. **Leaves** decurrent (whip) and scale-like. Foliage green but turning reddish-purple in winter. Leaf margins entire (20 X and 40 X). scale-leaf tips apiculate, mostly overlapping, both whip and scale leaves growing along the branchlets. **Seed cones** (1-2)3-4 seeded, blue-black to blue-brownish when ripe, borne on generally curved peduncles, mostly maturing in 2 yrs., 5-7 mm. **Seeds** 4-5 mm. 2n= 22 (Hall, Mukherjee and Crowley, 1979). **Pollen shed** spring. **Habitat** sand dunes, sandy and gravelly soils, prairies, slopes and along stream banks; sea level to 1000 m. **Uses** none known. **Dist.**: Canada: all provinces. AK, MT, WY, ND, SD, NB, MN, IA, WS, IL, MI, NY, VT, MA, ME. **Status**: this taxon is common and reproducing. Not threatened.

J. horizontalis appears to be a sibling species to *J. virginiana and J. scopulorum*. However, it cp DNA is very similar to that of the Caribbean junipers (Fig. 1.16.5), suggesting a more complex history.

    It hybridizes with both species (Adams, 1983; Fassett, 1945a,b; Palma-Otal, et al, 1983). *J. horizontalis* x *J. scopulorum* hybrid has been named *J. scopulorum* var. *patens* Fassett (= X *J. fassettii* Boivin).

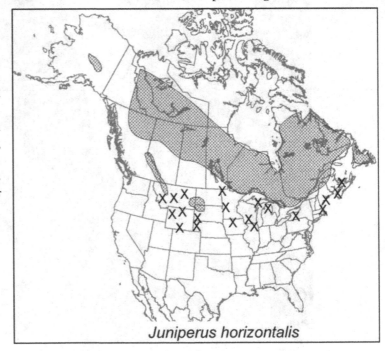

Juniperus horizontalis

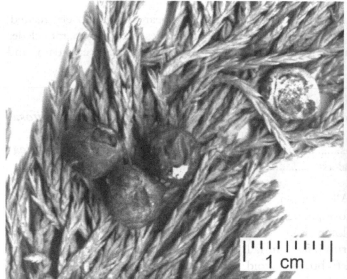

J. horizontalis leaves and seed cones

J. horizontalis bark

J. horizontalis, on the bank of the Saskatchewan River, Saskatoon, Saskatchewan, Canada. cf. *Adams 7096-7098*

Juniperus indica

    Juniperus indica and Rushforth's Bhutan juniper (*J. rushforthiana*) appear to be closely related. However, the new DNA sequencing (Fig. 1.3, 1.15, 1.15.1, Chpt. 1) show these taxa in separate clades and *J. rushforthiana* 12 MEs from *J. recurva* and 20 MEs from *J. indica* var. *indica* (Fig. 15.1, ex Adams and Schwarzbach, 2013c).

Key to *J. indica* varieties (plus *J. rushforthiana* that is often merged with *J. indica*).
1a. Low shrubs 50-100 cm tall, foliage very dense, seed cones 5- 8 mm long...*J. indica* var. *caespitosa*
1b. Shrubs or trees, foliage loosely set, seed cones 9-12 mm long
    2a. Monoecious, shrubs or shrubby trees...*J. indica* var. *indica*
    2b. Dioecious, trees with a strong central axis..*J. rushforthiana*

Juniperus indica var. **caespitosa** Farjon. Aljos Farjon, A mongraph of Cupressaceae and Sciadopitys, Kew Press, UK, p. 313 (2005), the caespitose juniper, Type: Nepal. Dhaulagiri Himal: Dolpo, Mugu Karnali, S. of Mugu, *S. Miehe 99-030001* (holotype K).
Dioecious. Decumbent to ascending shrub 0.5 - 1 m tall. **Branches** (nearly erect, very dense with short branchlets. **Seed cones** (sub) globose to broadly ovoid, (4.5-)5-8 x 4-6.5 mm, blue-black. **Pollen shed** spring? **Habitat** near timberline, on rocky slopes. **Uses** none known, possibly for incense. **Dist.** Farjon (2005) reports it in NW Nepal, S. Xizang (Tibet), Bhutan, incompletely known. **Status** unknown, Farjon (2005) lists IUCN as DD.

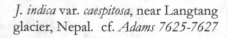

J. indica var. *caespitosa*
leaves and seed cones
cf *Adams 7625*

J. indica var. *caespitosa*, near Langtang glacier, Nepal. cf. *Adams 7625-7627*

Juniperus indica Bertol. var. *indica* Bertoloni, Misc. Bot. 23: 16, t. 1 (1862). Dian zang fang zhi bai, Indian juniper. Type: Farjon (p. 313, 2005) designated the illustration in Bertoloni as the lectotype).

> *Juniperus wallichiana* Hook. f. & Thomson ex E. Brandis, Forest Fl. N.W. & Central India: 537 (1874) Farjon (2005) concluded that Bertoloni based his illustration on the same specimens that Brandis used to describe *J. wallichiana*. Farjon concluded that if *J. wallichiana* was based on the same specimens, the name could not be accepted.
>
> *J. wallichiana* Hook. f. & Thomson ex E. Brandis var. *meionocarpa* Hand.-Mazz., Akad. Wiss. Wien, Math.- Naturwiss. Kl., Anz. 61: 107 (1924)
>
> *J. pseudosabina* auct., *non* Fisch. & Mey. (1842)
>
> *S. wallichiana* (Hook. f. & Thomson ex E. Brandis) var. *meionocarpa* (Hand.-Mazz.) W. C. Cheng and L.K. Fu, Chinese Dendrol. 1: 322 (1972)
>
> *Sabina wallichiana* (Hook. f. & Thomson ex E. Brandis) W.C. Cheng & L.K. Fu, Fl. Reipub. Pop. Sin. 7: 367 (1978)
>
> *Sabina indica* (Bertol.) L.K. Fu & Y. F. Yu, Higher Plants of China 3: 90 (2000)

Dioecious or Monoecious. **Shrubs** erect or procumbent, rarely small trees. **Trunk bark** dark brown, exfoliating in interlaced strips. **Branches** ultimate branchlets densely arranged, mostly straight, usually 4-angled, sometimes terete. **Leaves** both scale-like and decurrent; decurrent leaves usually present on young trees, in whorls of 3, ascending, 3-8 mm, apex acuminate; scale leaves with an oval to elliptical gland with a narrow extension almost to the leaf tip, decussate or sometimes in whorls of 3, closely appressed, rhombic, 1.2-2 mm, abaxial gland central, or basal in decurrent leaves, oblong or linear, depressed, leaf apex obtuse. **Seed cones** erect, black-brown when ripe, subglobose to turbinate, 9 -12 mm, maturing in 2 yrs. **Seeds** 1 (rarely 2) seeds per cone, ovoid, slightly flattened, 5-6 x ca. 4 mm, smoothly or obscurely ridged. **Pollen shed** late winter-early spring. **Habitat** rocky areas, forests or thickets on mountain slopes, 2600 - 5100 m. **Uses** incense. **Dist.**: E and S Xizang, NW Yunnan, Bhutan, N India, Kashmir, Nepal, Sikkim. **Status**: although it occurs over a large area, it does not seem common anywhere and is used for incense in Buddhist temples. The species should be considered threatened.

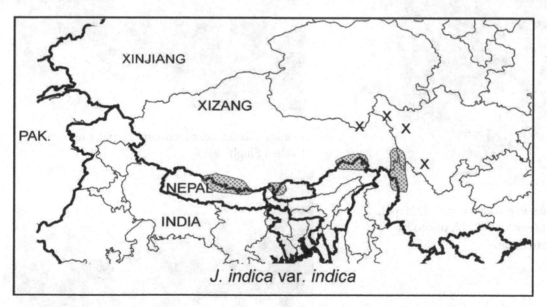

J. indica var. *indica*

J. indica var. *indica* leaves and seed cones, the 2 smaller cones are 1 year old and brown. The 2 larger cones are 2 years old and blue-black.

J. indica var. *indica*, bark

J. indica habitat, near Jomsom, Nepal, photo by Lakhan Singh

J. indica var. *indica*, Jomson, Nepal, cf. *Adams 8775-8777*, photo by Lakhan Singh

J. indica habitat, near Jomsom, Nepal, photo by Lakhan Singh

Juniperus jackii (Rehder) R. P. Adams. Type: Siskiyou Mtns., on the road from Waldo, Oregon to Crescent City, CA, 3000 ft., 25 Aug., 1904, *J. G. Jack (A. Rehder) s. n.*, lectotype: A!, designated by Farjon (2005). Named in honor of J. G. Jack.

Juniperus communis var. *jackii* Rehder Mitt. Deutsch. Dendrol. Ges. 1907 (16): 70 (1907).

Dioecious. Shrub procumbent, to 20 cm. **Trunk bark** thin, brown, exfoliating in strips. **Branches** procumbent, elongated. **Leaves** acicular, in whorls of 3, ascending, curved, (6) 7 - 9 (10) x 1-2 mm, concave adaxially with a single white stomatal band 3 or 4 times as wide as the green marginal bands, keeled abaxially, base jointed, not decurrent. **Seed cones** dark blue when ripe, glaucous, elongated, subglobose, 7-9 mm long. **Seeds** 1 (2) per cone, 3-4 mm. **Pollen shed** spring. **Habitat** serpentine and lava (ultramafic rocks). **Uses** none known. **Dist.**: Serpentine rock in OR and lava (Mt. Hood, OR) occasionally in granite (Trinty Alps, CA). **Status**: this taxon grows in very specific areas and thus could become threatened by habitat degradation.

The type locality is on serpentine, but var. *jackii* grows on a variety of ultramafic rock including high elevation lava (Mt. Hood, OR). The type locality is from south-most corner on the CA-OR boundary.

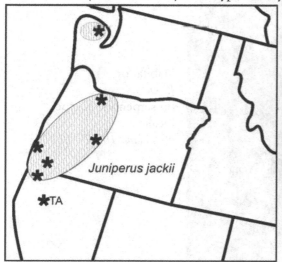

J. jackii at type locality on serpentine
cf. *Adams 10287*

Distribution of *J. jackii*. Asterisks (*) mark locations confirmed by DNA sequencing data. The *TA in CA is in the Trinity Alps on granite.

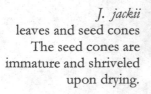

J. jackii
leaves and seed cones
The seed cones are
immature and shriveled
upon drying.

1 cm

J. jackii
on lava at timberline
on Mt. Hood, OR
cf. *Adams 10300*

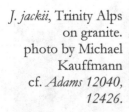

Habitat of
J. jackii
on serpentine at type
locality.
cf. *Adams 10287-10291*

J. jackii, Trinity Alps
on granite.
photo by Michael
Kauffmann
cf. *Adams 12040,
12426.*

Juniperus jaliscana Martinez, Anales. Inst. Biol. Univ. Nac. Mexico 17(1): 69(1946). Cedro, enebro, Jalisco juniper. Type: Mexico, Jalisco, Talpa de Allende Mun., Cuale. *M. Martinez* (with *Gonzalez) 7002* (holotype: MEXU!).

Dioecious. **Trees** (to 10 m) with straight stems, branches ascending to erect, forming a conic or round crown. **Trunk bark** stem bark 10-20 mm thick of fibrous, interconnected strips, grayish-brown surface over cinnamon brown inner bark. **Branches** terminal whip branches erect to strict, tips of branches very flexible and curved, but not recurved, bark reddish-brown and scaly; angle of branching of ultimate twig 50-60 degrees. **Leaves** decurrent(whip) and scale-like, scale leaves opposite, (rarely ternate), ovate or ovoid, obtuse, appressed tips with resin gland on center of leaf. 0.7-1.0 mm long, margins finely toothed, green. **Seed cones** female cones with soft, fleshy pulp, irregular and gibbous, reddish-brown, with a light coat of bloom, 7-8 mm. diameter, peduncles straight. **Seeds** (2-)4-9(-11) per cone, subovoid, angular, with resin grooves, 3-4 mm long, 2 mm wide; hilum about one-half length of seed. **Pollen shed** spring. **Habitat** Pine-Oak forests. **Uses** none known. **Dist.**: At El Puerto de las Viboras, Bosques de San Francisco, Pueblo Nuevo, Durango at 2670 m in the pine-oak forest of the Sierra Madre Occidental (Martinez, 1946) and in the pine-oak-juniper forest at 1335-1600 m on the hills at Cuale, Jalisco, Mexico. **Status**: EN, B1+2c. This species apparently has only a local distribution and may be considered rare in Mexico. It is likely threatened by cutting for fuel, posts, and pasture improvement.

Juniperus jaliscana

J. jaliscana bark

J. jaliscana
leaves and seed cones

J. jaliscana, 19 km e of MEX 200 on road to Cuale, Jalisco, Mexico, cf. *Adams 6846-6848*

Juniperus komarovii Florin, Acta Horti. Gothob. 3: 3 (1927). Ta zhi yan bai, Komarov's juniper.
Type: China, Sichuan, Sangarmai, 4000m, *H. Smith 4402* (holotype GB, isotypes LE; S!)

   J. glaucescens Florin, Acta Horti Gothob. 3: 5 (1927)

   Sabina komarovii (Florin) W. C. Cheng & W. T. Wang, [Chin. title; see Fl. Reipubl. Pop. Sin.
7:374 (1978)] 1: 261 (1961)

Monoecious or Dioecious. Trees to 20 m. **Trunk bark** brownish gray or gray, exfoliating in
interlaced thin strips. **Branches** drooping to pendulous; branchlet systems tapering and gradually
becoming shorter from base to apex of system; branchlets loosely arranged, ascending, straight or
slightly curved, terete or 4-angled, thick, ultimate ones 1.2-1.5 mm in diam. **Leaves** decurrent and
scale-like, decussate, occasionally in whorls of 3 on leading branches, scale-like, ovate-triangular or
triangular-lanceolate, 1.5-3.5(-6) mm, without cuticular wax, abaxial gland near base, ovate or elliptic,
leaf apex acute, rarely obtuse, slightly incurved but free. **Seed cones** erect, brownish black or black
when ripe, slightly glaucous, lustrous, ovoid to turbinate, 8-10(-12) mm, mature in 2 yrs. **Seeds** 1 per
cone, ovoid, rarely obovoid, 6-8.5 mm, obtusely ridged, narrowed by resin pits toward base. **Pollen
shed** spring. **Habitat** forest on high mountains; 3000-4000 m. **Uses** incense in temples. **Dist.**: S
Qinghai, NW Sichuan, China. **Status**: LRnt. This species is cut for incense and areas may be cleared
for pasture. It is likely threatened.

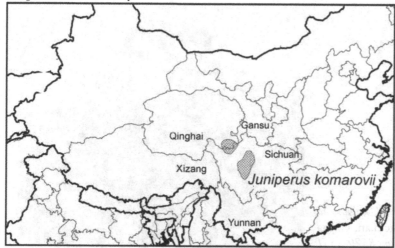

J. komarovii, bark exfoliating in
thin, narrow strips.

J. komarovii habitat
in deep soil on a river
terraces with Spruce,
near Songpan, Sichuan
cf. *Adams 8529-8531*
The 1998 field trip to
Sichuan was organized and
led by Prof. Yang Fei
(deceased), Kunming
Botanical Garden,
Kunming.

J. komarovii, seed cones and leaves. Notice the apiculate tipped ends of the seed cones.

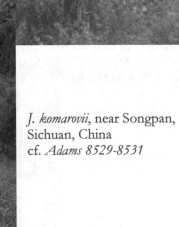

J. komarovii, near Songpan, Sichuan, China cf. *Adams 8529-8531*

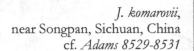

J. komarovii, near Songpan, Sichuan, China cf. *Adams 8529-8531*

Juniperus macrocarpa Sibth. & Sm., Fl. Graeca. Prodr. 2: 263 (1816). Large fruited juniper. Type: Greece. *J. Sibthorp s. n.* (specimen annotated by John Edward Smith) (lectotype OXF, dup at BM, UPS)

  J. communis L. var. *macrocarpa* (Sibth. & Sm.) Spach, Ann. Sci. Nat.Bot., ser. 2, 16: 290 (1841)

  J. lobelii Guss., Fl. Sicula Syn. 2:635 (1844)

  J. biasolettii Link, Flora 29:579 (1846)

  J. umbilicata Gordr. in Grenier & Godron, Fl. France 3:158 (1855)

  Sabina phoenicea (L.) Antoine var. *lobelii*(Guss.) Antoine, Cupress.-Gatt.: 43 (1857)

  J. willkommii Antoine, Cupress.-Gatt.: 9, t.7 (1857)

  J. sphaerocarpa Antoine, Cupress.-Gatt.: 11, t.10 (1857)

  J. macrocarpa Sibth. & Sm. var. *lobelii* (Guss.) Parl. in Candolle, Prodr. 16 (2): 477 (1868)

  J. oxycedrus subsp. *macrocarpa* (Sibth. & Sm.) Neilr., Verh. Zool.-Bot. Ges.Wein 1868: Veget Croat. 52(1868)

  J. macrocarpa Sibth. & Sm. var. *globosa* Neilr., Verh. Zool. Bot. Ges. Wien 19:780 (1869)

  J. oxycedrus subsp. *macrocarpa* (Sibth. & Sm.) Ball. Bull. J. Linn. Soc. London (Bot.) 16:670 (1878)

  J. major Dioscorides ex Bubani, Fl. Pyren. *1:44 (1897)*

  J. oxycedrus var. *macrocarpa* (Sibth. & Sm.) Silba, Int. Conifer Preserv. Soc. 7: 25 (2000)

Dioecious. Tree to 14 m. **Trunk bark** brownish red, exfoliating in strips. **Branches** erect. **Leaves** acicular, 12-20 mm long, 2 - 2.5 mm wide, in groups of three, acuminate, adaxial surface with 2 distinct stomatal bands, abaxial side keeled. **Seed cones** berry-like, purplish-red when ripe, globose, 12 - 18 mm long, maturing in 2 yrs. **Seeds** usually 3 per cone. **Pollen shed** winter. **Habitat** along sea coasts of the Mediterranean. **Uses** none known. **Dist.:** S. Portugal eastward to Cyprus. **Status:** the taxon is widespread in coastal areas of the Mediterranean and does not appear to be threatened. *Juniperus macrocarpa* is, morphologically, very distinct from *J. oxycedrus*, so it is surprising that it is often treated as part of *J. oxycedrus*.

Taxonomic note: Figs. 1.3, 1.11.1 (Chpt. 1) show that it is quite distinct in its DNA, although it is only 9 MEs removed from *J. oxycedrus* (Fig. 1.11.1), it is also quite distinct in its morphology. In this treatment, *J. macrocarpa* is recognized as a distinct species.

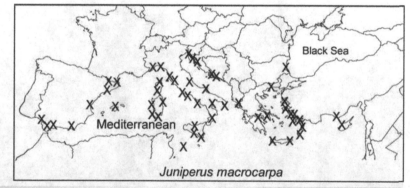

Juniperus macrocarpa

J. macrocarpa
Tarifa sand dunes,
Spain,
cf. *Adams 7205-7207*

J. macrocarpa leaves and seed cones. Note the cone scale impressions on the seed cone. These are similar to *J. drupacea* seed cones.

J. macrocarpa, bark exfoliating in thin, broad strips.

J. macrocarpa, Tarifa sand dunes, Spain, cf. *Adams 7205-7207*

Juniperus maderensis (Menezes) R. P. Adams, Phytologia 92(1): 53 (2010). Madeira Juniper, Type: Madeira, Serra do Faial, Curral das Freiras, Portugal. Basionym: *Juniperus oxycedrus* L. subsp. *maderensis* Menezes, Bull. Acad. Int. Geogr. Bot. 18 (No. 227-227): xii. 1908.

J. cedrus Webb & Berthel. subsp. maderensis (Menezes) Rivas Mart. et al., Itinera Geobot.
　15 (2): 703 (2002). Type: Madeira, Serra do Faial, Curral das Freiras.
Juniperus grandifolius Link, in Buch, Phys. Beschr. Canar. Ins.: 159. 1825. Farjon (2005) noted
　that this is a *nom. inval.* under Art. 34.1.

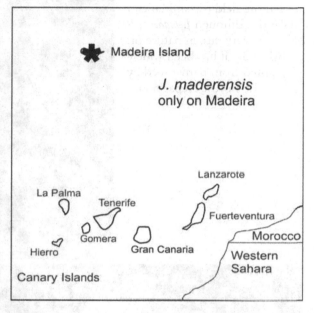

Dioecious. Trees to 30 m , or shrub at high altitudes. **Trunk bark** brown, peeling in coarse vertical strings, reddish-purple beneath. **Branches** level, upcurved toward tips, with pendulous branchlets. **Leaves** tips usu. mucronate but not elongated, acicular, straight, in whorls of three, 1-2 mm wide, two stomatal bands either side of the midrib on the adaxial surface, approx. as wide as the midrib, leaf midrib distinct, green, not covered with bloom. **Seed cones** red-brown, glaucous bloom, globose, 8-12 mm, approx. the same length as leaves, green when immature, mature in 2nd year. **Seeds** 3 per cone. **Pollen shed** fall. **Habitat** volcanic rocks and soils. **Uses** the wood of *J. maderensis* has long been used in carpentry and was even used in some of historic buildings of Funchal such as the Cathedral (1514) and the Old Customs House (1477). Apparently the trees were quite large as the wood was also used for beams. **Dist.**: Endemic to Madeira on the highest peaks in rocky areas. Nobrega collected it from Pico Ferreiro in 1988 and found 16 juniper plants on Pico das Torres in 1990. **Status**: It is extremely rare and endangered in nature, but the seeds have been collected and it is now widely cultivated on Madeira.

　　Recent DNA data (see Figs. 1.10, 1.11.1, Chpt. 1 and Adams et al. 2010b, 2012a) revealed that *J. maderensis* is a bit closer to *J. oxycedrus* (11 MEs) than to *J. cedrus* (13 MEs, Fig. 1.11.1). The taxon previously called *J. cedrus* in Madeira was recognized as a new species, *J. maderensis*.

Distinguishing characteristics that separate *J. cedrus*, *J. maderensis* and *J. oxycedrus*.

	J. cedrus	*J. maderensis*	*J. oxycedrus*
Leaf tips	blunt to acute rarely mucronate tipped	usu. with a mucronate tip but not an elongated tip	elongated, mucronate tips
Stomatal bands	wider than midrib.	approx. as wide as midrib	approx. as wide as midrib
Leaf midrib	indistinct, usu. covered with bloom	distinct, green not covered with bloom	distinct, green not covered with bloom
Seed Cones (mature)	bright red, little or no bloom	brown-red, with bloom	brown-red to bright red, little bloom
Seed cones (mature)	larger than leaf length	~same as leaf length	smaller than leaf length

J. maderensis, cultivated from native seed, Madeira Island, cf. *Adams 11496,* with Dr. Suzana S. Fontinha, who arranged and led the field trip in Madeira Island. Although *J. maderensis* is very rare in nature on Madeira, it has been widely planted from native seed by the Forest Department.

1 cm

J. maderensis, seed cones and leaves.

J. maderensis, bark.

Juniperus mairei Lemee & H. Leveille Monde Pl. 2(16): 20 (1914). Maire's juniper, Type: Yunnan, Jong-tohouan, *J. mairei* Lemee & H. Lev., *E. E. Maire s. n.*, (holotype A! barcode *#38339*, isotype E?)

> *Juniperus formosana* var. *mairei* (Lemee and H. Lev.) R. P. Adams and C-F. Hsieh, Biochem. Syst. Ecol. 30: 239 (2002).
>
> *J. chekiangensis* Nakai
>
> *J. formosana* f. *tenella* Handel-Mazzetti

Dioecious. **Trees** small. **Trunk bark** brown. **Branches** erect, dense. **Leaves** acicular, usually with raised oval or elliptical gland on the persistent branch (brown) leaf sheath, in whorls of 3, linear-lanceolate or linear-needlelike, 1.2-2 cm x 1.2-2 mm, slightly concave adaxially, with 2 white, broad stomatal bands separated by a narrow, green midvein, green and obtusely keeled abaxially, base jointed, not decurrent, apex sharply pointed. **Seed cones** 3 seeds per cone, 5-8 mm diam., glaucous, green when immature, turning reddish blue when mature in 2 yrs. **Pollen shed** summer. **Habitat** forests 200- 3400 m. **Uses** none known **Dist**.: S. Anhui, W. Fujian, E. Gansu, Guizhou, W. Hubei, S Hunan, S. Jiangsu, Jiangxi, NE Qinghai, S Shaanxi, Sichuan, S Xizang, Yunnan, Zhejiang. **Status**: scattered in large areas, probably not threatened.

Taxonomic notes: Recent DNA sequencing (Adams and Schwarzbach, 2012a) has shown (Figs. 1.3, 1.10) that *J. mairei* is in a clade with *J. jackii* (North America), and not in the clade with *J. formosana* as previously thought (Adams 2011). This gives support for the recognition of *J. mairei* at the specific level.

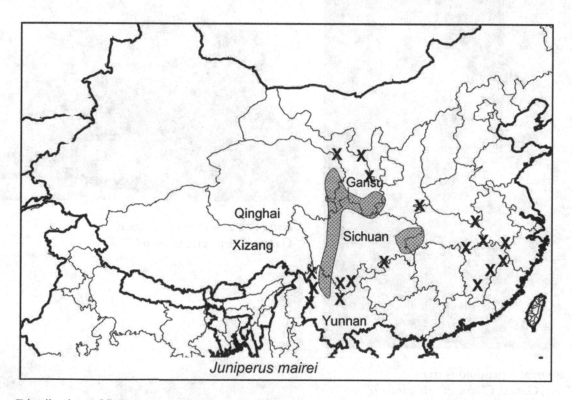

Distribution of *Juniperus mairei*.

J. mairei seed cones and leaves

J. mairei bark exfoliating
in thin, narrow strips.

J. mairei as a 2 m shrub near Jone,
Gansu, China. cf. *Adams 6772-6774*

J. mairei as a pendulous tree
(5 m), Gansu, China. cf. *Adams 6792*

J. mairei as a shrub in Gansu,
China with Guo Shou Jun.
cf. *Adams 6772-6774.*

(Prof. Ge-Lin Chu, not shown,
led the field trip)

Juniperus maritima R. P. Adams. Phytologia 89(3): 278 (2007). seaside juniper, Type: Canada, BC, Vancouver Island, Brentwood Bay, Lat 48° 34.794' N; Long 123° 20.211' W, elev. 5 m., 29 May 2006, *R. P. Adams 11056* (holotype BAYLU, isotype V).

Recent DNA sequencing shows *J. maritima* is in a well supported clade with *J. virginiana* and *J. scopulorum* but quite distinct (Figs. 1.3, 1.16.1, 1.16.5, Chpt. 1). This species is similar to *J. scopulorum* but differs in that the seed cones mature in 1 year (14-16 months), seeds are usually exserted from the cone, and the scale leaf tips are obtuse.

Morphological comparison of *J. maritima*, *J. scopulorum* and *J. virginiana*:

	J. maritima	*J. scopulorum*	*J. virginiana*
seed cones mature	1 yr (14-16 mos.)	2 years	1 year
seed cone diam.	6-8 mm	6-9 mm	3-6(7) mm
seed cone shape	globose to reniform	globose to reniform	ovoid
seeds per cone	(1) 2	(1) 2 (3)	1-2 (3)
exserted seeds	very common	rare	rare
scale leaf overlap	< 1/5 length	< 1/5 length	> 1/4 length
scale leaf tips	obtuse	acute to obtuse	acute
branchlets (6-15mm diam.)	smooth, reddish-brown	smooth, bright reddish-brown	brown with persistent old leaves

Dioecious. Trees single stemmed to 15 m or more, pyramidal to round crown. **Trunk bark** brown, exfoliating in thin strips. **Branches** foliage erect or occasionally lax, green but turning reddish-brown in the winter, twigs (3-5 mm diam.) with persistent dead scale leaves, bark on twigs (6-15 mm diam.) smooth reddish-brown beneath. **Leaves** both decurrent (whip) and scale. Whip leaves growing only at branchlet tips (on mature trees), with an elliptical or elongated gland. Scale leaves overlapping by less then 1/5 length), tips obtuse. Scale-leaf margins entire (20 X and 40 X). **Seed cones** globose to reniform, blue-black to blue brownish, maturing in 14 to 16 mos., borne terminally, 6-8 mm in diam., (1) 2 seeded. **Seeds** tan to brown, 2-4 mm long, commonly exserted due to insect damage. **Pollen shed** March-April. **Habitat** near the seashore on s and w exposed rock, on sand, on rock in the rain shadow of Mt. Olympia. **Uses** none known **Dist.**: Puget Sound and Strait of Georgia, WA and BC See Adams et al. (2010d) for notes on distribution on the Olympic Peninsula. **Status**: this taxon has very limited distribution and grows in areas of prime development, so it may become threatened.

See Adams (2009c) for a recent comparison of the leaf essential oils of *J. maritima* with *J. horizontalis*, *J. scopulorum* and *J. virginiana*.

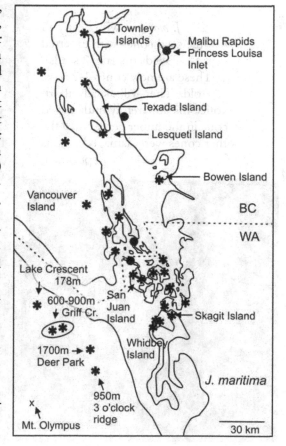

J. maritima leaves and normal seed cones

J. maritima, bark exfoliating in narrow thin strips.

J. maritima with nearly all the seed cones with excerted seeds (usually 2 seeds). These are new cones 1-2 mos. old. They will usually abort. Notice the apparently undamaged cone in the upper center. All the other cones were damaged on this specimen.

Author and the largest known tree of *J. maritima* in the USA. Skagit Island, WA. *Adams 11077*

Wade Calder coring a *J. maritima* tree on Lesqueti Island, BC

J. maritima, Anacortes, Fidalgo Island, WA, cf. Adams 11076

J. maritima, Brentwood Bay, Victoria, BC, Canada cf. *Adams 11056*, holotype

Wind shaped, *J. maritima* on sand dune, Whidbey Island St. Pk., WA, cf. *Adams 11075*

J. maritima (center and lower right), Squitty Bay, Lesqueti Island, BC. cf. *Adams 11065*

Juniperus martinezii Perez de la Rosa, Phytologia 57: 81 (1985) Martinez juniper Type: Mexico, Jalisco, Sierra Cuatralba, El Cuarento, 4 km e of village, *J. A. Perez de la Rosa, 661*. (holotype IBUG, isotypes F, K).

Juniperus flaccida Schltdl. var. *martinezii* (Perez de la Rosa) Silba, Phytologia 58: 367(1985).

Dioecious. Trees 4-8 m tall, crown round to broadly conical. **Trunk bark** grayish-brown, exfoliating in interlaced strips. **Branches** ashen color, terminal branchlets flaccid or pendulous. **Leaves** usually opposite and not imbricate (1.5-)2(-2.5) mm long, 1.0 (-1.5) mm wide, ovate to elongate-ovate, acuminate, mucronate, with dorsal gland, hyaline margin barely dentate (40x). **Seed cones** ovoid (5-)6(-8) mm long by (5-)6(-9) mm wide, brownish, gibbous. **Seeds** 1-2(-3), 3-5 mm long by 3-4 mm wide. **Pollen shed** late spring. **Habitat** rocky areas. **Uses** none known. **Dist.**: Mexico, see map. **Status**: LRnt. known only from a few populations, likely to become threatened due to range land improvement.

Recent DNA sequencing (Adams and Schwarzbach, 201a,b) revealed that *J. martinezii* is more closely related to *J. durangensis* than to *J. flaccida* (see Fig. 1.14, Chpt. 1, and Fig. 5.22.3). There is strong support for it recognition at the specific level. In this treatment, it is treated as a distinct species, ***J. martinezii***.

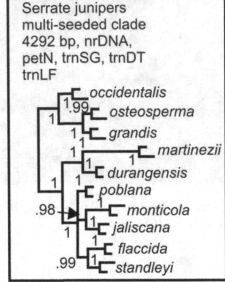

Figure 5.22.3. Clade containing *J. martinezii* and related species.

J. martinezii with bark exfoliating in interlaced narrow strips.

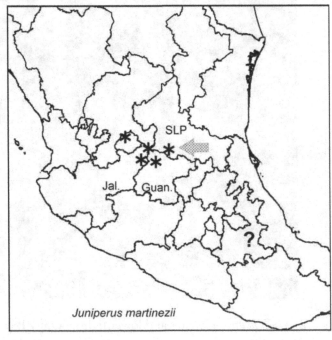

J. martinezii
leaves and seed cones

J. martinezii, north of Lago de Moreno,
La Quebrada Ranch, Mexico, cf. *Adams 8709*

J. martinezii, north of Lago de
Moreno, La Quebrada Ranch,
Mexico, cf. *Adams 8709*
with Glenn Harris and a local
rancher.

220 Robert P. Adams

Juniperus microsperma (Cheng & L.K. Fu) R. P. Adams. Biochem. Syst. Ecol. 28: 540 (2000). Xiao zi yuan bai, small seeded juniper. Type: China, Tibet, Songzong, Shnowa, (old name = Sangru), 3500 m. *(First) Expedition to Xizang (Tibet) 10019* (holotype PE!)

    Sabina convallium (Rehder & E. H. Wilson) W. C. Cheng & W. T. Wang var. *microsperma* W.
      C. Cheng & L.K. Fu. Acta Phytotax. Sin. 13: 86 (1975)
    S. microsperma (W.C. Cheng & L.K. Fu) W.C. Cheng & L.K. Fu. Xizangica 1: 390 (1983)
    J. convallium Rehder & E. H. Wilson var. *microsperma* (W.C. Cheng & L.K. Fu) Silba,
      Phytologia Mem. 7: 33 (1984)

Dioecious or monoecious. Trees to 10 m with strong central axis. **Trunk bark** gray to cinnamon, exfoliating in thin strips. **Branches** erect. **Leaves** decurrent (whip) and scale-like, scale leaves with an elliptic, ovate, or orbicular, convex gland abaxially. **Seed cones** bluish brown without noticeable bloom (glaucousness), irregular globose, 5-7 x ca. 5 mm. **Seeds** 1 per cone, flattened ovoid, ca 4 x 3 mm. **Pollen shed** April-May. **Habitat** high mountains; 3200-4000 m. E. Xizang. **Uses** none known. **Dist.**: China: Xizang (Tibet), 3500 to 4000 m. **Status**: DD. At this time it is known from a very limited area. It is likely threatened.

    Juniperus microsperma has been thought to be very closely related (even conspecific) with *J. convallium*, but recent DNA sequencing (Figs. 1.16.1, 1.16.2, Chpt. 1) have shown *J. microsperma* is in a clade with *J. erectopatens* and not with *J. convallium* (Fig. 5.22.4). Its leaf oil is quite unique (Adams, Mao and Liu, 2013).

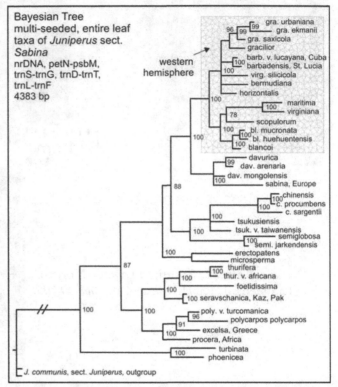

Figure 5.22.4. Bayesian tree of multi-seeded, smooth leaf taxa of section *Sabina* of the eastern hemisphere.

type specimen of *J. microsperma*
(First) Expedition to Xizang (Tibet)
10019 (holotype PE!)

J. microsperma, leaves and seed cones
from near the type locatity.
cf. *Liu, J-Q (Jian-Quan Liu)-QTP-2011-201*
Lab Acc. *Robert P. Adams 13633*

J. microsperma, habitat near
Song Zong, Tibet, near type locality
~29° 58' N; 94° 52' E., elev. ca 2650 - 2700 m
photo by Kangshan Mao

Juniperus monosperma (Engelmann) Sargent. Silva N. Amer. 10: 89 (1896). One-seeded juniper, cherry-stone juniper. Type: USA, Colorado, Fremont Co., Canon City, *G. Englemann s. n.* (lectotype MO, designated by T. Zanoni, Feb. 1992)

> *J. occidentalis* Hook. var. *monosperma* Engelm., Trans. St. Louis Acad. Sci. 3: 590 (1878)
>
> *J. californica* Carriere var. *monosperma* (Engelm.) Lemmon, Handb. W. Amer. Conebearers, ed. 2: 17 (1892)
>
> *J. occidentalis* Hook. var. *gymnocarpa* Lemmon, Handb. W. Amer. Cone-bearers, ed. 3: 80 (1895)
>
> *J. occidentalis* Hook. f. *gymnocarpa* (Lemmon) Rehder, J. Arnold Arbor. 7: 239 (1926)
>
> *J. mexicana* Schiede ex Schltdl. & Cham. var. *monosperma* (Engelm.) Cory, Rhodora 38: 183 (1936)
>
> *J. gymnocarpa* (Lemmon) Cory, Rhodora 38: 184 (1936)
>
> *Sabina monosperma* (Engelm.) Rydberg, Bull. Torrey Bot. Club 32: 598 (1905)

Dioecious. **Shrub** or small tree, 2-7 (-12) m, usually with stems branching near the ground. **Trunk bark** thin, gray to brown, exfoliating in thin strips revealing cinnamon color. **Branches** ascending to erect, with an ashy-white peeling bark. **Leaves** both decurrent (whip) and scale-like. Ultimate branchlets approx. 2/3 as wide as scale leaf length, square or six-sided but not terete. Whip- and scale-leaf margins denticulate (20 X). Scale leaves acute to acuminate. Whip leaf gland 3/4 as long as the leaf, adaxial (inner) leaf surface glaucous. Scale leaves 1-3 mm, ovate, acute to acuminate, green. Few (less than 1/5) whip-leaf glands ruptured and with a white crystalline exudate (visible without a lens). **Seed cones** 6-8 mm, soft and juicy pulp, globose to ovoid, reddish-blue to brownish-blue, white glaucous, 1(2-3) seeded, the hilum scar approx. 1/3 as long as seed. **Seeds** 4-5 mm long. **Pollen shed** late winter—early spring. **Habitat** common shrub in dry rocky soils and slopes; 1000-2300 m. **Uses** not rot resistant, so not used for fence posts. **Dist**.: Ariz., Colo., N. Mex., Okla., Tex. Often reported from Mexico, but these plants should be referred to *J. angosturana* or *J. coahuilensis*. True *J. monosperma* has not been seen in Mexico by the author. **Status**: This species is the dominant plant on millions of hectares in New Mexico, USA. It is reproducing and is considered a pest (weed) in pastures. It is not threatened.

The synonym, *J. occidentalis* Hook. var. *gymnocarpa* Lemmon, came from a term 'gymnocarpa' that seems to be have been introduced by Lemmon in 1895. It is common to see the seeds exerted from the seed cones on most *Juniperus* species. This is apparently due to insect damage to the immature cone. See the figure to the left of excerted seeds in *J. saltillensis*.

Hybridization between *J. monosperma* and *J. pinchotii* (Hall and Carr, 1968) has been negated using numerous chemical and morphological characters (Adams, 1969; 1975). Note also that pollination for *J. monosperma* is in late winter - early spring whereas *J. pinchotii* is in fall. Hybridization with *J. coahuilensis*, a sibling species, does appear likely and is under investigation.

Gymnocarpy (in *J. saltillensis*) Gymnocarpy is common in many juniper species in the southwestern US and Mexico.

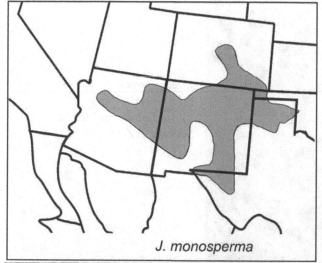

J. monosperma, bark exfoliating
in thin, narrow strips.

J. monosperma leaves and seed cones.

1 cm

J. monosperma, on limestone, 7 mi. e of Gruver,
Texas, USA. cf. *Adams 7789*

J. monosperma near the canyon rim in Palo Duro
canyon, Texas.

J. monosperma
rare mutation with
twisted limbs and trunk,
Sedona, AZ.

J. monosperma
Reserve, NM

J. monosperma
on volcanic pumice
north of Flagstaff, AZ

Juniperus monticola Martinez

Recent DNA sequencing has shown that *J. monticola f. compacta* from Sierra Potosi should not be included in *J. monticola*, but is a new species, **J. zanonii** (see Figs. 1.14, 1.14.1, Chpt. 1).

See note by Adams and Zanoni (1993) that the correct name is *J. monticola* not *J. sabinoides*.

Key to forms:

1a. Mature twigs 5-10 mm long, angle of branching of ultimate twig 50-60 degrees, foliage very dense and tightly compacted, prostrate shrubs (less than 1 m), alpine, sub-alpine habitat
....***J. monticola f. compacta***

1b. Mature ultimate twigs 10-40 mm long; angle of branching of ultimate twig 40-60 degrees; foliage not tightly compacted; spreading, low shrub to tree to 10m

   2a. Mature ultimate twigs 10-15(-20) mm long, angle of branching of ultimate twig 40-50 degrees, foliage not tightly compacted, spreading, tortuous-stemmed shrub to tree to 10 m. ....***J. monticola f. monticola***

   2b. Mature ultimate twigs 20-40 mm. long, angle of branching of ultimate twig 55-60 degrees, foliage not tightly compacted, shrubs to 1.5 m. ......***J. monticola f. orizabensis***
   Forma *orizabensis* is scarcely distinct. Additional analyses may negate the use of forma *orizabensis*.

Juniperus monticola Martinez f. **compacta** Martinez, Anales Inst. Biol. Univ. Nac. Mexico 17: 87 (1946). Compact mountain juniper, Type P!, Mexico: Vera Cruz de Perot. Humboldt and Bonpland s.n.

   Cupressus sabinoides H.B.K, Nova. Gen. et Sp. Pl. 2: 3 (1817).

   J. mexicana Sprengel, Syst. Veg. 3: 309 (1826), *nom. superfl. illeg.*

   J. sabinoides (Kunth) Nees. Linnaea 19: 706 (1847), *non* Griseb., Spic. Fl. Rumel. 2: 352 (1846).

   J. sabinoides Humb. (erroneously attributed) in Lindley and Gordon, J. Hort. Soc. 5: 202 (1850).

   J. monticola Martinez var. *monticola* f. *compacta* Martinez, Bol. Soc. Bot. Mexico 7: 19 (1948).

   J. compacta R. P. Adams, Phytologia 89: 368 (2007).

Dioecious. **Prostrate shrubs** with twisted branches forming mats (to 1 m high). **Trunk bark** cinnamon brown, exfoliating in thin strips. **Branches** procumbent. **Leaves** decurrent (whip) and scale-like, foliage very densely compacted; angle of branching of ultimate twig 50-60 degrees mature ultimate twigs 5-10 mm long. **Seed cones** female cones with soft, fleshy pulp, globose or gibbous, dark bluish-black, with a light coat of bloom. 5-9(-10) mm diameter, peduncles usually curved. **Seeds** (2-)3-7(-9) per cone, usually angular and grooved, hilum about 0.5 to 0.6 length of seed. **Pollen shed** fall. **Habitat** at edge of pine forest treeline (about 3000 m elevation); or above treeline (alpine) in mountain grasslands of *Calamagrostis* and *Festuca*, usually in volcanic rocky crevices, or on volcanic rocky outcrops at 3000-4300 (-4500) m elevation. **Uses** none known. **Dist.**: Cerro Pelado and Ajusco, Distrito Federal: Nevado de Colima, Jalisco; Popocatepetl, Iztaccihuatl, Tlaloc and Nevado de Toluca, Mexico; Malinche, Tlaxcala; and Cofre de Perote, Vera Cruz; Mexico. **Status**: This form is found in very scattered, alpine areas. Because it is a small shrub of no economic importance and these areas are not suited for pasture, it is likely not threatened.

Juniperus monticola f. compacta

J. monticola f. *compacta*, bark.

J. monticola f. *compacta*, leaves and seed cones.

J. monticola f. *compacta* (shrub)
Malinche, Tlaxcala
(photo by Alicia
Mastretta).

J. monticola f. *compacta*
Nevada de Toluca
(photo by Tom A.
Zanoni).

Juniperus monticola Martinez f. ***monticola***, Anales Inst. Biol. Univ. Nac. Mexico 17(1): 79 (1946). Cedro, sabino, sabino de castillo, tlascal (Hidalgo); mountain juniper. Type: Mexico, Hidalgo, Mineral del Monte. *C. A. Eherenbrg s.n.* (holotype B, lost, lectotype: MO)

    Cupressus sabinoides Kunth in Humboldt & al., Nov. Gen. Sp. Pl. 2(5): 3 (1817)

    J. tetragona Schltdl., Linnaea 12: 495 (1838), *non* Moench, Methodus 699 (1794)

    J. mexicana Spreng., Syst. Veg., ed. 16, 3: 909 (1826)

    J. sabinoides (Kunth) Nees, Linnaea 19: 706 (1847), *non* Griseb. (1846)

    J. sabinoides Humb. ex Lindl. & Gordon, J. Hort. Soc. London 5: 202 (1850), *non* Griseb. (1846)

    J. sabinoides (Kunth) Nees f. *monticola* (Martinez) M. C. Johnst. Taxon 34(3): 506 (1985)

    Sabina tetragona (Schltdl.) Antoine, Cupress.-Gatt.: 40 (1857)

Dioecious. **Trees** or spreading shrub with tortuous branches, if a tree, then to 10 m with crown flattened to broadly conic. **Trunk bark** stem bark 5-10 mm thick, gray to grayish-brown, exfoliating in fibrous, longitudinal strips. **Branches** terminal whip branches spreading and tortuous to ascending, tips occasionally curved, bark reddish-brown to grayish-brown, distichous. **Leaves** decurrent (whip) and scale-like, scale leaves usually opposite, often thick and appearing as a string of beads on the ultimate twig, rounded or obtuse, appressed tips, 1.0-2.0 mm long, margins finely denticulate, grayish-green to green, resin gland sometimes obvious on surface of leaf, angle of branching of ultimate twig 40-50 degrees, foliage not tightly compacted. **Seed cones** with soft, fleshy pulp, globose or gibbous, dark bluish-black, with a light coat of bloom (glaucous), 5-9(-10) mm diameter, peduncles usually curved. **Seeds** (2-)3-7(-9) per cone, usually angular and grooved, hilum about 0.5 to 0.6 length of seed. **Pollen shed** fall. **Habitat** rocky areas in subalpine forests of oak-juniper, pine or *Abies*. **Uses** none known. **Dist.**: Mostly subalpine forests of oak-juniper, pine, or *Abies*, as understory shrub at 2400 m elevation or higher, or at above treeline on Nevada de Colima, Jalisco, in rocky crevices to 4300 m elevation, Mexico. **Status**: occurs in several areas and habitats, not threatened.

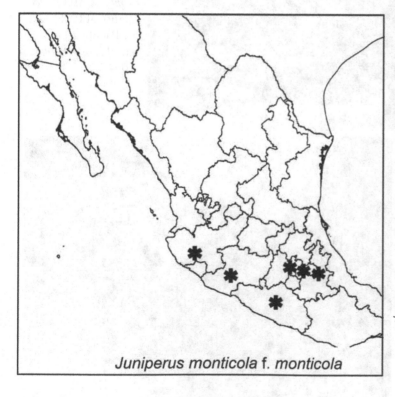

Juniperus monticola f. *monticola*

J. monticola f. monticola
bark

228 Robert P. Adams

J. monticola f. monticola
leaves and seed cones.

1 cm

J. monticola f.
monticola, El Chico
National Park, east
of Pachuca, Mexico,
cf. *Adams 10544-
10547*

J. monticola f. *monticola*
El Chico National Park,
east of Pachuca, Mexico,
cf. *Adams 10544-10547*

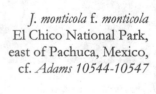

Juniperus monticola f. ***orizabensis*** Martinez. Anales Inst. Biol. Univ. Nac. Mexico 17: 91 (1946). Orizaba mountain juniper Type: Mexico, Vera Cruz, Pico de Orizaba. *I. de B. 162* (holotype Farjon(2005) says at MEXU, but I have not been able to locate it, lectotype: *M. Martinez 8001.* MEXU!)

> *Juniperus monticola* Martinez var. *monticola* f. *orizabensis* Martinez, Anales Inst. Biol. Univ. Nac. Mexico 17(1): 91 (1946)
>
> *J. sabinoides* (Kunth) Nees var. *sabinoides* f. *orizabensis* (Martinez) M. C. Johnston, Taxon 34(3): 506 (1985)

Dioecious. Shrubs to 1.5 m, with twisted branches. **Trunk bark** light brown, exfoliating in this strips or plates. **Branches** spreading. **Leaves** decurrent (whip) and scale-like, foliage not tightly compacted; angle of branching of ultimate twig 55-60 degrees; mature ultimate twigs 20-40 mm long. **Seed cones** female cones with soft, fleshy pulp, globose or gibbous, dark bluish-black, with a light coat of bloom. 5-9(-10) mm diameter, peduncles usually curved. **Seeds** (2-) 3-7(-9) per cone, usually angular and grooved, hilum about 0.5 to 0.6 length of seed. **Pollen shed** fall. **Habitat** alpine habitats at treeline and above treeline. **Uses** none known. **Dist.:** in rocky crevices on Pico de Orizaba, Sierra Nevada, and Cofre de Perote, Vera Cruz; Pena Nevada, Tamaulipas at 3700-4500 m elevation; also in oak forest near Mineral Catoree, San Luis Potosi at 2850 m elevation, Mexico. **Status:** known from only a few sites, but these alpine sites are not heavily disturbed and because it is a shrub, it is of no economic use. It is not threatened.

Juniperus monticola f. orizabensis

1 cm

J. monticola f. *orizabensis* leaves and seed cones

J. monticola f. *orizabensis*, Pico Orizaba, Mexico
(photo T. A. Zanoni)

Juniperus morrisonicola Hayata, Gard. Chron. ser. 3, 43:194 (1908). Mt. Morrison juniper (Taiwan)
Type: Taiwan, Nantou: Chia-i Pref., Mt. Morrison (= Yu Shan = Yushan or Mt. Jade), 23.5° N, 121.0°
E, *S. Nagasawa 585* (holotype TI, isotype BM).

 J. squamata sensu Kanehira, Form. Trees rev. ed. 60. f.22. 1936, *non* Lambert
 Sabinella morrisonicola (Hayata) Nakai, Chosen Sanrin Kaiho 165: 14 (1938)
 J. squamata Buch.-Ham. ex. D. Don var. *morrisonicola* (Hayata) H. L. Li & H. Keng, Taiwania
  5: 81. t. 28. (1954)

Dioecious or monoecious. Tree or shrub, procumbent, up to 5-10 m in height. **Trunk bark**
brown, exfoliating in thin plates. **Branches** long, decumbent, terete, the branchlets ascending. **Leaves**
only decurrent leaves, mucronate tips, 3-5 mm long, 1 mm broad, densely and ternately arranged,
concave and whitish above, grooved and green beneath. **Seed cones** greenish brown to tan when
immature, then purplish black when mature in 1 yr., subtended by several bracts at the base globose or
ellipsoid, 6-8 mm long. **Seeds** 1 per cone, globose or ovoid. **Pollen shed** spring. **Habitat** rocky cliffs
and in bamboo thickets. **Uses** none known. **Dist.**: Taiwan, 3000- 4000 m. **Status**: this species is not
widely distributed, but it is in protected areas of Taiwan in rocky areas that not likely to be burned nor
cut over.

  Juniperus morrisonicola has traditionally been treated as a part of *J. squamata* (Farjon, 2005) but
both terpenoids (Fig. 3.4, Chpt. 3) and DNA fingerprinting (Adams, 2000c) revealed (Fig. 3.5, Chpt. 3)
it to be so different that it was recognized at the specific level. This decision has now been validated by
DNA sequencing (Schwarzbach et al., in prep.), see figure 1.13 which clearly shows that *J. morrisonicola*
is in a clade with *J. prezewalskii* and well separated from *J. squamata*.

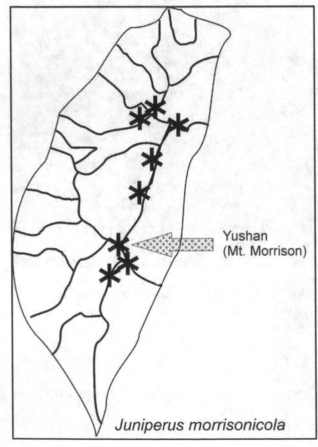

Yushan
(Mt. Morrison)

Juniperus morrisonicola

J. morrisonicola bark, exfoliating in thin plates.

J. morrisonicola
leaves and seed cones

J. morrisonicola
Hehuanshan (Mt. Hohuan) area, Taiwan
with Chang-Fu Hsieh
cf. *Adams 9068-9070*

J. morrisonicola
Hehuanshan
(Mt. Hohuan)
area, Taiwan, with
Chang-Fu Hsieh
cf. *Adams 9068-
9070*

Juniperus. navicularis Gand., Bull. Soc. Bot. France 57: 55 (1910).   Sea sands juniper, fastigiate juniper Type: Portugal, Alemtejo, Sado, Troia, *J. Daveau 1346* (lectotype P, designated by Farjon (p. 335, 2005).

   J. oxycedrus subsp. *rufescens auct. lusit.*
   J. oxycedrus subsp. *transtagana* Franco, Fedde's Repert.Sp. Nov. Regni Veg. 68(3):166 (1963)
   J. oxycedrus L. var. *transtagana* (Franco) Silba, Phytologia 7: 35 (1984)

Dioecious. **Shrubs** up to 2 m, branching at the base (fastigiate). **Trunk bark** brown, exfoliating in thin plates. **Branches** somewhat pendulous. **Leaves** acicular, 4-12 x 1 - 1.5 mm, obtuse and mucronate, with two glaucous bands above. **Seed cones** female maturing in 2 years, globose to pyriform, yellowish when unripe, maturing as red or reddish-copper, 7-10 mm. **Seeds** free, usually 3 per cone. **Pollen shed** summer? **Habitat** on Pliocene sands with *Pinus* near the seacoast. **Uses** none known. **Dist**.: Pliocene sands, W. Portugal. **Status**: common in a few areas of Portugal, but these areas do not seem to be utilized, so it is likely not threatened.

   Farjon (2005) treats this taxon as a subspecies of *J. oxycedrus*, but the new sequence data (Fig. 1.3, 1.10, 1.11.1, Chpt. 1, Adams and Schwarzbach, 2013e) clearly places *J. navicularis* in a clade with *J. brevifolia*, not with *J. oxycedrus*. The sequencing data re-confirms the terpene and RAPDs data (Adams, 2000) that *J. navicularis* is a distinct species.

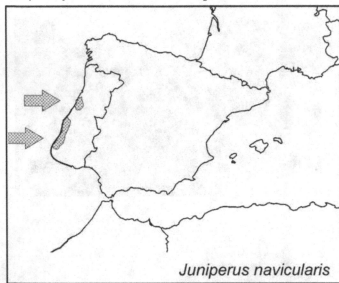

J. navicularis bark exfoliating in smooth plates.

J. navicularis on Pliocene sand with *Pinus*, 1 km se of Apostics, s of Lisbon, Portugal. cf. *Adams 8239-8243*

J. navicularis
leaves and seed cones
(immature, so they wrinkled
upon drying)

J. navicularis on Pliocene sand with *Pinus*, 1 km se of Apostics, s of Lisbon, Portugal.
cf. *Adams 8239-8243*

Juniperus occidentalis

Recently, Adams (2012) recognized a new forma *J. occidentalis* f. *corbetii* that is similar to *Juniperus occidentalis* but differing habit, being a shrub with compact foliage (see below).

Juniperus occidentalis Hook., Fl. Bor. Amer. 2(10): 166 (1838). Western juniper, Sierra juniper. Type: USA, Washington, Columbia River, *D. Douglas s. n.* (holotype K!)

Chamaecyparis boursieri Decne., Bull. Soc. Bot. France 1: 70 (1854)
J. andina Nutt., N. Amer. Sylva 3: 95, t.110 (1849)
J. pseudocupressus Dieck, Neuheit.-Off. Nat.-Arb. Zoschen 1899: 8 (1899)
J. californica var. *siskiyouensis* L.F. Henderson, Rhodora 33: 203 (1931)
J. occidentalis f. *robinsoni* O. V. Matthew
Sabina occidentalis (Hook.) Antoine, Cupress.-Gatt.: 64 (1857)

Monoecious / Dioecious approx. 50% of the plants monoecious (Vasek, 1966). **Trees** to 20 m, 1 (-3) trunks. **Trunk bark** brown. **Branches** ascending. **Leaves** decurrent (whip) and scale-like, both kinds with visible glands with exudate that turns dark brown to black. **Seed cones** blue to blue-black, with resinous pulp, maturing in 2 yrs., 7-10 mm long (avg. 8.3). **Seeds** 1-2(3) per cone (1.6 avg.). **Pollen shed** late spring. **Habitat** dry rocky foothill and mtn. slopes; (near sea level) to 1500-3000 m; Calif., Idaho, Ore., Nev., Wash. Map: Vasek, 1966. **Uses** fence posts (but not very rot resistant). **Dist.**: Sierra Nevada of northern California, Oregon and Washington into Nevada. **Status**: common and reproducing. Considered a pest (weed) on pastureland in Oregon. Not threatened.

Vasek (1966) reported hybridization with *J. osteosperma* in n. w. Nevada. Terry et al. (2000) confirmed hybridization between *J. occidentalis* and *J. osteosperma* using cp and nuclear DNA markers. Recent DNA sequencing (Fig. 1.12) shows that *J. occidentalis* var. *australis* is in a separate clade from *J. occidentalis* var. *occidentalis*, which supported the recognition (Adams et al., 2006) of *J. occidentalis* var. *australis* at the specific level (*J. grandis* R. P. Adams).

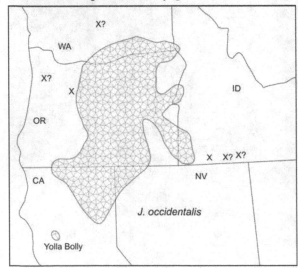

J. occidentalis, bark exfoliating in narrow strips.

J. occidentalis branch cross section showing rotted core of heartwood. Rotted heartwood is also common in *J. grandis* and *J. osteosperma*.

J. occidentalis
leaves with glands,
pollen cones
and seed cones
(monecious)

J. occidentalis forest
with sage (*Artemisia*)
in the foreground.
Bend, Oregon, USA.

cf. Adams 8725-8734.

J. occidentalis, 19 km WSW of Susanville, CA
cf Adams 12342-12346

Juniperus occidentalis forma **corbetii** R. P. Adams. Phytologia 94(1) 29 (2012). TYPE: United States, Oregon, Deschutes Co., 32 km E of Bend, OR on OR 20, shrubs, 0.5 - 1m tall, 43° 53.922'N, 120° 59.187'W, 1274 m, OR; 4 Aug 2009, *Robert P. Adams 11949* (HOLOTYPE: BAYLU, PARATYPES: *Robert P. Adams 11950, 11951*, BAYLU).

Similar to *Juniperus occidentalis* but differing habit, being a shrub with compact foliage.

The typical variety, with a strong central axis and pyramidal crown, grows on a nearby hillside, whereas f. *corbetii* grows along a dry wash on a mix of lava and sand. No female cones were found in this population.

Mark Corbet with the shrubby form of *J. occidentalis, J. o.* f. *corbetii*.
32 km east of Bend, OR (cf. *Adams 11949-11951*).

Juniperus osteosperma (J. Torrey) Little, Leafl. Western Bot. 5:125 (1948). Utah juniper. Type: USA, Arizona, Coconino Co., Bill Williams Mt., (probable location according to E. L. Little, Jr.), *J. M. Bigelow s. n.* (lectotype: NY!, see Little, 1948; isolectotype MO).

    J. tetragona Schltdl. var. *osteosperma* Torr., Pacif. Railr. Rep. 4(5):141 (1857)

    Sabina osteosperma (Torr.) Antoine, Cupress.-Gatt.: 51 (1857)

    J. californica var. *utahensis* Engelm., Trans. St. Louis Acad. Sci. 3:588 (1878)

    J. californicus var. *utahense* Vasey

    J. occidentalis Hook. var. *utahensis* (Engelm.) Kent, Veitch's Man. Conif.: 289 (1881)

    J. utahensis (Engelm.) Lemmon, (Cone Bearers Calif.) Calif. State Board Forest. Bienn. Rep. 3:183 (1890)

    J. knightii A. Nelson, Bot. Gaz. (Crawfordsville) 25:198 (1898)

    J. monosperma (Engelm.) var. *knightii* (A. Nelson) Lemmon, Handb. W. Amer. Cone-bearers, ed. 4:114 (1900)

    J. utahensis (Engelm.) Lemmon var. *cosnino* Lemmon, Bull. Sierra Club 4:123, t. 62 (1902)

    S. knightii (A. Nelson) Rydberg, Bull. Torrey Bot. Club 32: 598 (1905)

    S. utahensis (Engelm.) Rydberg, Bull. Torrey Bot. Club 32: 598 (1905)

    J. megalocarpa Sudw., Forestry & Irrig. 13:307 (1907)

    S. megalocarpa (Sudw.) Cockerell, Muhlenbergia 3:143 (1908)

    J. utahensis (Engelm.) Lemmon var. *megalocarpa* (Sudw.) Sargent, Bot. Gaz. (Crawfordsville) 67:208 (1919)

    J. californica Carriere subsp. *osteosperma* (Torr.) E. Murray, Kalmia 12:21 (1982)

    J. californica Carr. var. *osteosperma* (Torr). E. Murray, Kalmia 12:21 (1982)

Monoecious or rarely Dioecious (10%). Shrubs multi- (seldom one) stemmed shrub or tree, 3-6 (-12) m. with round crown. **Trunk bark** exfoliating in thin gray-brown strips. Bark on twigs (5-10 mm diam.) brown or gray, not exfoliating in scales or flakes. **Branches** erect. **Leaves** decurrent (whip) and scale-like, foliage light yellow-green. Whip- and scale-leaf margins denticulate (20 X). Leaf glands not conspicuous and not ruptured (embedded in the leaf, therefore not visible). **Seed cones** fibrous, bluish-brown, with white glaucous, often almost tan beneath the glaucous, (6-) 8-9(-13) mm. Maturing in 1-2 yrs. **Seeds** 1(2) avg. 1.07 per cone, 4-5 mm long. **Pollen shed** spring. **Habitat** dry, rocky soil and slopes; 1300-2600 m. **Uses** none known, not rot resistant. Trunks of living trees often with rotted heartwood. **Dist.**: Ariz., Calif., Colo., Idaho, Mont., Nev., N. Mex., Utah, Wyo., USA. **Status**: abundant in Utah and adjacent states. Considered a weed, not threatened.

    The dominant juniper of Utah. Reported to hybridize with *J. occidentalis* (see *J. occidentalis* above).

J. osteosperma

J. osteosperma bark exfoliating in thick interlaced strips.

J. osteosperma
leaves and seed cone

1 cm

J. osteosperma
habitat
Red Bluffs, Arizona

J. osteosperma, near Fremont,
Utah, USA with Bill Hadley.
cf. *Adams 5552*

Juniperus ovata (R. P. Adams) R. P. Adams, Phytologia 95(2) 175 (2013). oval gland juniper, TYPE: U. S. A., Texas, Crockett Co., 5 km w. Ozona, 6 Dec. 1994, *R. P. Adams 7463* (holotype: BAYLU, Paratypes: *R. P. ADAMS 7664, 7465, 7466, 7467* (BAYLU). Other specimens examined: MEXICO, Coahuila, *Adams 1066-1076*. U.S.A., Texas, Crockett Co., Ozona, *Adams 7424-42* (BAYLU), Coryell Co., TX, *Adams 7463-82* (BAYLU!).

Juniperus ashei var. *ovata* R. P. Adams, oval gland ashei, Phytologia 89(1): 17 (2007).

Recent DNA sequencing (Figs. 1.3, 1.1.14, 1.14.1, Adams and Schwarzbach, 2013a,b) found *J. ovata* is in a clade with *J. saltillensis* and *J. zanonii*, not with *J. ashei* as previously thought.

Dioecious. **Trees** with broad, bushy rounded or irregularly open crown, to 15 m, with a single trunk branching at 1-3 m or occasionally branching at the base. **Trunk bark** exfoliating in thin brown strips. **Branches** brown but usually with a grey-white fungus. **Leaves** both whip and scale-like. Whip leaves with a raised, oval or elliptical glands (not prominent on scale leaves, but round on scale leaves). Whip- and scale-leaf margins denticulate (20 X). **Seed cones** ovoid to subglobose, maturing in one year, dark blue and glaucous, (5) 6 (8) mm., seeds (1) 2 (avg. 1.7) per cone. **Seeds** 4-6 mm long. 2n = 22 ? **Pollen shed** Dec. -Feb. **Habitat** Limestone glades and bluffs, 150-600 m. **Uses** fence posts. **Dist.**: Tex., northern Mexico. **Status**: abundant on limestone in central/ west Texas. The range is expanding, and it is regarded as a weed in Texas.

Key to *J. ashei* and *J. ovata*:
1. Glands on whip leaves hemispheric; female cones (8) 9 (10) mm in
    diameter; seeds 16-27 mm², 1 (rarely 2, avg. 1.01) per cone..............*J. ashei*
1. Glands on whip leaves oval to elliptical; female cones (5) 6 (8) mm
    diam.; seeds 13-16 mm², 2 (avg. 1.7), per cone....................................*J. ovata*

The whip leaf glands are illustrated in figure 5.22.5. Notice hemispherical glands on *J. ashei* (Fig. 5.22.5, left) and the raised, oval to elongated glands on *J. ovata* (Fig. 5.22.5, right). It should be noted that a few nearly hemispherical glands are present on whip leaves of *J. ovata*. This is informative, as these characters can be used to distinguish *J. ovata* from *J. ashei*, yet exclude other nearby juniper species such as *J. monosperma* (Englem.) Sarg., *J. pinchotii* Sudw. and *J. coahuilensis*. (Mart.) Gaussen ex R. P. Adams.

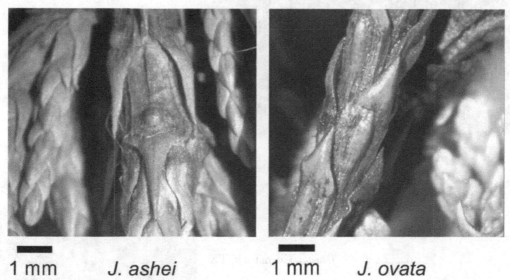

1 mm   *J. ashei*   1 mm   *J. ovata*

Figure 5.22.5. Comparison of whip leaf glands for *J. ashei* and *J. ovata* (= *J. ashei* var. *ovata*).

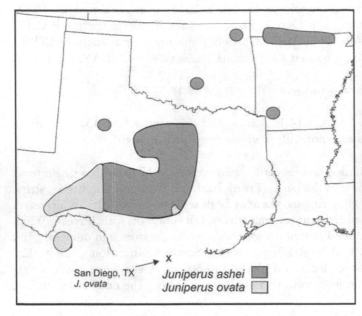

Heartwood of *J. ovata*, New Braunfels, TX, showing rotted center, indicative of a low a concentration of oil in the heartwood.

J. ovata, New Braunfels, TX

J. ovata, Comstock, TX

Juniperus oxycedrus L.

Many authors map *J. oxycedrus* as occurring across the Mediterranea region into Turkey, but recent DNA sequencing data clearly shows (Figs. 1.3, 1.10) that *J. deltoides* is in a clade with *J. brevifolia* and *J. navicularis*, not with *J. oxycedrus*. *Juniperus deltoides* is separated by 40 MEs from *J. oxycedrus* and 32 MEs from *J. brevifolia* (Fig. 5.23) giving clear support for its recognition (see also detailed discussion under *J. deltoides*).

In the previous edition (Adams, 2011) I recognized *J. oxycedrus* var. *badia* H. Gay. However, new DNA sequencing indicates (Fig. 5.23) that *J. o.* var. *badia* is separated by only 4 MEs from *J. oxycedrus*. These few DNA differences and the lack of morphological differences suggest that var. *badia* is not sufficiently differentiated to merit its recognition.

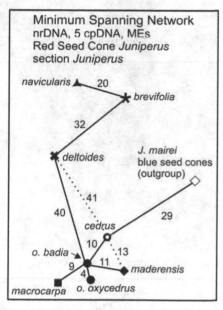

Fig. 5.23. Min. spanning network

Juniperus oxycedrus L., Sp. Pl. 2:1038 (1753). Prickly juniper; Russian: mozhzhevel'nik krasnyi (red); Georgian: gvia, gia, tvia; Armenian: gikhi; Tataric; dyshi-ardysh; German: spitzblattriger oder Cederwacholder; French: genevrier cade; Italian: ginepro rosso; Croatian: smirk; Serbian: srvena fen'a. Type: Country and locality unknown, leg. ign. *LINN 1198.9* (holotype LINN!)

J. rufescens Link, Atti della Setima Adun. Scienz. Ital. Nap. 878 (1846) *nom. illegit*

J. wittmanniana Fisch. & Lindley (1850) invalid

J. oxycedrus var. *wittmanniana* hort. ex Carriere, Traite Gen. Conif. 14(1855), invalid

J. marschalliana Stev., Bull. Soc. Imp. Naturalistes Moscou. 29: 244 (1856)

J. rhodocarpa Steven, Bull. Soc. Imp. Naturalistes Moscou 30: 397 (1857)

J. tenella Antoine, Cupress.-Gatt. 20, t.27-29 (1857)

J. oxycedrus var. *brachyphylla* Loret in Billot, Annot. Fl. France & Allem. 282(1865)

J. oxycedrus L. var. *rufescens* (Link) Carriere, Traite Gen. Conif. ed. 2, 2:13 (1867)

J. bieberstieniana hort. ex K. Koch Dendrol. iiII. 112 (1873) nom. nudium

J. heterocarpa Timb.-Lagr. ex Loret & Barrandon, Pl. Montpellier 610 (1876)

J. oxycedrina St.-Leg., Ann. Soc. Bot. Lyon 7:128 (1880)

J. oxycedrus L. subsp. *rufescens* (Link) Asch. & Graebn., Syn. Mitteleurop. Fl. 1: 248 (1897)

J. oxycedrus var. *badia* H. Gay, Assoc. Fran. Adv. Sci. Compt. Rend. 1889:501 (1889).

J. oxycedrus subsp. *badia* (H. Gay) Debeaux, Fl. Kabylie 411(1984)

Dioecious. Trees or shrub to 6 m tall and to 1 m in circumference. **Trunk bark** light gray, smooth. **Branches** no young branches yellowish-brown; branches erect, spreading or ascending; branchlets green, short, obtusely 3-angled. **Leaves** acicular, linear, terminating in a long spiny points, 15-20 mm long, with two white bands, prominently keeled, eglandular. **Seed cones** brownish-red to reddish-purple, glaucous, solitary, axillary, 6-12 mm long, globose. **Seeds** (1-)2-3(-4) per cone, broadly ovate, slightly trigonous. **Pollen shed** spring. **Habitat** dry slopes, in the open or in thin woods, rising in the mountains up to 1000 m, often on the margins of woods in stony soil. **Uses** Wood reddish, with white sapwood, very durable; used for construction timber and carpentry; pencil-making. Fruits have stimulant and diuretic action, containing 1/5% of essential oil. As with other species, dry distillation of the wood yields a brown oily fluid, oleum eupyreumaticum cadinum (Cade oil), which is used as an anthelminthic and is applied against dermatitis. An excellent ornamental tree for gardens of the dry South. 2n = 22. **Dist.**: N. Africa, Portugal, Spain, France. **Status**: common and abundant, not threatened.

J. *oxycedrus* leaves and seed cones.

J. *oxycedrus* bark

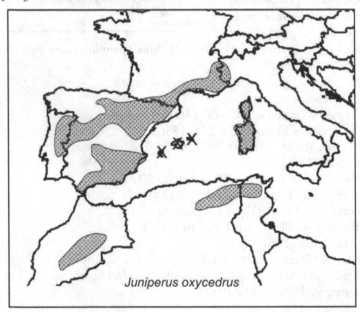

Juniperus oxycedrus

J. *oxycedrus* El Penon, Spain cf. *Adams 7080*

J. *oxycedrus*, Ruidera, Spain, cf. *Adams 9053*

Juniperus phoenicea

Juniperus phoenicea has been treated as having two varieties (var. *phoenicea*, var. *turbinata*) by Adams (2011) and Farjon (2005, 2010). However, new DNA sequencing data shows that not only is *J. phoenicea* not part of the serrate junipers of the western hemisphere (Fig. 1.3, Chpt. 1), but there is 100% support that *J. phoenicea* and *J. p.* var. *turbinata* are in separate clades (Fig. 1.16.1, Adams and Schwarzbach, 2013d). In addition, the taxa arc differ by 23 MEs (Fig. 5.23) and from the *chinensis* group by 66 MEs. It is clear from these data that the taxa are as dissimilar as other recognized species (cf. *virginiana - davurica*, Fig. 5.24) and this led Adams and Schwarzbach (2012c) to treat *J. phoenicea* var. *turbinata* as a distinct species, *J. turbinata*.

More recently, Adams et al. (2013) analyzed DNA from through the range of *J. phoenicea* (and var. *turbinata*) and found that *J. phoenicea* was present only on the Iberian peninsula (Fig. 5.24.1), whereas var. *turbinata* (*J. turbinata*) was found throughout the Mediterreanea area. Even the very distant populations in the Sinai and Madeira Island are in the group with other var. *turbinata* (*J. turbinata*) populations.

The differences are seen in the minimum spanning network (Fig. 5.24.2) where the *J. phoenicea* OTUs differ by 13 MEs from the Sinai OTUs. Notice (Fig. 5.24.2) that most of the var. *turbinata* (*J. turbinata*) OTUs differ by only 1 or 2 MEs with a few differing by 4 MEs.

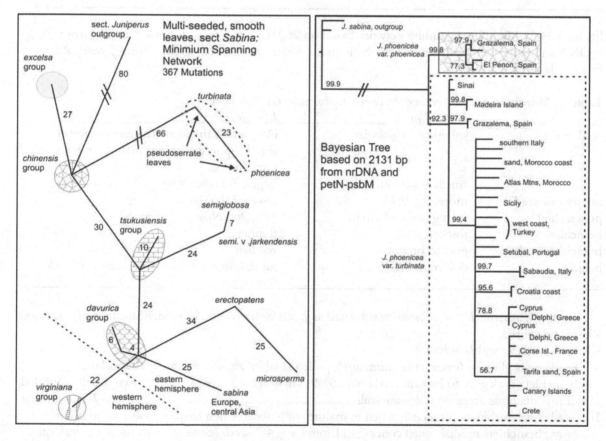

Fig. 5.24. Min. spanning network, multi-seeded, smooth leaves, sect. *Sabina*, from Adams et al. (2013d).

Fig. 5.24.1. Bayesian tree of *J. phoenicea* and *J. turbinata* (var. *turbinata*), from Adams et al. (2013).

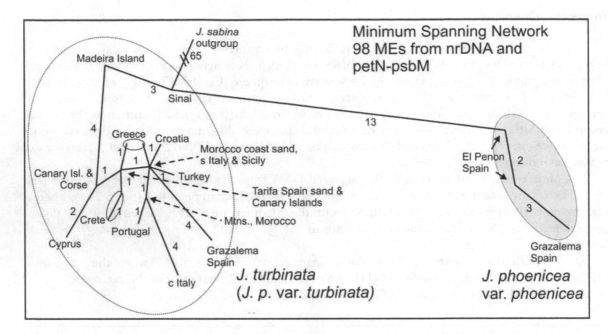

Figure 5.24.2. Minimum spanning network based on 98 MEs (base substitutions + indels) from nrDNA and petN-psbM sequence data. Note that *J. phoenicea* differs by 13 MEs from *J. turbinata*. From Adams et al. (2013).

Table 1. Morphological differences between *J. phoenicea* and *J. turbinata*.

	J. phoenicea	*J. turbinata*
seed cones	spherical to globular	elongate or turbinate (when immature) almost spherical (when mature) in some populations.
	smaller, 5-9 mm long	larger, 7-11mm long
number of seeds	more, 3(7-9)13	fewer, 3(4-7)10
pollen shed	spring (Feb. - March)	fall (Oct.-Nov.)
branchlets	thicker	thinner
branchlets bark	gray to brown	reddish
habitat	dolomitic soil	sand, Cambrian limestone, volcanic rock

In this treatment, *J. p.* var. *turbinata* is treated as *J. turbinata*. As an aid to identification of these taxa a key is presented:

Key to *J. phoenicea* and *J. turbinata*:

1a. Seed cones globose (even when immature), pollen shed in late winter-spring (Feb-Mar), branchlets bark gray to brown, seed cones 5 - 9 mm long, 7-9 seeds/ cone, crown usually rounded, in mountainous areas on dolomite soils...***J. phoenicea***

1b. Seed cones turbinate (especially when immature) or globose when mature, pollen shed in Oct.-Nov., branchlets reddish, seed cones 7 - 11 mm long, 4-7 seeds/cone, crowns more conical, on coastal sand dunes and on limestone (in mountains)...***J. turbinata***

Juniperus phoenicea L., Sp. Pl. 2:1040 (1753). Phoenician Juniper. Type: France, Languedoc, Herault, Montpellier, *leg. ign.* (lectotype LINN!), see Farjon (p. 337, 2005) for comments.

    J. lycia L., Sp. Pl. 2:1039 (1753)

    J. tetragona Moench, Methodus 699 (1794)

    J. terminalis Salisb., Prodr. Chap. Allerton 397 (1796)

J. phoenicea L. var. *lobelii* Guss., Pl. Rariores 370, t. 62 (1826)

J. phoenicea L. var. *malacocarpa* Endl., Syn. Conif. 30 (1847)

J. phoenicea L. var. *sclerocarpa* Endl., Syn. Conif. 30 (1847)

J. bacciformis Carriere, Traite Gen. Conif. 56 (1855)

Sabina bacciformis (Carriere) Antoine, Cupress.-Gatt. 69 (1857)

S. phoenicea (L.) Antoine, Cupress.-Gatt. 42 (1857)

S. lycia (L.) Antoine, Cupress.-Gatt. 44 (1857)

J. phoenicea L. var. *prostrata* Willk., Suppl. Prodr. Fl. Hispan. 4 (1893)

Sabinella phoenicea (L.) Nakai, Chosen Sanrin Haiho 165: 14 (1938)

J. phoenicea L. subsp. *phoenicea* f. *prostrata* Debreczy & Racz, Studia Bot. Hung. 29: 87 (1999)

Monoecious or dioecious. Trees small up to 8 m, or shrub. **Trunk bark** brown, exfoliating in strips. **Branches** twigs 1 mm in diameter, terete, scaly. **Leaves** decurrent (whip) and scale-like, decurrent (juvenile) leaves 5-14 x 0.5-1 mm, patent, acute, mucronate, with two stomatiferous bands both above and beneath; adult leaves 0.7-1 mm, scale-like, ovate-rhombic, closely appressed, obtuse to somewhat acute, with an oblong furrowed gland on the back and a distinct scarious border, sometimes with minute serrations near the base (40X) but not serrated as the serrate junipers of the western hemisphere. **Seed cones** 5-9 mm, ripening in the second year, globose to ovoid, blackish when very young, later green or yellowish and not or slightly pruinose, dark red when ripe. **Seeds** 3 (7-9) 13 seeds per cone, free. 2n=22. **Pollen shed** late winter - early spring. **Habitat** mountains on dolomite soils. **Uses** none known. **Dist.:** Mediterranean region, Algeria, Morocco, and Portugal. **Status**: common and reproducing, not threatened.

J. phoenicea bark

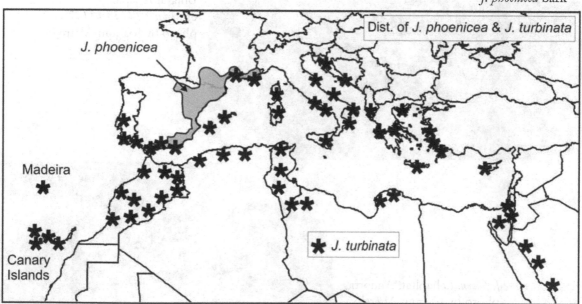

Distributions of *J. phoenicea* and *J. turbinata*, see Adams et al. (2013) for details.

J. phoenicea, leaves and seed cones.
Bujaraloz, Spain,
cf. *Adams 14134*

J. phoenicea as a shrub, west of
Bujaraloz, Spain
cf *Adams 14133-14137*.
photo by Joaquin Altarejos.

J. phoenicea habitat, Andorra,
photo by Joaquin Altarejos.

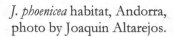

Juniperus pinchotii Sudworth, Forest. and Irrig. 11: 204 (1905). Copper berry juniper, Pinchot juniper, red-berry juniper. Type: USA, Texas, Palo Duro Canyon, 'Palodura Canyon', *G. L. Clothier s. n.* (holotype US!) (not Brewster Co., Big Bend Natl. Park, as listed in Farjon (p. 340, 2005). Sudworth (p.206, 1905) clearly states that the range of this juniper is not known, but it "has been found growing abundantly in Palodura (sic) Canyon in Briscoe, Randall and Armstrong counties, Texas". Of course, Palo Duro Canyon is found in those Texas counties, so Clothier surely collected his specimen from that area and called Sudworth's attention to it.

J. erythrocarpa Cory, Rhodora 38:186 (1936)

J. monosperma (Engelm.) Sarg. var. *pinchotii* (Sudw.) Melle, Phytologia 4:29 (1952)

J. texensis Melle, Phytologia 4:26 (1952)

J. pinchotii Sudw. var. *erythrocarpa* (Cory) Silba, Phytologia Mem. 7:35 (1984)

Dioecious. **Trees,** shrub to small shrubby tree, 1-6 m, usually multi-stemmed at the base and forming broad shrubs. **Trunk bark** thin, ashy-gray, exfoliating in long strips. **Branches** stiff, erect or spreading, the bark in long, narrow scales. **Leaves** both decurrent (whip) and scale-like. Whip- and scale-leaf margins denticulate (20 X), leaves yellow-green. Adaxial leaf surface not glaucous. Many glands ruptured and with a white, crystalline (mostly camphor) exudate, both whip and scale leaf glands elliptical to elongate. **Seed cones** copper to copper-red, not glaucous, globose to ovoid, 6-8 (-10) mm; soft and juicy, sweet pulp, 1(2) seeded, the hilum scar approx. ½ as long as the seed. **Seeds** 4-5 mm long. **Pollen shed** fall. **Habitat** 300-1000(-1700) m; gravelly soils on rolling hills and ravines, limestone, gypsum. **Uses** occasionally used as fence posts, but it is not rot resistant. **Dist.**: N. Mex., Okla., Tex.; northeastern Mexico. **Status**: this species is abundant in its range and is a weed. Not threatened.

Hybridization with *J. coahuilensis* (see above). No hybridization with *J. ashei* (see above) nor *J. monosperma* (see above). Farjon (p. 340, 2005) has confused the species description by stating the seed cones are orange-red, rose-red, usually with whitish-glaucous bloom. The absence of bloom on the mature seed cones of *J. pinchotii* is a key character separating it from *J. coahuilensis*. In addition, the seed cones of *J. pinchotii* are never rose-red, as this is found in *J. coahuilensis*. This accounts for Farjon (p. 343, 2005) showing *J. pinchotii* records far south into Mexico. These are records should be referred to *J. coahuilensis*.

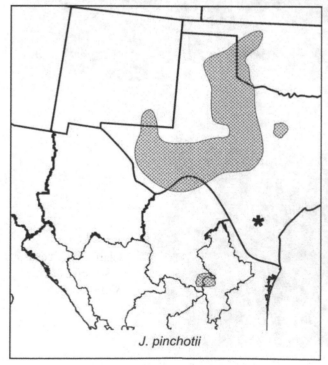

J. pinchotii

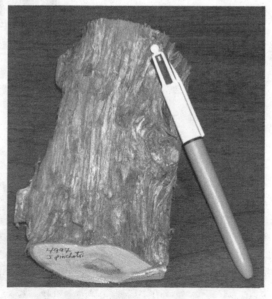

J. pinchotii, bark exfoliating in thin, narrow strips.

J. pinchotii
leaves and seed cones

J. pinchotii,
with Janice E. Adams and
Sankyo Nguyen,
Sonora, TX, USA

J. pinchotii,
Clarendon, Texas,
USA. cf. *Adams
8736-8745*

Juniperus pingii

In 2005, Farjon recognized several varieties of *J. pingii*.
J. pingii var. *wilsonii* (Rehder) Silba. Phytologia Mem. 7:36 (1984).

Farjon selected a lectotype from the 2 sheets of *E. H. Wilson 985* at E, choosing "the left hand specimen on sheet II" (see below). However, Farjon's illustration (p. 345, plate 94.2, 94.3) of *J. pingii* var. *wilsonii* (*J. squamata* Lamb. f. *wilsonii* Rehder) was based on G. Miehe 98-0504 and G. Miehe 9800401, not from E. H. Wilson 985, the lectotype. Farjon's illustration is very different from any for the specimens on E. H. Wilson 985 at A!, E! and K!. Farjon's drawing appears with clasping scale leaves (plate 94.5). All the Wilson 985 specimens at A, E and K, clearly have leaf tips that are free, straight and diverge at about at an 20-35° angle and appear to be *J. squamata* f. *wilsonii* (see leaves photo below). The *J. squamata* f. *wilsonii* specimens of E. H. Wilson 985 (several per sheet at A, E, and K) are scarcely distinct from the lectotype of *J. squamata* (W. L. Webb, 6043c, Bhutan, K!), confirming E. H. Wilson's keen eye to detect minor variation in the field.

J. pingii var. *wilsonii*, lectotype chosen by Farjon (p. 347, 2005),
left hand specimen on sheet II, of *E. H. Wilson 985* at E.
= *J. squamata* f. *wilsonii* Rehder
Notice the leaves are not clasping the stem, but have free tips.

J. squamata, lectotype chosen by Farjon (p. 382, 2005) from
W. S. Webb W 6043c at K.
Note the long leaves are not clasping the stem, but have free tips (as seen in *J. squamata* f. *wilsonii* above).

Rehder (p. 192, 1920) in his description of *J. squamata* f. *wilsonii*, thrice cited "E. H. Wilson (No. 985, in part)" collected from three different locations. Rehder comments that "cultivated at the Arboretum it looks very different from the typical form.....the leaves are shorter and broader than in the type and more closely imbricate". My material (*Adams 5521*) obtained from the living *J. squamata* f. *wilsonii* at Arnold Arboretum resembles *J. squamata* more than *J. pingii*.

   J. pingii var. *wilsonii* is treated as **J. squamata** var. **wilsonii** (Rehder) R. P. Adams in this monograph.

   Farjon (2005) treated *J. pingii* var. *carinata* Y. F. Yu & L. K. Fu as a synonym of *J. pingii* var. *wilsonii*. Based on the description of *J. pingii* var. *carinata*, my samples fit that taxon, not *J. squamata* f. *wilsonii* (= *J. pingii* var. *wilsonii*).

   Juniperus carinata has generally been recognized as a variety of *J. pingii*, but recent DNA sequencing (Adams and Schwarzbach, 2013c) shows that *J. pingii* and *J. carinata* form a clade in which they are about as distinct as some other species pairs (Figs. 1.15, 1.15.1, Chpt. 1). In the present treatment, *J. carinata* is treated as distinct species. See **Juniperus carinata** for discussion.

J. pingii W. C. Cheng var. **chengii** (L. K. Fu & Y. F. Yu) Farjon (p344, 2005)
  Basionym: **J. chengii** L. K. Fu & Y. F. Yu. Novon 7(4): 443 (1998). Type: China, Yunnan: NW Yunnan, Zhongdian, *T. T. Yu 10960* (holotype PE, isotypes A, BM).
Type is similar to *J. pingii* except the leaves are shorter and opposite. Treated herein as **J. pingii**.

J. pingii var. **miehei** Farjon, Type: China, Xizang (Tibet), Zangbo River, Upper Zangbo basin, s of Saga, along road to Gyirong, *G. Miehe (with S. Miehe) 9532-01* (holotype K, isotype GOET). Procumbent shrub.
Examination of the type, revealed it to be very similar to **J. pingii** var. **chengii**, except the seed cones are globose and smaller (5-7 mm diam.).

Key to **J. pingii** varieties in this treatment:
1a. Seed cones globose, 5-7 mm diam., leaves predominately opposite, branchlets quadrangular, procumbent shrubs..**J. pingii** var. **miehei**
1b. Seed cones subglobose, 9-11 mm. long, leaves generally terete, branchlets round, trees with strong central axis in boreal forests, leaves with abaxial keel and narrow grooves next to the keel (or a raised oil gland in some cases)...**J. pingii** var. **pingii**

Juniperus pingii W. C. Cheng var. **miehei** Farjon, A monograph of Cupressaceae and *Sciadopitys*. Kew Press. p. 346 (2005). type: China, Xizang (Tibet), Zangbo River, S. of Saga, along road to Gyirong, *G. Miehe and S. Miehe 9532-01* (holotype K, isotype GOET).

Monecious? **Procumbent shrub**. **Foliage** dense, erect. **Leaves** short, scale-like. **Seed cones**: globose, 5-7 mm long. Otherwise as *J. pingii*. **Habitat** open alpine steppe, Tibetan Plateau **Uses** none known. **Dist.**: China: Xizang (Tibet), known only from a locality in the upper Zangbo River basin. **Status**: unknown due to lack of data.

Leaves and seed cone from the type of *J. pingii* var. *miehei*, *G & S Miehe 9532-01*, K.

Juniperus pingii W.C. Cheng ex Ferre, Bull. Soc. Hist. Nat. Toulouse 79:76 (1944). Chui zhi xiang bai. Type: Illustration in Bull. Soc. Hist. Nat. Toulouse 79:76 (1944) chosen as lectotype by Farjon (p.346, 2005), but there is a specimen noted as type at PE!

J. baimashanensis Y.F.Yu & L.K. Fu, Novon 7:443-444(1997)

J. chengii Y.F.Yu & L.K. Fu, Novon 7:443-444(1997)

Sabina pingii (W.C. Cheng ex Ferre) W.C. Cheng & W. T. Wang in Cheng & Fu, eds., Fl. Reipubl. Pop. sin. 7:355 (1978)

J. pingii subsp. *polycarpos* (K. Koch) Takht; Fl. Erevana, p. 53 (1972)

Monoecious. **Trees** to 30 m. **Trunk bark** cinnamon, exfoliating in thin plates. **Branches** branchlets pendulous, prominently 6-angled. **Leaves** all decurrent, in whorls of 3, 3-5(-7) x 1-1.5 mm, glaucous, concave and with a faint green midvein adaxially, keeled and without longitudinal grooves abaxially, mucronate tips. **Seed cones** axillary, brownish green when immature, black when ripe (in 2nd yr.), lustrous, ovoid or subglobose, 7-9 mm, 1 seeded. **Seeds** ovoid or subglobose, 5-7 mm, with prominent resin pits, base rounded, apex obtuse. **Pollen shed** spring. **Habitat** forest on mountain slopes. **Uses** none known. **Dist.**: 2600-3800 m. SW Sichuan NW Yunnan. **Status**: LRnt. Limited distribution, it may become threatened.

J. pingii var. *pingii*
leaves and seed cones
Notice the long, sharp leaves
and the presence of young and
more mature seed cones.

J. pingii bark,
exfoliating
in wide, thin
plates.

J. pingii var. *pingii*
Deqin Co., Yunnan, China
cf. *Adams 8506-8507*

Juniperus poblana (Mart.) R. P. Adams, Phytologia 88: 239 (2006). Poblano juniper (Poblano is Spanish for Pueblo, the type locality). Type: Mexico, Puebla, Amozoc at 2300 m., *M. Martinez 507* (holotype: MEXU!).

    Juniperus flaccida var. *poblana* Martinez, Anales Inst. Biol. Univ. Nac. Mexico 17:31 (1946).
    Cupressus thurifera Kunth in Humboldt & al., Nova Gen. Sp. Pl. 2(5): 3 (1817)
    Cupressus thurifera Schltdl., Linnaea 12: 493 (1838), *non* Kunth (1817)
    Chamaecyparis thurifera (Kunth) Endl., Syn. Conif.: 62 (1847)

Dioecious. **Trees** to 8 m, branched with round crowns. **Trunk bark** brown exfoliating in wide strips. **Branches** flaccid only at the tips, otherwise erect, flattened (distichous). **Leaves** scale and decurrent, scale leaves on ultimate twigs with sharp, acute, divergent tips (appearing decurrent). **Seed cones** spherical, glaucous, bluish brown, 9-15 mm, mature cones usually show suture lines from fusion of cone-scales, appearing like a soccer ball. **Seeds** (4-)6-10(-13) per cone. **Pollen shed** spring. **Habitat** usually on dry, calcareous slopes, in pure stands or in mixed forests, at 1200-2300 m elevation. **Uses** none known. **Dist.:** from Jalisco E to Oaxaca, Mexico. **Status:** limited distribution in areas that may be cleared for ranching, so it may become threatened in the future.

    Juniperus poblana was thought to be a part of *J. flaccida*, but recent sequencing (Adams and Schwarzbach, 2011) has shown that *J. poblana* is part of a well supported clade with *J. jaliscana* and *J. monticola*, not with *J. flaccida* (Fig. 5.25). See also Figs. 1.3, 1.14, 1.14.1, Chpt. 1.

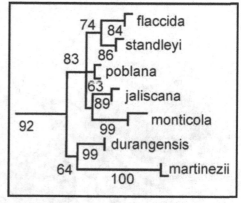

Figure 5.25. Clade from serrate junipers tree based on nrDNA, petN-psbM, trnD-trnT, trnL-trnF, and trnS-trnG data.

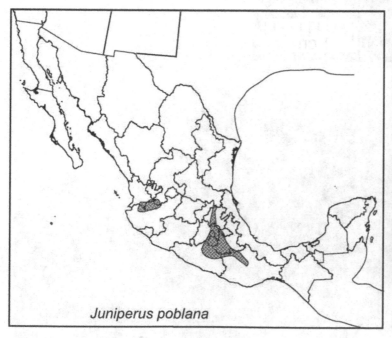

Juniperus poblana

J. poblana, with bark exfoliating in narrow, interlaced strips.

J. poblana, at KM 62,
62 km s of Oaxaca,
on MEX 190, Oaxaca, Mexico
cf. *Adams 6871*

1 cm

J. poblana
leaves and seed cones

J. poblana
on MEX 190, near San
Dionisio Ocotopec
turnoff, Oaxaca, Mexico
cf. *T. A. Zanoni 2674*
photo by T. A. Zanoni
Author and Walt Kelley at
lower right.

Juniperus polycarpos K. Koch, Linnaea 22:303 (1849).

This is a very variable species, often included as a subspecies or variety of *J. excelsa* (see Farjon, 2005). However, its DNA is clearly different Fig. 1.3, Chpt. 1). Recently, Adams (2004b) recognized *J. turcomanica* as a variety of *J. polycarpos*. DNA sequencing and SNPs (Adams et al. 2008) revealed that *J. excelsa*, *J. polycarpos* var. *polycarpos*, *J. p.* var. *seravschanica* and *J. p.* var. *turcomanica* are each these taxa are distinct in their DNA with each taxon is separated by from 7 to 9 bp. (Fig. 5.26).

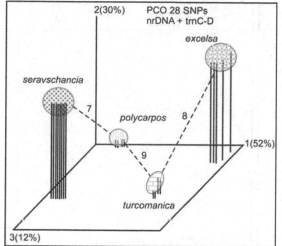

Figure 5.26. PCO based on 20 SNPs of nrDNA + trnC-D resolved by from 7 to 9 bp. (from Adams et al., 2008).

Figure 5.27 shows the patterns among these taxa based on morphology, terpenoids, RAPDs, nrDNA and cp DNA sequences. It is apparent that evolution has proceeded unevenly between the cp trnC-trnD region and the nrDNA (Fig. 5.4.1).

Of particular interest is that the nrDNA of *excelsa* (E) and *turcomanica* (T) is almost identical, whereas the cp DNA is very different. The cp DNA (trnC-trnD) is nearly identical between *turcomanica* and *polycarpos* (Fig. 5.4.1). This suggests that as some point in time, perhaps *turcomanica* is of hybrid origin between *excelsa* and *polycarpos* (or their ancestors).

The Bayesian tree for the multi-seeded, smooth leaf, section *Sabina* junipers (Fig. 1.16.1, 5.28) shows *J. polycarpos* and *J. p.* var. *turcomanica* in a well supported clade and *J. excelsa* in another clade. *J. seravschanica* in a well supported clade with *J. foetidissima* and *J. thurifera*.

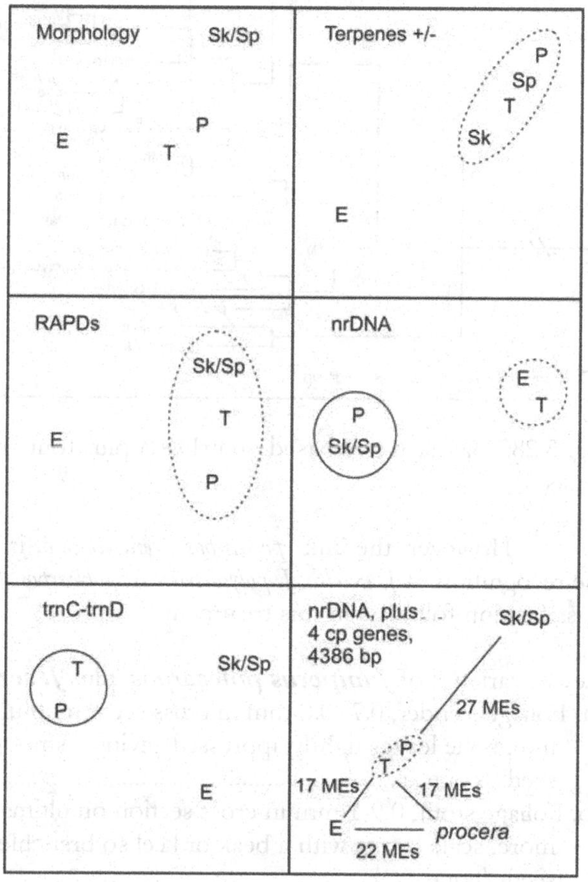

Figure 5.27. Graphic summaries of morphology, terpenes, RAPDs, nrDNA, trnC-trnD, and combined nrDNA + trnC-trnD data. E=*J. excelsa*, P = *J. p.* var. *polycarpos*, T = *J. p.* var. *turcomanica*, Sk = *J. p.* var. *seravschanica*, Kazakhstan, Sp = *J. p.* var. *seravschanica*, Pakistan. Adams et al. (2008, 2013d).

The magnitude of the differences among these taxa is shown in the minimum spanning network (Fig. 5.28.1). *Juniperus p.* var. *turcomanica* is separated by 17 MEs (mutational events = nucleotide substitutions + indels) from *J. excelsa* and *J. procera*. The link *polycarpos - turcomanica* is 8 MEs.

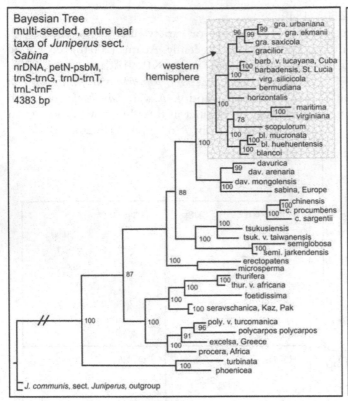

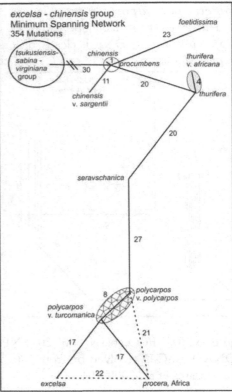

Fig. 5.28. Bayesian tree based on nrDNA plus four cp genes.

Fig. 5.28.1. Min. spanning network. #s on lines are Mutational Events (MEs).

However, the link *polycarpos - seravschanica* is 27 MEs (Fig. 5.28.1). These DNA data support the recognition of *J. excelsa*, *J. polycarpos* var. *polycarpos*, *J. p.* var. *turcomanica*, and *J. seravschanica*. That is the classification followed in this treatment.

Key to varieties of *Juniperus polycarpos* (plus *J. seravschanica* for convenience)

1a. Foliage slender, 0.7- 0.8 mm in cross section on ultimate branchlets, seed cones 7-9(-10) mm, scale leaves tightly appressed, giving a smooth branchlet, (1-)2-3(-4) seeds/cone..*J. polycarpos* var. *turcomanica*

1b. Foliage stout, 0.9-1 mm in cross section on ultimate branchlets, seed cones 9-11 mm or more, scale leaves with a beak or keel so branchlet appears as a string of beads, 3-6 seeds/cone

    2a. Seed cones 9-11mm, at least some scale leaves with very narrow, elongated, brown glands, not ruptured..*J. polycarpos* var. *polycarpos*

    2b. Seed cones 8-10 mm, scale leaves with clear, ellipsoid glands, often ruptured, with a clear exudate...*J. seravschanica*

Juniperus polycarpos K. Koch var. **polycarpos**, Linnaea 22:303 (1849). Common Names:Russian: mazhzhevel'nik mnogoplodnyi or vostochnyi (many-fruited or eastern juniper); kara-archa. Type: Turkey, Gumushane, Anadolu, Daglari, Taltaban, *P. E. E. Sintensis 5520* (isotype B, Destroyed in 1943, neotype L, see Farjon 1993, isoneotypes BM, E, K)

    Sabina polycarpos (K. Koch) Antoine, Cupress.-Gatt.: 47, t. 63 (1857)

    J. excelsa Wall., sensu Aitch., J. Linn. Soc., Bot. 18:97 (1880)

    J. polycarpos K. Koch var. *pendula* Mulk., Dokl. A.N. Armen. S.S.R. 45 (2): 86 (1967)

    J. excelsa M.-Bieb. subsp. *polycarpos* (K. Koch) Takht. Fl. Yerev.: 53 (1972)

    J. excelsa M.-Bieb. var. *farreana* P.N. Mehra, Nucleus 19 (2):135 (1976). *nom. inval.*, Art. 36.1

J. excelsa var. *polycarpos* (K. Koch) Silba, Phytologia Memoirs 7: 34 (1984)
J. excelsa subsp. *polycarpos* var. pendula (Mulk.) Imkhan., Bot. Zurn. 75: (3): 407 (1990)
J. excelsa M.-Bieb. subsp. *polycarpos* (K. Koch) Takht. var. *pendula* (Mulk.) Imkhan., Bot. Zurn. 75 (3): 407 (1990)

Dioecious. **Tree** to 6-7 m tall or a low shrub with a dense crown. **Trunk bark** light brown, exfoliating in thin wide strips. **Branches** leafy branchlets short, firm, rather stout. **Leaves** decurrent (whip) and scale-like, ovate or deltoid, long-pointed, on branchlets obtuse to rhombic or ovate-rhombic, acute to subobtuse, slightly keeled on the back so branchlet appears as a string of beads, with an ovate inflated gland, at least some scale leaves with very narrow, elongated, brown gland, not ruptured. **Seed cones** short-stalked, solitary or in groups, blackish-blue when mature, brownish-blue when immature, globose, 9-11mm diam. **Seeds** 4-6 per seed cone, ovate-oval, ribbed, brownish. **Pollen shed** late spring. **Habitat** between 300 and 2,500 m, as solitary trees or shrubs, sometimes together with other juniper species, forming juniper forests, tolerant of dry stony soils. **Uses** wood suitable for building and carpentry. **Dist.**: Western Himalayas to Caucasus. **Status**: var. *polycarpos* is in areas with goat grazing, so it is threatened.

J. polycarpos var. *polycarpos* habitat, Armenia cf. *Adams 8761-8763*   *J. polycarpos*, bark

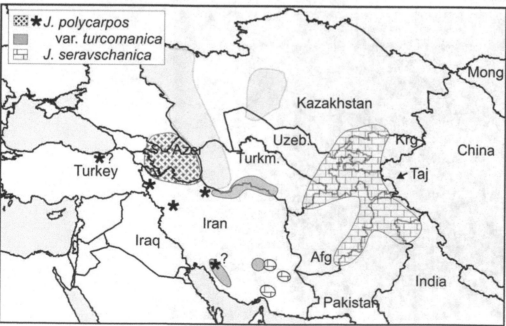

Distribution of *J. polycarpos* and *J. p.* var. *turcomanica* plus *J. seravschanica*. ?= uncertain *polycarpos*.

J. polycarpos var. *polycarpos*
n. side of Lake Sevan, Armenia
cf. *Adams 8761-8763*

Dr. Vrezh Manakyan organized
the expedition and Jessica D.
(Adams) Black assisted with
logistics of the trip.

J. polycarpos var. *polycarpos*
leaves and seed cones

1 cm

J. polycarpos var. *polycarpos,,* n. side of Lake Sevan, Armenia with Dr. Vrezh Manakyan, a true gentleman,
who led the field trip in 1999. cf. *Adams 8761-8763*

Juniperus polycarpos K. Koch var. ***turcomanica*** (B. Fedtsch.) R. P. Adams, Phytologia 86: 52 (2004). Turkman juniper. Type: Turkmenistan, Kopet Dag (Mts.), Dschalilu. *D. P. Gedevanov (with D. A. Dranitsyn) 148* (lectotype LE!, see Imkhanitskaya, 1990)

 J. turcomanica B. Fedtsch. in Fedtschenko & al., Fl. Turkmenii 1:14 (1932)
 Sabina turcomanica B. Fedtsch). Nevski, Trudy Bot. Inst. Akad. Nauk S.S.S.R. ser. 1, Fl. Sist. Vyss Rast. 4:218 (1937)
 J. excelsa M.-Bieb. subsp. *turcomanica* (B. Fedtsch.) Imkhan., Bot. Zurn. 75 (3):408 (1990)

Dioecious. Tree to 6 m, with spreading crown. **Trunk bark** reddish, exfoliating in scales. **Branches** leafy terminal branchlets slender (0.7- 0.8 mm in cross section on ultimate branchlets). **Leaves** decurrent (whip) and scale-like, dorsal gland nearer the base, round or rarely ovate, short scale leaves tightly appressed giving a smooth branchlet. **Seed cones** fruits borne on short leafy stalks, globose, black, pruinose, 7-9(-10) mm in diam. **Seeds** (1-2-)3 (-4) per seed cone, brownish, oval-oblong, lustrous, ca 6 mm long and 2.5 mm broad, some seeds deviate greatly in shape. **Pollen shed** March-April. **Habitat** in open groves or isolate groups on dry mountain slopes. **Uses** none known. **Dist.:** Elburz and Kopet Mts. of Iran and Turkmenistan. **Status**: this taxon has a limited distribution in areas subject to goat and sheep grazing and is threatened.

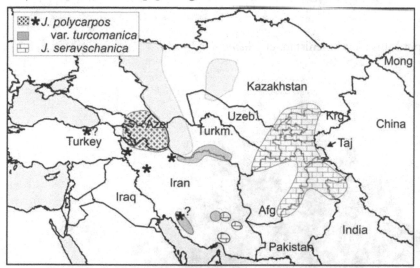

J. polycarpos var. *turcomanica* bark in thin, broad strips.

Distribution of *J. polycarpos* and *J. p.* var. *turcomanica* plus *J. seravschanica*. ?= uncertain *polycarpos*.

J. polycarpos var. *turcomanica* leaves and seed cones

J. polycarpos var. *turcomanica*, Kopet Mtns., Turkmenistan, cf. *Adams 8757-8760*

J. polycarpos var. *turcomanica*, Kopet Mtns. Turkmenistan, with Jessica D. (Adams) Black, cf. *Adams 8757-8760*

Juniperus procera Hochst. ex Endl., Syn. Conif., 26 (1847). Tall juniper, east African juniper. Type: Ethiopia, Gonder Prov., Amhara, Adda Marim, near Enschedcap, *G. H. W. Schimper 537* (lectotype L, see Farjon, 1993, isolectotypes BM, K, LE, M).

   J. hochstetteri Antoine, Cupress.-Gatt. p.s.n., t. 33 (1857)

   Sabina procera (Hochst. ex Endl.) Antoine, Cupress.-Gatt.: 36, t. 47 (1857)

   J. abyssinica hort. ex K. Koch, Dendrol. 2 (2): 132 (1873)

Dioecious. **Tree** to 40 m high. **Trunk bark** outer bark of trunk thin, grey brown, with shallow longitudinal fissures, exfoliating in thin papery strips. **Branches** adult branchlets 0.5-1 mm diameter with opposite, decussate scale leaves, acute, hooded and keeled at the tip, narrow translucent margins and elliptic oil gland on the back near the base. **Leaves** decurrent and scale-like, decurrent (juvenile) leaves in 3's, linear, spine tipped, to 10 mm. long margins rounded, gland linear, extending 3/4 length of leaf and decurrent on stem. Intermediate leaves paired, grading into the adult form. **Seed cones** 3-4 pairs of scales, mature in 1 yr., reddish brown to purplish black with blue bloom, subglobose to irregular, 6-8(10) mm diameter. **Seeds** 2 - 5 per cone, flattened or triangular seeds. **Pollen shed** summer? **Habitat** upland dry evergreen forest, often dominant at 1350-3100 m. **Uses** source of cedarwood oil in the past, furniture, woodworking. **Dist.**: Ethiopia, Kenya, Tanzania, Saudi Arabian Peninsula. **Status**: this taxon is cut for lumber and fuel and is likely threatened.

Juniperus procera

J. procera bark exfoliating in thin, broad strips.

J. procera
along road to from Kijabe
to Rift Valley Academy, Kenya.
cf. *Adams 6007-6009*

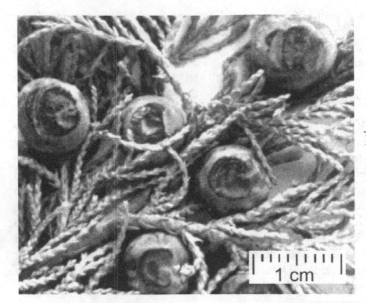

J. procera
leaves and seed cones

J. procera, near Guder, 64 km w of Addis Ababa, Ethiopia, cf. *Adams 6184-6188*

Juniperus przewalskii Komarov, Bot. Mater. Gerb. Glavn. Bot. Sada RSFSR 5: 28 (1924). Qi lian yuan bai. Przewalski juniper. Type: China, Gansu, Qilian Shan, Daban Shan, 13,000 ft., *N. M. Przewalski 318* (lectotype LE! , isolectotype PE).

Sabina przewalskii (Kom.) W. C. Cheng & L. K. Fu, Fl. Reipubl. Pop. Sin. 7: 375 (1978)

Monoecious. **Trees** to 20 m. rarely shrubs. **Trunk bark** brown, exfoliating in strips. **Branches** branchlet systems not tapering and subequal in length throughout system, branchlets loosely arranged, straight or slightly curved, terete or 4-angled, thick ultimate ones 1.2-1.5 mm in diam. **Leaves** both decurrent and scale-like on mature trees, only decurrent (juvenile) leaves on young trees, decurrent leaves in whorls of 3, spreading, free part 4-8 mm, scale-like leaves decussate, usually glaucous, rhombic-ovate, 1.2-3 mm, usually with cuticular wax, abaxial gland basal, orbicular, ovate, or elliptic, convex, leaf apex acute, free. **Seed cones** tan when immature, bluish black or black when mature in 2 yrs., ovoid or subglobose, 0.8-1.3 cm. **Seeds** 1 per seed cone, turbinate to subglobose, 7-12 x 6-10 mm obscurely or prominently ridged, with resin pits. **Pollen shed** fall. **Habitat** forests on mountain slopes; 2600-4300 m. **Uses** incense, furniture. **Dist.**: Gansu, Qinghai, N. Sichuan (Songpan). **Status**: LRnt. common in some areas and reproducing, not threatened.

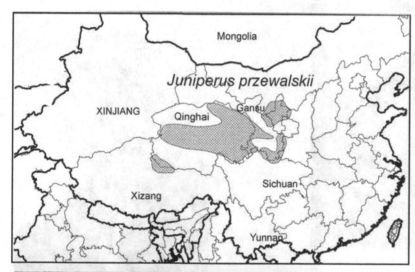

J. przewalskii, bark exfoliating in narrow strips

J. przewalskii
Qinghai, China
cf. *Adams 10336-10341*

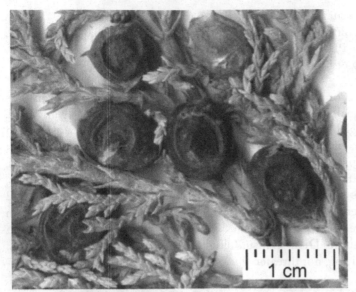

J. przewalskii
leaves and seed cones

J. przewalskii, Qinghai, China

J. przewalskii, on Tao River,
25 km w of Jone, Gansu.
cf. *Adams 6775-6777*, field trip
led by Prof. Ge-Lin Chu.
in 1991.

J. przewalskii, river bank, Qinghai,
China, cf. *Adams 10336-10341*
field trip arranged by Dr. Jinquan
Liu in 2004.

Juniperus pseudosabina Fischer & C.A. Meyer, Index Sem. Hort. Petrop. 8 (1841) 24, 65 (1842). Turkestan juniper, Xin jiang fang zhi bai. pseudo sabina juniper. Type: Kazakhstan, Dzhungarskiy Altau, Tarbagatay Mts., *A. G. von Schrenk s. n.* (holotype LE!, lectotype K, designated by Farjon, p. 355, 2005).

> *J. sabina* Pall., Reise Russ. Reich. 3(1): 273 (1776), *non* L. (1753)
>
> *Sabina fischeri* Antoine, Cupress.-Gatt.: 53 (1857)
>
> *J. centrasiatica* Kom., Bot. Mater. Gerb. Glavn. Bot. Sada RSFSR 5:27 (1924)
>
> *J. turkestanica* Kom., Bot. Mater. Gerb. Glavn. Bot. Sada RSFSR 5:26 (1924)
>
> *S. pseudosabina* (Fisch. & C.A. Mey.) W. C. Cheng & W. T. Wang, Acta Phytotax. Sin. 13(4): 75 (1975)
>
> *S. centrasiatica* (Kom.) W. C. Cheng & L.K. Fu, Fl. Reipubl. Pop. Sin 7:370 (1978)
>
> *S. pseudosabina* (Fisch. & C.A. Mey.) W. C. Cheng & W. T. Wang var. *turkestanica* (Kom.) C. Y. Yang, Fl. Reipubl. Pop. Sin. 7:369 (1978)
>
> *J. pseudosabina* Fisch. & C. A. Mey., var. *turkestanica* (Kom.) Silba, Phytologia Mem. 7:36. (1984)

Dioecious or rarely monoecious. **Shrubs** erect or procumbent, or small trees to 12 m. **Trunk bark** cinnamon brown, exfoliating in thin strips. **Branches** ultimate branchlets densely arranged, mostly straight, 4-angled or sometimes ± terete. **Leaves** both scale-like and decurrent (whip), decurrent leaves usually present on seedlings and young trees, with oval or elliptical gland but seldom with a thin extension almost to the tip, decussate or in whorls of 3, ascending, 4-8 mm, apex acuminate, scale leaves with scarious border, appressed or with free apex, rhombic, obtuse, 1.5-2 mm abaxial gland central, or basal in decurrent leaves, oblong or broadly linear, depressed. **Seed cones** green when immature, bluish black or brownish black when mature in 2 yrs., often glaucous, turbinate, 8 - 10 mm. **Seeds** 1 per cone, ovoid or ellipsoid, slightly flattened, 6-7 x 4-6 mm, ridged, base rounded or pointed, apex blunt. **Pollen shed** winter. **Habitat** thickets on mountains, 2000-3300 m. **Uses** none known. **Dist.**: Afghanistan, Kazakhstan, Kyrgyzstan, Mongolia, Pakistan, Tajikistan, Uzbekistan, Xinjiang. **Status**: grows in areas where goats and sheep graze, it may be threatened.

Adams & Turuspekov (1998) demonstrated that *J. centrasiatica* Kom. and *J. turkestanica* Kom. are conspecific with *J. pseudosabina*..

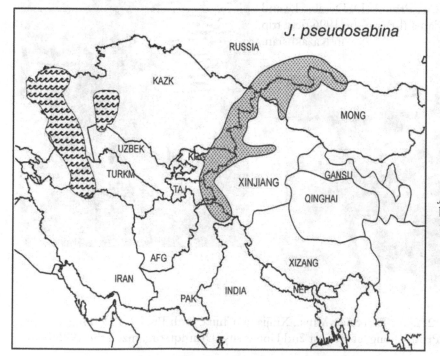

J. pseudosabina
bark

J. pseudosabina
leaves and seed cones

J. pseudosabina, (foreground) with
J. sabina (background), 30 km n of
Jarkent (Paniflor), Kazakhstan
ca. 2000 m. cf. *Adams 7808-7810*

Dr. Yerlan Turuspekov , Prof. A. D. Dembitsky and
the author, R. P. Adams (l-r), on the 1996 field trip
in Kazakhstan.

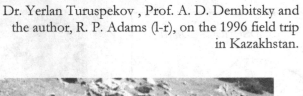

J. pseudosabina, Heaven Lake, 2120m, Tian Shan Mtns., Xinjiang, China, with Prof. J. P. Zhong and his
wife, cf. *Adams 7833-7835*. Prof. Zhong, at August 2nd University, Urumqi, organized the 1996 field
trip.

Juniperus recurva Buch.-Ham. ex D. Don, Prodr. Fl. Nepal. (2): 55 (1825). Drooping juniper, recurved juniper, Chu zhi bai. Type: Nepal, Unknown, Naraianghet? ["Narainhetty"], *F. Buchanan-Hamilton s. n.* (holotype BM!).

Juniperus butanensis Wender., Pfl. Bot. Gart. 1, Conif.: 36 (1851)

Sabina recurva (Buch.-Ham. ex D. Don) Antoine, Cupress.-Gatt. 67 (1857)

Juniperus recurva Buch.-Ham. ex D. Don var. *typica* Patschke, Bot. Jahrb. Syst. 48: 776 (1913),
nom. inval., Art. 24.3

Sabinella recurva (Buch.-Ham. ex D. Don) Nakai, Chosen Sanrin Kaiho 165: 14 (1938)

Monoecious or rarely dioecious. Trees shrubs or trees. **Trunk bark** light grayish brown or brown; crown conical or broadly pyramidal. **Branches** ascending in apical part of plant and spreading toward base, branchlets pendulous, curved. **Leaves** all decurrent (juvenile), mucronate tips, in whorls or 3, loosely appressed, greenish white or slightly glaucous adaxially, 5 mm. **Seed cones** axillary, slightly glaucous when young, maturing purplish black and not glaucous, ovoid, 6-12 x 5-9 mm. **Seeds** 1 per seed cone, ovoid or conical-ovoid, 5-9 x 3-6 mm. 2n=22 (Merha, 1988). **Pollen shed** summer. **Habitat** rocky areas, high mountains. **Uses** wood is burned as incense. **Dist.**: Afghanistan, Himalayas, China: Tibet (3000-4000 m). **Status**: common in areas but often cut for incense, may become threatened.

Recent DNA sequencing has placed *J. recurva* in a clade with *J. indica* rather than with *J. pingii* in spite of the similar nature of their leaves (see Fig. 1.3, Chpt. 1).)

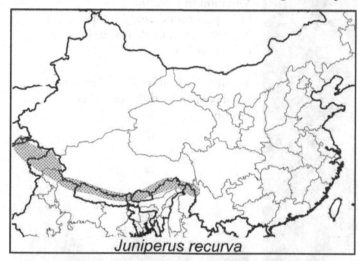

Juniperus recurva

J. recurva, bark

J. recurva
Sing Gompa, Nepal
cf. *Adams 7215-7217*

J. recurva, leaves and seed cones
cf. *Adams 7215*

Variation in *J. recurva*
in Nepal around Sing Gompa
Note variation in shapes and colors.
cf. *Adams 7215*

Thanks to Dr. Ram Chaudary who
organized and led the expedition to Sing
Gompa.

J. recurva
Sing Gompa, Nepal
cf. *Adams 7215-7217*

Juniperus rigida Siebold & Zucc.

Key to varieties of ***Juniperus rigida***: (see Figs. 1.3, 1.11.2, Chpt. 1 for phylogenetic information).

1a. Trees or erect shrubs; seed cones 6 - 9 mm, leaves 12 -25 mm long, dry areas in mountains
....***Juniperus rigida*** var. ***rigida***

1b. Procumbent shrubs, seed cones 8 - 12 mm, leaves 8 - 15 mm long, sandy seashores.............
....***Juniperus rigida*** var. ***conferta***

Juniperus rigida Siebold & Zucc. var. ***conferta*** (Parl.) Patschke, Bot. Jahrb. Syst. 48: 678 (1913). Shore juniper, crowded or compact juniper. Type: Japan, Ryukyu Islands?, *C. Wright s. n.* (holotype K!)

    J. conferta Parl., Conif. Nov.: 1 (1863)

    J. litoralis Maxim., Bull. Acad. Imp. Sci. Saint-Petersbourg 12:230 (1868)

    J. rigida Siebold & Zucc. subsp. *litoralis* (Maxim.) Urussov, Bjull. Glavn. Bot. Sada 122: 56
      (1981)

    J. coreana Nakai, Bot. Mag. (Tokyo) 40: 161 (1926)

Dioecious. Procumbent shrubs. Trunk bark brown, exfoliating in strips. **Branches** procumbent. **Leaves** acicular, 8-15 mm. long, 1-2 mm across, usually with a short point, straight or slightly incurved, rigid with a white groove above. **Seed cones** globose, 8-12 mm across, brown to bluish black and slightly glaucous. **Seeds** 2 -3 per seed cone. **Pollen shed** spring? **Habitat** sandy seashores. **Uses** widely cultivated for ground cover. **Dist.**: Hokkaido, Honshu, Kyushu, Sakhalin. **Status**: in protected areas, not threatened.

J. rigida var. *conferta* leaves and seed cones

J. rigida var. *conferta*,
cultivated, Royal Bot. Gardens,
Kew, UK.

Juniperus rigida var. conferta

Juniperus rigida var. ***rigida***, Abh. Math.-Phys. Cl. Konigl. Bayer. Akad. Wiss. 4(3): 233 (1846). Du song. Type: Japan, Honshu, *J.(?) Burger (comm. P. F. von Siebold 1842) s. n.* (lectotype M, designated by Farjon, p. 365, 2005).

    J. communis Thunb., Fl. Japon: 264 (1784), *non* L. (1753)
    J. seoulensis Nakai, Bot. Mag. (Tokyo) 31: 23 (1917)
    J. utilis Koidz., Bot. Mag. (Tokyo) 44: 99 (1930)
    J. utilis Koidz. var. *modesta* Nakai, Chosen Sanrin Kaiho 158: 26 (1938)
    J. rigida Siebold & Zucc. f. *modesta* (Nakai) Y. C. Zhu, Pl. Med. Chinae Bor.-orient.
      (ed. Z. Y. Chang): 59 (1989)

Dioecious. **Trees** to 10 m or erect shrubs; crown pyramidal or cylindrical. **Trunk bark** brown, exfoliating in strips. **Branches** ascending, branchlets pendulous, 3-angle when young. **Leaves** acicular, in whorls of 3, green abaxially, linear-needlelike, thick, "V"-shaped in cross section, 1-2.3 cm x ca. 1 mm, rigid, deeply grooved with a narrow, white stomatal band adaxially, prominently keeled abaxially, apex sharply pointed. **Seed cones** axillary, light brownish blue or bluish black when ripe, usually glaucous, globose, 6-8 mm in diam. **Seeds** 3 per seed cone, often subovoid, ca. 5 mm, indistinctly 4-ridged, apex obtuse or rounded. 2n=44. **Pollen shed** spring? **Habitat** dry areas in mountains, below 2200 m. **Uses** none known. **Dist.**: Gansu, N. Hebei, Heilongjiang, Jilin, Liaoning, Nei Mongol, Ningxia, Qinghai, Shaanxi, Shanxi (Japan, Korea). **Status**: not threatened.

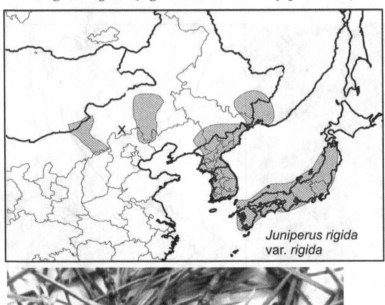

Juniperus rigida var. rigida

J. rigida bark

J. rigida, Beijing Botanic Garden, China
cf. *Adams 6797-6799*

1 cm

Juniperus rushforthiana (R. P. Adams) R. P. Adams. Phytologia 94(3) 401 (2012). Rushforth's juniper, Type: Bhutan, Soe, at Soe Tajitang campsite, tree 15 m, 11,400 ft., *Rushforth 0802 (= Adams 8140)* (holotype: BAYLU).

Juniperus indica Bertol. var. R. P. Adams, Phytologia 90(2): 244 (2008).

Note: Farjon (2005) includes *J. rushforthiana* in *J. indica*, but see Figs. 1.3, 1.15.1, Chpt. 1, for recent DNA sequencing analysis.

Dioecious. Trees to 20 m. **Trunk bark** dark appearing black, orange underneath, exfoliating in strips. **Branches** erect, ascending. **Leaves** decurrent (whip) and scale-like, dark green, whip leaves usually present on young trees, in whorls of 3, ascending, 3 - 8 mm, apex acuminate, scale leaves with an oval to elliptical gland with a narrow extension almost to the leaf tip, decussate or sometimes in whorls of 3, closely appressed, rhombic, 1.2 - 2 mm, abaxial gland central, or basal in decurrent leaves, oblong or linear, depressed, leaf apex obtuse. **Seed cones** erect, subglobose to turbinate, 10 - 13 mm, brown when young, black-brown when ripe, maturing in 2 yrs. **Seeds** 1 per cone. **Pollen shed** April-May. **Habitat** dry open forests, 2800- 4600 m. **Uses** wood is utilized for fuel and incense in Buddhist temples. 2n = 44 (Mehra, 1988). **Dist.:** Known only from Bhutan, but expected to occur in neighboring Xizang, Sikkim and eastern Nepal(?). **Status:** because it is used for incense and in areas of deforestation, it is likely threatened.

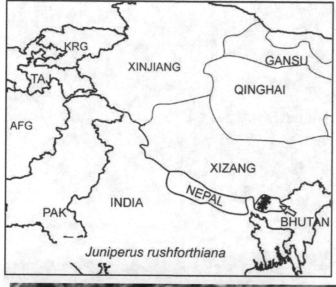

J. rushforthiana
bark on young tree.
(not typical of adult trees)

J. rushforthiana
leaves and seed cones,
from holotype *Rushforth 0802* at E.

1 cm

Juniperus rushforthiana, Bhutan, Soe, at Soe Tajitang campsite, 11,400 ft.,
Rushforth 0802 (photo Keith Rushforth).

Juniperus sabina

Juniperus sabina is a variable species that spans a very large geographic range. Adams and Schwarzbach (2013e) sequenced samples of *J. sabina* and *J. davurica* from an extended range and found *J. sabina* from Switzerland in a clade with *J. davurica* (Fig. 5.29). However, *J. sabina* is separated by 25 MEs from *J. davurica* (Fig. 5.30). In this treatment, only *J. sabina* (var. *sabina*) is recognized.

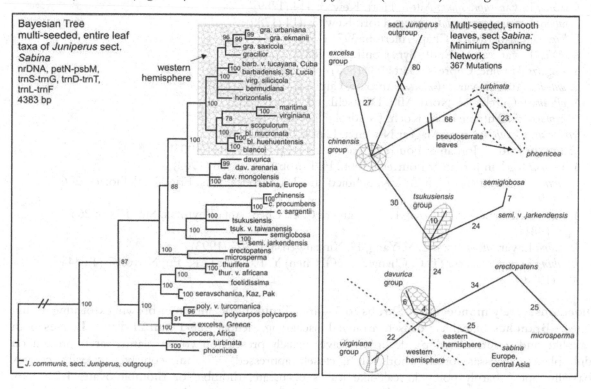

Fig. 5.29. Bayesian tree. Note the clade with *J. sabina* and *J. davurica*.

Fig. 5.30. Min. spanning network. *J. sabina* is separated by 25 MEs from *J. davurica*.

Key to **Juniperus sabina** and **J. davurica** (as an aid in identifying these taxa):

1a. Both scale-like and whip- (decurrent with free tips) leaves found on mature plants, interspersed on a single branchlet, not just at the ends of rapidly growing branchlets, occasionally, one leaf type predominates..**J. davurica**

1b. Whip (decurrent) leaves found only on juvenile plants or at the ends of rapidly growing branchlets, otherwise all leaves scale like

    2a. Shrubs to 1 m; growing on rocky areas, seeds with an acute tip that tapers into a globose shaped base, 1-3 seeds per globose cone..**J. sabina**

    2b. Prostrate to shrubs to 1 m; with long lateral branchlets, growing on sand, seeds with an obtuse (occasionally acute when only one seed per cone) tip on an ellipsoid to globose shaped base, 1-4 seeds per cone

        3a. Most seed cones bi-lobed, seeds (1) 2 (3-4) per cone, seeds an elongated ellipsoid with an acute tip that resemble a duck bill, ultimate branchlets radially distributed around lateral branches.....................................**J. davurica** var. **arenaria**

        3b. Most seed cones globose, seeds 2-4 per cone, seeds a flattened globose with an obtuse tip, ultimate branchlets growing from the top (upper) side of long lateral branches.....................................**J. davurica** var. **mongolensis**

Juniperus sabina L., Sp. Pl. 2:1039 (1753). Savin, Cha zi yuan bai. Type: [country and locality unknown], *leg. ign. LINN 1198.4* (holotype LINN!).

  Sabina vulgaris Ant. (used in Fl. China)
  J. lusitanica Mill., Gard. Dict., ed. 8: Juniperus No. 11 (1768)
  J. lycia Pall., Fl. Ross. I, 1:14 (1788)
  J. sabina L. var. *cupressifolia* Aiton, Hort. Kew. 3: 414 (1789)
  J. sabina L. var. *tamariscifolia* Aiton, Hort. Kew. 3: 414 (1789)
  J. humilis Salisb., Prodr. Chap. Allerton: 397 (1796)
  J. sabina L. var. *vulgaris* Endl., Syn. Conif.: 22 (1847)
  S. vulgaris Antoine, Cupress.-Gatt.: 58 (1857)
  S. vulgaris Antoine var. *arborescens* Antoine, Cupress.-Gatt.: 58 (1857)
  S. officinalis Garcke, Fl. Nord.-Mitt. Deutschl., ed. 4: 387 (1858)
  S. cupressifolia Antoine ex K. Koch, Dendrol. 2 (2): 125 (1873)
  Juniperus x *kanitzii* Csato, Magyar Novenyt. Lapok 10: 145 (1886)
  S. alpestris Jord. in Jordan & Fourreau, Icon. Fl. Europ. 3: 5, t. 369 (1903)
  S. villarsii Jord. in Jordan & Fourreau, Icon. Fl. Europ. 3: 4, t. 368 (1903)
  J. sabina L. var. *lusitanica* (Mill.) C. K. Schnied. in Silva-Tarouca, Uns. Friel.-Nadelholzer: 206 (1913)
  S. vulgaris Antoine var. *yulinensis* T. C. Chang & C. G. Chen, Acta Phytotax. Sin. 19 (2): 263 (1981)
  J. sabina L. var. *monosperma* C. Y. Yang, Fl. Xinjiangensis 1: 305 (1993)
  J. sabina L. var. *yulinensis* (T. C. Chang & C. G. Chen) Y. F. Yu & L. K. Fu, Novon 7 (4): 444 (1998)

Dioecious, rarely monoecious. Shrubs to 3 – 4m. **Trunk bark** cinnamon-brown, exfoliating in thin plates. **Branches** branchlets densely arranged, ascending, slender, 0.8-1 mm in diam. **Leaves** both scale-like and decurrent (whip), decurrent leaves usually present on young plants, rarely present on adult plants, decussate or in whorls of 3, closely appressed, 3-7 mm, concave adaxially, convex abaxially, apex sharply pointed, scale-like leaves decussate, rhombic or rhombic-ovate, 1-2.5 mm. abaxial gland central, prominent, elliptic. **Seed cones** light brownish green, brown when young, purplish blue, or black when mature in 2 yrs., often glaucous, usually irregularly globose, 5-8 x 5-9 mm. **Seeds** (1-) 2 seeds per seed cone, ovoid, slightly flattened, 4-5 mm, ridged, with resin pits, apex blunt or slightly pointed. **Pollen shed** late winter-spring. **Habitat** forests or thickets on rocky mountain slopes and sand dunes, 1000-3300 m. **Uses** extremely widely cultivated for landscaping. **Dist.**: Europe, central Asia, w China. e. China? **Status**: widely cultivated, common in nature, not threatened.

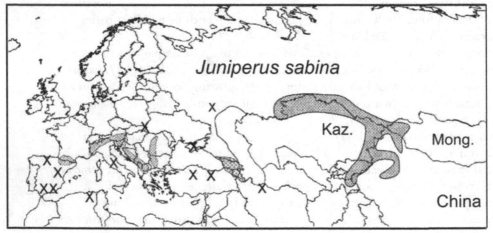

Distribution of *J. sabina*. X represents an outlying location. See *J. davurica* for locations previously mapped as *J. sabina* in China and Mongolia.

J. sabina, bark exfoliating in thin, broad plates.

J. sabina leaves and seed cones.

J. sabina
Pyrenees, Spain
cf. *Adams 7573-7577*

J. sabina, 30 km n of Jarkent (Paniflor), Kazakhstan cf. *Adams 7811-7813*, Kazakhstan 1996 field trip organized by Prof. A. D. Dembitsky and Dr. Yerlan Turuspekov.

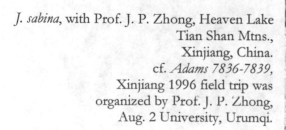

J. sabina, with Prof. J. P. Zhong, Heaven Lake Tian Shan Mtns., Xinjiang, China. cf. *Adams 7836-7839*, Xinjiang 1996 field trip was organized by Prof. J. P. Zhong, Aug. 2 University, Urumqi.

Juniperus saltillensis M. T. Hall, Fieldiana Bot. 34(4): 45 (1971). Saltillo juniper. Type: Mexico, Coahuila: 30 km s of Saltillo on MEX HWY 57, *Hall 66305* (holotype: F!, isotypes MEXU!, MO, NY).
    J. ashei Buch. var. *saltillensis* (M.T. Hall) Silba, Phytologia Mem. 7: 32 (1984)

Dioecious. Trees to 7m or broad shrubs, high, trunk branching at or just above base. **Trunk bark** thick, light ashy-gray, longitudinally divided into narrow strips, bark gray smooth not peeling. **Branches** erect or strict, terminal whips recurved at tips, angle of branching of ultimate twigs about 60 degrees. **Leaves** decurrent (whip) and scale-like, scale leaves mostly opposite on ultimate twig, triangular-ovate, tips appressed, obtuse to rounded, 0.5-1.5 mm long, margins finely denticulate, usually with resin gland protruding like a bead on the dorsal surface, gland same color as older light grayish-green leaves. **Seed cones** ovulate cone with soft juicy resinous pulp, slightly ellipsoid to globose, dark bluish-black with heavy white bloom appearing bright light blue, 4.5-8 mm diameter. **Seeds** mostly 1(-2, rarely 3) per cone, ovoid, obtuse, dark brown, 4-5 mm. long, hilum light brown to bark brown, one third length of seed. **Pollen shed** Jan. – Feb. **Habitat** at the edge of the *Bouteloua* grasslands and border of pine woodlands at 1550-2900 m. elevation in the W foothills and adjacent grasslands. **Uses** fence posts. **Dist.**: Chihuahua, Coahuila, Zacatecas and Nuevo Leon, Mexico. **Status**: the taxon is in areas that are subject to clearing for pasture improvement and may become threatened.

    Adams et al. (1980) considered *J. saltillensis* to a sibling species to *J. ashei*, and recent DNA analysis (Adams, 2011a) has *J. saltillensis* in a clade with *J. ovata* (*J. ashei* var. *ovata*)(Figs. 1.14), with *J. zanonii*.

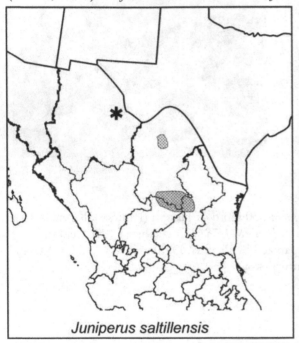

Juniperus saltillensis

J. saltillensis bark exfoliating in thin, broad strips.

J. saltillensis, with Walt Kelley, s of Agua Nueva, near Carneros Pass, 3 km e of MEX 54, Coahuila, Mexico cf. *Zanoni 2777* (photo by T. A. Zanoni)

J. saltillensis
leaves and seed cones

J. saltillensis, habitat, on limestone, with pinyon pine, on MEX 60, 14 km e of San Roberto Junction, Nuevo Leon, Mexico, cf. *Adams 6886-6890*

J. saltillensis, on MEX 60, 14 km e of San Roberto Junction, Nuevo Leon, Mexico, cf. *Adams 6886-6890*

Juniperus saltuaria Rehder & E. H. Wilson in Sargent, Pl. Wilson. 2: 61 (1914). Fang zhi bai. Type: China, Sichuan, Peling Shan, *W. Purdom 1911* (holotype A!)

  Sabina saltuaria (Rehder & E. H. Wilson) W. C. Cheng & W. T. Wang. [Chin. title; see Fl. Reipubl. Pop. Sin. 7:366 (1978)] 1: 257 (1961)

Monoecious. **Trees** to 20 m, rarely shrubs. **Trunk Bark** brown, exfoliating in narrow strips. **Branches** spreading or ascending, branchlets usually curved, 4-angled, 1-1.7 mm in diam. **Leaves** both scale-like and decurrent (whip), decurrent leaves present on young plants, in whorls of 3, 4.5-6 mm, ridged abaxially, apex sharply pointed, scale-like leaves decussate, 4-ranked, closely appressed, triangular-rhombic, gibbous, 1-2 mm, abaxial gland basal, inconspicuous, orbicular or ovate, slightly depressed leaf apex obtuse. **Seed cones** erect, black or bluish black when ripe, turbinate or subglobose, 4-8(-10) mm. **Seeds** 1 seed per cone, irregularly ovoid-globose, 3.5-7 x 3-5 mm, ridged and pitted. **Pollen shed** spring. **Habitat** forests or thickets on mountains, 2700-4600 m. **Uses** none known. **Dist.**: S. Gansu, SE Qinghai, W. Sichuan, E Xizang, NW Yunnan, China. **Status**: common in areas, not threatened.

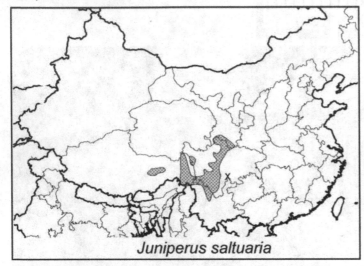

Juniperus saltuaria

J. saltuaria bark exfoliating
in narrow strips.

J. saltuaria
leaves and seed cones

J. saltuaria, Deqin Co., Yunnan, China cf. *Adams 8494-8496*

J. saltuaria, on Duoer River, ca. 23 km se of Forestry Station, 3100 m, Gansu, China, cf. *Adams 6789-6791*

J. saltuaria habitat, Deqin Co., Yunnan, China cf. *Adams 8494-8496*. The 1998 Yunnan trip was organized by Prof. Fei, Kunming Botanic Garden, who is now deceased.

Juniperus scopulorum Sargent, Gard. & Forest 10:420, f. 54 (1897). Rocky mountain juniper. Type: not designated by Sargent; USA, Wyoming, Yellowstone Natl. Pk, Mammoth Hot Springs, *C. S. Sargent, s. n., 8 July 1906* (lectotype: A!, designated by T. Zanoni, 1978).

    J. excelsa Pursh, Fl. Amer. Sept. 2:647 (1814), *non* M.-Bieb. (1800)

    J. virginiana L. var. *montana* Vasey, (Cat. Forest Trees U.S. 30) Rep. U.S. Dept. Agric. 1875: 47 (1876)

    J. occidentalis Hook. var. *pleiosperma* Engelm., Trans. St. Louis Acad. Sci. 3: 590 (1878)

    J. virginiana L. var. *scopulorum* (Sarg.) Lemmon, Handb. W. Amer. Cone-Bearers, ed. 4: 114 (1900)

    Sabina scopulorum (Sarg.) Rydberg, Bull. Torrey Bot. Club 32: 598 (1905)

    J. scopulorum var. *patens* Fassett, Bull. Torrey Bot. Club 72: 46 (1945) [= X *fassettii* Boivin (*horizontalis* x *scopulorum*)]

    J. scopulorum Sarg. var. *columnaris* Fassett, Bull. Torrey Bot. Club 72:482 (1945)

    J. scopulorum Sarg. f. *columnaris* (Fassett) Rehder, Bibliogr. Cult. Trees Shrubs: 63 (1949)

    J. fassettii A. Boivin, Naturaliste Canad. 93:372 (1966)

    J. scopulorum var. *columnaris* Fassett (environmentally induced by gases from burning coal, see Adams, 1982 for analysis and discussion)

    J. virginiana L. subsp. *scopulorum* (Sarg.) E. Murray, Kalmia 13:8 (1983)

Dioecious. **Trees** single (rarely multi-) stemmed tree to 20 m. pyramidal to occasionally round crowns. Twigs (3-5 mm diam.) with smooth bark, twigs (6-15 mm diam.) with bark exfoliating in plates, reddish-copper beneath. **Trunk bark** brown, exfoliating in thin strips. Foliage light to dark green but often blue and blue-gray due to glaucousness. **Branches** erect to occasionally pendulous at the tips. **Leaves** both decurrent (whip) and scale. Whip leaves growing only at branchlet tips (on mature trees). Scale leaves not overlapping, or, if so, then not by more than 1/5 the length, obtuse to acute, margins entire at 20 X (and 40 X). **Seed cones** maturing in 2 yrs., globose to 2 lobed, appearing light blue when with heavy glaucous, but dark blue-black beneath glaucous (when mature). [Note: cones may appear tan beneath the glaucous when immature], 6-9 mm, borne on mostly straight peduncles. **Seeds** (1)2(3) per cone, 4-5 mm long. 2n = 22 (Hall, Mukherjee and Crowley, 1973). **Pollen shed** March-April. **Habitat** rocky soils, and slopes, eroded hillsides, sea level (Vancouver Isl., Puget Sound), otherwise 1200-2700 m. **Uses** fence posts. **Dist.**: Canada: Alberta, B.C., USA: Ariz., Colo., Idaho, Mont., Neb., N.D., N. Mex., Nev., Ore., S.D., Tex., Utah, Wash., Wyo. **Status**: abundant and increasing, considered a weed in rangelands. Not threatened.

    Juniperus scopulorum hybridizes with its eastern sibling species, *J. virginiana* in the zones of contact in the Missouri R. Basin (Comer, Adams and Van Haverbeke, 1982; Flake, Urbatsch and Turner, 1978; Van Haverbeke, 1968). Relictual hybridization seems to be present with *J. blancoi* (Adams, 2011b) in northern Mexico and with *J. virginiana* in the Tex. Panhandle (Adams, 1983). It also hybridizes with *J. horizontalis* (see *J. horizontalis*, above).

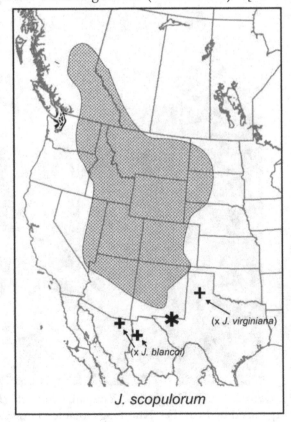

J. scopulorum

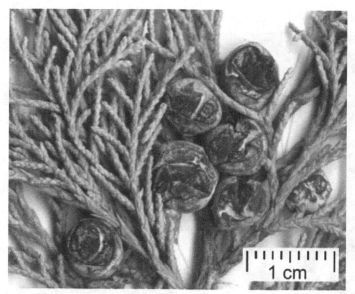

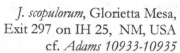

J. scopulorum bark exfoliating in thin plates.

J. scopulorum, leaves and seed cones.

J. scopulorum, Glorietta Mesa, Exit 297 on IH 25, NM, USA cf. *Adams 10933-10935*

J. scopulorum, 15 m tree, Charleston Peak, Nevada, USA. cf. *Adams 10277-10281*

Columnar trees of *J. scopulorum* produced by the environmental induction of burning coal gases from a nearby smelter at Butte, MT, USA (see Adams, 1982, for analysis and discussion). 16 km w of Butte, MT. cf. *Adams 10885-10889.*

Very glaucous tree of
J. scopulorum, Glorietta Mesa,
Exit 297 on IH 25, NM, USA
cf. *Adams 10933-10935.*
Glaucous trees of *J. scopulorum* are
very common in the southern
Rocky Mountains.

J. scopulorum (x *virginiana?*) trunk used for fence post on Curry Ranch, south of Amarillo, Texas, USA.
This trunk came from an old tree in
Palo Duro Canyon on the John Curry Ranch.

Juniperus semiglobosa

Previously (Adams, 2011) I recognized 2 varieties of J. semiglobosa, but new DNA sequencing has necessitated the recognition of 3 varieties (see Figs. 1.3, 1.16.2, Chpt. 1).

Key to varieties:

1a. Mature (ripe) seed cones very sweet.................................*Juniperus semiglobosa* var. *talassica*

1b. Mature (ripe) seed cones not sweet, but with turpentine flavor

    2a. Scale-leaves mostly flat, scarcely overlapping, giving the ultimate branchlets a smooth appearance, glands on whip-leaves not conspicuous, not generally raised..*J. semiglobosa*

    2b. Scale-leaves beaked, overlapping by about 1/4, giving the ultimate branchlets a rough appearance, glands on whip-leaves conspicuous, generally raised......

    ....*J. semiglobosa* var. *jarkendensis*

Juniperus semiglobosa Regel var. *semiglobosa*, Trudy Imp. S.-Peterburgsk. Bot. Sada 6(2): 487 (1879). Himalayan pencil juniper or cedar, Dhup. Kun lun duo zi bai. Type: Tajikistan, Zaravschan Range, Saratag Pass, Isfara, *B. A. Fedtschenko s. n.* (lectotype: LE!, by Farjon, 1993)

    J. excelsa auct., non M.-Bieb.: Wallich, Cat. 6041 (1832); Brandis, Forest fl. NW India: 538 1874)

    J. schugnanica Kom., Bot. Zurn. (Moscow & Leningrad) 17: 482 (1932)

    J. media V.D. Dmitriev, Trudy Sektora Agrolesomelior. Lesn. Khoz. Komiteta Nauk Uzbeksk. S. S. R. 4: 31 (1938)

    J. tianschanica Sumnev., Bot. Mater. Gerb. Inst. Bot. Zool. Akad. Nauk Uzbeksk. SSR 8: 24 (1946)

    J. drobovii Sumnev. Bot. Mater. Gerb. Inst. Bot. Zool. Akad. Nauk Uzbeksk. SSR 8: 22 (1946)

    J. semiglobosa Regel var. *drobovii* (Sumnev.) Silba, Phytologia 68: 33 (1990)

    Sabina semiglobosa (Regel) L.K. Fu & Y. F. Yu, Higher Plants of China 3:88 (2000)

Dioecious, rarely monoecious. **Trees** to 10 m tall, with a rather loose and narrow often weeping crown. **Trunk bark** brown, cinnamon, exfoliating in thin, long plates. **Branches** branchlets slender, pendulous. **Leaves** both scale and decurrent (whip) leaves, decurrent leaves terete, with a white median band above, lustrous, long-pointed, imbricated leaves rhombic, obtuse, with an oblong dorsal gland, tightly appressed to the branchlets. **Seed cones** black when mature in 2 yrs, bi-lobed (semiglobosa) or reniform, flatly truncate at the top, black, with scattered mealy bloom, 6-7 mm in diameter **Seeds** 2 - 3(-4) per seed cone, 4 - 5 mm long and 3 mm broad, flattish, cuneate or with convex abaxial surface and lateral grooves. **Pollen shed** spring. **Habitat** growing partly by itself, partly together with other juniper species, forming thin juniper woods on dry mountain slopes, moraine deposits, and lakeshore terraces. 2500-3300 m. **Uses** Wood reddish, used for pencil-making, furniture, fuel and charcoal,

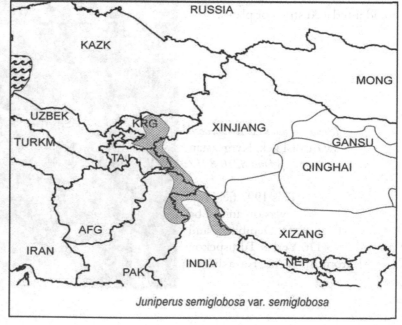

Juniperus semiglobosa var. *semiglobosa*

incense, female cone is medicinal. Resin content higher than in other juniper species. **Dist**.: SW Xinjiang, W Xizang, China; Afghanistan, N. India, Kashmir, Kazakhstan, Kyrgystan, Tajikistan, Uzbekistan. **Status**: occurs over a wide range, might be threatened by goats.

J. semiglobosa var. *semiglobosa* leaves and seed cones

1 cm

J. semiglobosa, bark exfoliating in thin, broad interlaced strips or plates.

J. semiglobosa var. *semiglobosa* sw of Bishkek, Kyrgyzstan cf. *Adams 8210-8212*

This 1997 field trip was organized by Prof. A. D. Dembitsky and Dr. Yerlan Turuspekov from Alma Ata, Kazakhstan.

Juniperus semiglobosa var. ***jarkendensis*** (Kom.) . R. P. Adams. Phytologia 94(3) 354 (2012). Jarkant (Yarkant) juniper. Type: China, Xinjiang, Kunlun Shan, Yarkant River, mtns. near Shache (Yarkant), *V. I. Robarovski 409* (holotype LE!)
   Juniperus jarkendensis Kom. Not. Syst. Herb. Petrop. 4: 181 (1923).
   J. sabina var. *jarkendensis* (Kom.) J. Silba, Phytologia
      68: 33(1990)
   Sabina vulgaris var. *jarkendensis* (Kom.) C. Y. Yang in
      Fl. Reipub. Pop. Sin. 7: 360 (1978)

Dioecious. Small Trees to 4 m or taller, with strong central axis, crown very open. **Trunk bark** brown, cinnamon beneath, exfoliating in thin plates. **Branches** some trees with tight foliage and other trees with drooping foliage. **Leaves** decurrent and scale-like, foliage light green, decurrent leaves only on juvenile growth. **Seed cones** 8-9 mm long, bilobed or ovoid with knobby tip, reddish brown to dark blue, glaucous. **Seeds** 2(-4) per cone, subovoid, angular. **Pollen shed** spring. **Habitat** high mountains with *J. pseudosabina*, *J. sabina* and Abies. **Uses** none known. **Dist.**: Xinjiang, China, 3000-4000 m. **Status**: It has a narrow distribution in an area that is subject to cutting for fuel. It is likely threatened.

       This species is often merged under *J. sabina*, but Adams (1999) found that it is more closely associated with *J. semiglobosa* in its terpenoids and RAPDs. DNA sequencing (Figs. 1.3, 1.16.1, 1.16.2, 5.31) indicates that *J. jarkendensis* appears to be best treated as *J. semiglobosa* var. *jarkendensis* (Adams and Schwarzbach, 2013d). *Juniperus sabina* is loosely linked to the *J. davurica* group (Fig. 5.31) and not at all allied with *J. semiglobosa*.

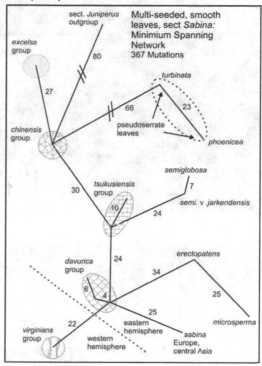

Figure 5.31 Minimum spanning network showing *J. semiglobosa* and *J. s.* var. *jarkendensis* separated by only 7 MEs. Note that *J. sabina* is far removed.

J. semiglobosa var. *jarkendensis* bark exfoliating in thin, broad strips or plates.

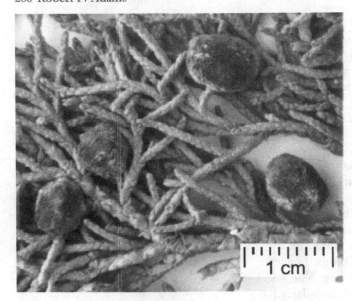

J. semiglobosa var. *jarkendensis*
leaves and seed cones

1 cm

J. semiglobosa var. *jarkendensis*, above Sasche (Yarkant), Xinjiang, China, cf. *Adams 7820-7825*. Notice the variation among adjacent trees (dark green foliage on tree on left and very pale green/silver foliage on the tree on the right). Terpenes and RAPDs were analyzed for both these trees and no differences were found. The field trip to Sasche, Xinjiang was organized and led by Prof. J. P. Zhong, Aug. 2 University, Urumqi.

Juniperus semiglobosa Regel var. ***talassica*** (Lipinsky) Silba, Phytologia 68: 34 (1990). Type: Kyrgyzstan, Talass-Alatoo Range, Sir-Darin Oblast, Karagoin Pass, *Z. A. von Minkwitz 1351* (holotype C?, n. v., isotype LE). Named after the type locality, Talas.

J. talassica Lipinsky in O.E.Knorring & Z. A. Minkvich, Rastit. Aulie-Atinsk. u.s. Dar'inskoi obl./Travaus d'expedition pour exploration des regions de Colonization Russe 2. Explor. Bot. 1909. 6:185-186, pl. (1912).

Although var. *talassica* is very similar to *var. semiglobosa*, it is reported that the fruits of var. *talassica* have a much higher sugar content (sweeter) than the fruits of var. *semiglobosa* . The fruits (seed cones) of var. *talassica* do seem sweeter than var. *semiglobosa* (Adams, pers. obs.). However, this variety appears to be dubious. Adams (1999) found that var. *semiglobosa* and var. *talassica* were very similar in their leaf terpenoids and only a little different in their RAPDs.

Dist.: Kyrgystan, Central Asia.

J. semiglobosa var. *talassica* bark, exfoliating, in thin, wide plates.

J. semiglobosa var. *talassica*, leaves and seed cones.

J. semiglobosa var. *talassica*
20 km s of Talas, Kyrgystan.
cf. *Adams 8220-8222*

Juniperus seravschanica Kom., Bot. Zurn. (Moscow & Leningrad) 17:481 (1932) Afghan juniper.
Type: Tajikistan, Zaravschan Range, Zaravschan Valley, Darch, *V. L. Komarov s. n.* (lectotype LE!)
Named after locality, Zaravschan.

    J. macropoda Boiss., Fl. Orient. 5:709 (1884)

    J. polycarpos var. *seravschanica* (Komarov) Kitamura, Fl. Pl. W. Pakist. Afghan Add. & Corr. Fl.
        Afghan.: 68 (1966).

    Sabina seravschanica (Kom.) Nevski, Trudy Bot. Inst. Akad. Nauk S.S.S.R., ser. 1, Fl. Sist.
        Vyss. Rast. 4:245 (1937)

    J. polycarpos K. Koch var. *seravschanica* (Kom.) Kitam., Add. & Corr. Fl. Afghan.: 68 (1966)

    J. excelsa M.-Bieb. subsp. *seravschanica* (Kom.) Imkhan., Bot. Zurn. 75 (3): 407 (1990)

Dioecious. Tree 5-10 m tall. **Trunk bark** reddish or reddish-gray bark, dense crown. **Branches** branchlets rather slender and relatively short, 1 - 1.5 mm thick, slightly moniliform or not so, yellowish-green or glaucescent. **Leaves** decurrent (whip) and scale-like, oblong, acute, scale leaves with a beak or keel so branchlet appears as a string of beads, with an oblong gland, scale leaves with clear, ellipsoid glands, often ruptured, with a clear exudate. **Seed cones** subglobose, strongly pruinose, very hard, with a woody pericarp layer, 8-10 mm diam. **Seeds** 3-6 per seed cone. **Pollen shed** fall-winter. **Habitat** forming juniper woods in lower and middle parts of the timber zone, in pure stands or mixed with *J. semiglobosa*. **Uses** fruits large, but not juicy and with very low sugar content. Wood useful for joinery and carpentry. An ornamental tree recommended for control of sliding and erosion of mountain slopes. **Dist**: Central Asia to Pakistan. **Status**: common and reproducing, not threatened. See Figs. 1.3, 1.16.1, 1.16.3 (Chpt. 1) for DNA sequencing data supporting specific status.

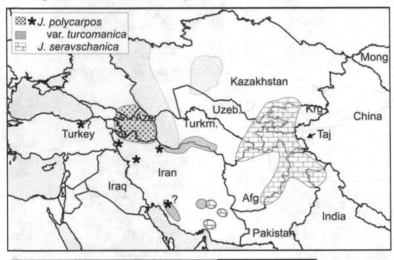

J. seravschanica
bark exfoliating in thin,
wide plates.

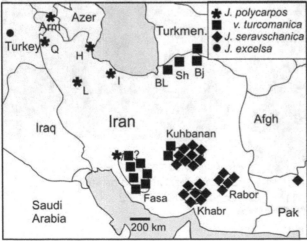

Distribution of *J. seravschanica* in Iran.
(from Adams, Hojjati and Schwarzbach, 2014)

The populations of *J. seravschanica* in southern Iran appear to have some DNA differences from populations in Pakistan and Kazakhstan.

J. seravschanica
leaves and seed cones.

J. seravschanica
2 km s of Dzhabagly, Kazakstan
cf. *Adams 8224-8226*

J. seravschanica, Ziarat Forest,
Pakistan (photo by William M. Ciesla)
cf. *Adams 8483-8486*.

Juniperus squamata Buchanan-Hamilton ex D. Don in Lambert.

In this treatment, 2 infra-specific taxa are recognized. *Juniperus squamata* and *J. pingii* from western China and the Himalayas are still in need of considerable study. See discussion of *J. pingii* and *J. squamata* f. *wilsonii* (= *J. pingii* var. *wilsonii*) under *J. pingii*. It may be that hybridization is involved. There is much variation in leaf types and foliage. It is clear from DNA and terpene data that *J. morrisonicola* is not conspecific with *J. squamata* (see Figs. 1.3, 1.11, 1.15.1 for DNA seq. discussion). Below is a key to these infra-specific taxa based on present, incomplete knowledge:

Note: for *J. squamata*, all leaves are decurrent (**not acicular as sometimes stated**). Each decurrent leaf is composed of a blade (free portion) and sheath (that clasps the stem).

1a. Leaf blades 6-7 mm long, seed cones 8-9 mm long..............***J. squamata*** var. **squamata**
1b. Leaf blades 4-5 mm long, seed cones 4-5 mm long.....…......….....***J. squamata*** var. **wilsonii**

Juniperus squamata Buch.-Ham. ex D. Don in Lambert var. **squamata**, Descr. Pinus 2: 17 (1824). Scaly-leaved Nepal juniper, Gao shan bai. Type: Bhutan, 'Alpibus', *W. S. Webb W 6043C* (lectotype K-W!, designated by Farjon (p.382, 2005).

J. squamata var. *meyeri* Rehder, J. Arnold. Arb. 3: 207 (1922), type A

J. squamata var. *prostrata* Hornibr., Dwarf. Conif. 77 (1923. ex cult. plant.

J. franchetiana Lev. ex Kom., Bot. Mater. Gerb. Glavn. Bot. Sada RSFSR 5: 30 (1924)

J. recurva Buch.-Ham. ex D. Don var. *densa* hort. ex Carriere, Traite Gen. Conif.: 27 (1855)

Sabina recurva (Buch.-Ham. ex D. Don) Antoine var. *densa* (Carr.) Antoine, Cupress.-Gatt.: 67 (1857)

S. squamata (Buch.-Ham. ex D. Don) Antoine, Cupress.-Gatt.: 66 (1857)

J. densa (Carriere) Gordon Pinetum Suppl.: 32 (1862)

J. excelsa var. *densa* Endl.

J. recurva var. *squamata* (D. Don) Parl. full ref.: P. Parlatore, in De Candolle Prodromus, Vol. 16 part 2, p. 482 (1868)

J. squamata Buch.-Ham. ex D. Don var. *hongxiensis* Y.F. Yu & L.K. Fu, Novon 7 (4):444 (1998)

J. squamata Buch.-Ham. ex D. Don var. *parvifolia* Y.F. Yu & L.K. Fu, Novon 7 (4): 444 (1998)

J. baimashaneusis Y.F. Yu & L.K. Fu, Novon 7 (4) 443 (1998)

Dioecious, occasionally monoecious. **Shrubs**. **Trunk bark** cinnamon-brown, exfoliating in thin, broad strips or plates. **Branches** ascending or horizontally spreading, branchlets densely arranged, straight or curved, usually short, not angled. **Leaves** both decurrent (whip) and scale-like, in whorls of 3, spreading or ascending, sometimes nearly appressed, straight or slightly curved, (2.5) 5-7 x 1.2-2.5 mm, slightly concave, with white stomatal bands adaxially, obtusely ridged with longitudinal, thin groove on ridge or at base abaxially, base decurrent, apex acute or acuminate. **Seed cones** black or bluish black when mature (in 2 years), ovoid or subglobose, 4-8 x 4-6 mm, 1-seeded. **Seeds** ovoid, 3.35-6 x 2-5 mm, ridged, with resin pits. **Pollen shed** spring. **Habitat** rocky areas in high mountains. **Uses** wood is used for fuel in high mtns. and for incense. **Dist.**: Himalayans, w. China. **Status**: common, not threatened at present. Note that on some specimens of *J. squamata*, it appears that the pollen was shed in the spring compared to fall for *J. fargesii*.

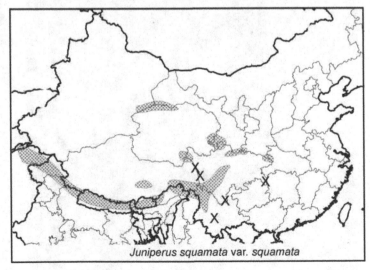

Juniperus squamata var. squamata

J. squamata var. *squamata* bark

J. squamata var. *squamata* leaves and seed cones

J. squamata cv. *meyeri*
Kew Gardens, UK

J. squamata cv. *meyeri*
Kunming Botanic Garden
Kunming, China.
cf. *Adams 6795-6796*

Juniperus squamata var. ***wilsonii*** (Rehder) R. P. Adams, Phytologia 94(3) 399 (2012). Xiang bai. Type: China, Sichuan, Tachien lu, 9000 – 14,000 ft., E. H. Wilson 985 (1908), A!

    J. squamata Buch.-Ham. ex D. Don f. *wilsonii* Rehder, J. Arn. Arb. 1: 191 (1920).

    J. pingii W.C. Cheng ex Ferre var. *wilsonii* (Rehder) Silba, Phytologia Mem. 7: 36 (1984)

    Sabina pingii (W. C. Cheng ex Ferre) W. C. Cheng & W. T. Wang var. *wilsonii* (Rehder) W.C. Cheng & L.K. Fu, Fl. Reipubl. Pop. Sin. 7: 356 (1978)

    S. squamata (Buch.-Ham. ex D. Don) Antoine var. *wilsonii* (Rehder) W. C. Cheng & L.K. Fu, [Chin. title; see Fl. Sichuan 2:177 (1983)] 1: 320 (1972)

    S. wilsonii (Rehder) W.C. Cheng & L.K. Fu, Forest Sci. Techn. 4: 455 (1981)

    J. wallichiana Hook. f. & Thomson ex E. Brandis var. *loderi* Hornibr., Gard. Chron., ser. 3, 85: 50 (1929)

    J. squamata Buch.-Ham. ex D. Don var. *loderi* (Hornibr.) Hornibr. in Chittenden, Rep. Cult. Conif.: 74 (1932)

Dioecious. **Shrubs** erect or procumbent, or small trees to 6 m tall. **Trunk bark** exfoliating in strips. **Branches** branchlets not pendulous, stout, prominently 6-angled. **Leaves** all decurrent, blades 4-5 mm long, blades free at 30° - 45° angle to stem. **Seed cones**. green when immature, black when mature in 2 yrs., 4-5 mm long, subglobose to elongate. **Seeds** 1 per cone. **Pollen shed** winter? **Habitat** thickets in mountain regions, 2600 - 4900 m. **Uses** none known. **Dist.**: S. Gansu, NW Hubei, S. Qinghai, S. Shaanxi, Xizang, Yunnan, China. **Status**: unknown.

Silba [Phytologia Mem. 7: 36 (1984)] placed *J. squamata* f. *wilsonii* as a variety of *J. pingii* but see discussion under *J. pingii*.

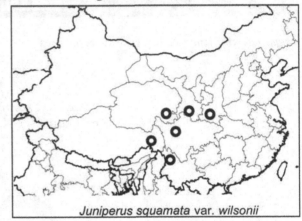

Juniperus squamata var. *wilsonii*

J. squamata var. *wilsonii*, *E. H. Wilson 985*, sheet II, left hand specimen.

J. squamata var. *wilsonii* cultivated from seeds obtained from the *E. H. Wilson 985* collections in China, grown at Arnold Arboretum. Acc. # 1010-64A, PI266613. cf. *Adams 5521*. Note the more open leaf arrangement in cultivation.

Juniperus standleyi Steyermark, Publ. Field Mus. Nat. Hist., Bot. Ser. 23 (1): 3 (1943). Cipres, cipres enano, chancel, huito, huitun, si-lop (Huehuetenango, Guatemala), Standley juniper. Type: Guatemala, San Marcos, Volcan de Tacana, *J. A. Steyermark 36137* (holotype F!).

Dioecious. Trees or shrubs, 0.3-15 m, crown irregularly shaped. **Trunk bark** stem bark of longitudinal, interconnected strips. **Branches** spreading to spreading ascending, main stem often contorted, terminal whips broadly spreading to ascending, tip of whip branches usually straight, reddish-brown, angle of branching of ultimate twig 40-50 degrees. **Leaves** decurrent (whip) and scale-like, scale leaves opposite or ternate, ovate to broadly ovate, with rounded or obtuse appressed tips, 1.5-2.0 mm long, margin finely toothed, yellowish to dark green, resin gland at base of scale leaf. **Seed cones** dark blue, with a light coat of bloom (glaucous), with soft fleshy pulp, ovoid-conic or irregular, 7-9 mm diameter, on short, curved peduncles. **Seeds** 3 - 5 (-6) per cone, 4-5 mm long, 2.5-3.7 mm wide, subconic, grooved, dark brown, hilum about 2/3 length of seed. **Pollen shed** Nov.-Jan. **Habitat** above timberline near the summit of Volcan de Tacana on the Chiapas, Mexico side, and Depto. San Marcos, Guatemala; in pine and pine-juniper forests on moist slopes and sheltered ravines, often on limestone, in the Sierra de los Cuchumatanes, Depto. Huehuetenango, Guatemala, at elevations of 3000 m or higher. **Uses** none known. **Dist.**: Mexico/Guatemala border. **Status**: EN, B1+2b. Although it has a small area of distribution, the alpine areas are not as likely to be disturbed, as areas where cropland and pasturing is expanding. It seems more likely that it is vulnerable than endangered.

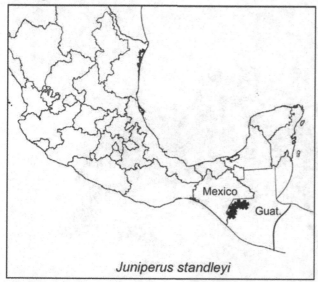

Juniperus standleyi

J. standleyi, bark exfoliating, in narrow strips. (photo T. A. Zanoni)

Small forest of *J. standleyi* trees, 26 km n of Huehuetenango on RN 9, Guatemala. cf. *Zanoni 2644*. (photo T. A. Zanoni)

J. standleyi
leaves and seed cones.

J. standleyi, Sierra de los Cuchumatanes, 26 km n of Huehuetenango on RN 9, Guatemala. cf. *Zanoni 2644*. (photo T. A. Zanoni)

Juniperus taxifolia

Previously, Adams (2004) treated *J. taxifolia* var. *lutchuensis* as a distinct species (*J. lutchuensis*) based on RAPDs data (Adams, et al., 2002) showing it is as quite different. However, the most recent sequence data (Adams and Schwarzbach, 2013e) has these taxa in the same clade (see Figs. 1.3, 1.16.4, Chpt. 1) quite removed from *J. chinensis*. At present, these taxa are treated as varieties.

1a. Leaves with one stomatal band or if 2, then midrib faint and not full length of leaf, leaves with acute to obtuse tips, boat-shaped, 8 - 12 (-14) mm; endemic to Bonin Islands...***J. taxifolia*** var. ***taxifolia***

1b. Leaves with 2 distinct stomatal bands, with midrib the runs full length of leaf, leaves acute to apiculate tips; boat-shaped, 7 - 10 mm and wide (1.1 - 1.5 mm), endemic to Ryukyu Islands and s. coast of Shizuoka Pref., Oshima (Island), Japan.. ....***J. taxifolia*** var. ***lutchuensis***

Juniperus taxifolia var. **lutchuensis** (Koidz) Satake Bull. Natn. Sci. Mus. Tokyo 6(2): 193 (1962) Type: Japan, Ryuku Islands, Okinawa, *T. Miyagi 17* (holotype KYO).

    J. lutchuensis Koidz, Bot. Mag., Tokyo 32:138 (1918).

    J. conferta sensu Wilson, J. Arn. Arb. 1: 97-121(1919), 171-186 (1952).

    J. conferta var. *maritima* Wilson ex Nakai, Bot. Mag. Tokyo 36: 105 (1922) *nom. nud.*

    J. lutchuensis var. *maritima* (Wilson) Iwata & Kusaka, l.c., ed 2: 195 (1954). sugimoto, l.c. 59 (1961).

Dioecious. Prostrate shrub to somewhat upright 1-3 m tall. **Trunk bark** brown, exfoliating in thin strips. **Branches** procumbent or horizontally spreading; branchlets dull brown, ascending, glabrous. **Leaves** acicular, 7-14 mm long, 1-1.5 mm wide ternate, awl-shaped, base jointed, apex soft and obtuse, with one ridged green midrib and two white stomatiferous bands on upper surface(some leaves on a branch may lack the midrib, showing only 1 stomatiferous band), light green and slightly ridged on lower surface. **Seed cones** purple brown when mature, tan-brown when immature, on short stalks, densely clothed with small scale leaves, globose, green, ca 2 mm in diam., consisting of 3 coalescent ovuliferous scales enveloping 3 ovules disposed alternately with the scales; cones ripening January to February of the third year, berry-like, globose, 8-9 mm in diameter. **Seeds** 3 per cone, ovoid, keeled, triangular in cross section, brown, ca. 5 mm long. Flowers February to March, solitary in axils of previous year's shoots. **Pollen shed** spring. **Habitat** grows from near sea level to 300 m elevation, in sunny stony places. **Uses** none known. **Dist.**: Ryukyu Islands, s. coast of Shizuoka Pref., Oshima (Island), Japan. **Status**: a rather rare taxon with limited distribution. It may become threatened.

Juniperus taxifolia
var. *lutchuensis*

J. taxifolia var. *lutchuensis* bark

J. taxifolia var. *lutchensis*
leaves and seed cones

J. taxifolia var. *lutchensis*
as a prostrate creeping plant
on rock, Shizuoka Pref., Japan
cf. *Adams 8538-8540*

Prof. Jin Murata arranged the
1998 field in Japan.

J. taxifolia var. *lutchuensis*,
as a prostrate plant on
rock, on the seashore,
Oshima Island, Japan.
cf. *Adams 8541-8543*

Juniperus taxifolia Hook. & Arn., Bot. Beechey Voy. 6: 271 (1838). Shima-muro, Hide. Type: Japan, Bonin Islands, Ogasawara Group, *G. T. Lay (with A. Collie) s. n.* (holotype K!).

Dioecious. Small trees 1-3(-13) m tall, or erect shrubs. **Trunk bark** light brown, exfoliating thin, narrow strips. **Branches** ascending or horizontally spreading, branchlets dull brown, ascending, glabrous. **Leaves** acicular, 7-14 mm long, 1-1.5 mm wide ternate, awl-shaped, base jointed, apex soft and obtuse, with one ridged green midrib and two white stomatiferous bands on upper surface, light green and slightly ridged on lower surface. **Seed cones** on short stalks, densely clothed with small scale leaves, globose, greenish copper color when immature, ca 2 mm in diam., consisting of 3 coalescent ovuliferous scales enveloping 3 ovules disposed alternately with the scales, cones ripening to purple brown in January to February of the third year, berry-like, globose, 8-9 mm in diameter. **Seeds** 3 per seed cone, ovoid, keeled, triangular in cross section, brown, ca. 5 mm long. Flowers February to March, solitary in axils of previous year's shoots. **Pollen shed** winter? **Habitat** grows from near sea level to 300 m elevation, in sunny stony places. **Uses** none known. **Dist.**: Japan, Bonin Islands. **Status**: unknown.

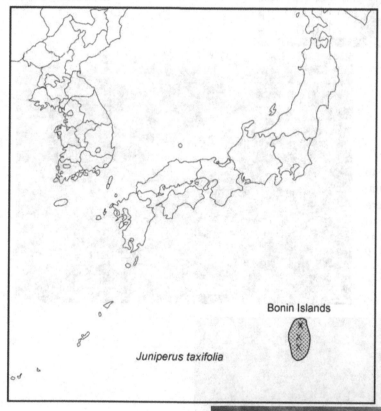

Bonin Islands

Juniperus taxifolia

J. taxifolia var. *taxifolia*,
Bonin Islands, Japan
(photo Prof. Jin Murata)

J. taxifolia var. *taxifolia*
leaves and seed cones

 1 cm

J. taxifolia var. *taxifolia*,
Bonin Islands, Japan
(photo Prof. Jin Murata)
Murata's collections were
sent to the author and
relabeled as
Adams 8448-8449

J. taxifolia var. *taxifolia*,
Bonin Islands, Japan
(photo Prof. Jin Murata)
cf. *Adams 8448-8449*

Juniperus thurifera L.

Two varieties are recognized in this treatment. Adams et al. (2003) found the leaf essential oils in *J. thurifera* var. *africana* (from Morocco) to be higher in sabinene, in contrast, var. *thurifera* (from Spain, and France) was higher in 2-carene and limonene. RAPDs showed the Morocco plants (var. *africana*) to be barely distinct from var. *thurifera*. Terrab et al. (2008) using AFLPs found var. *africana* to be distinct in Morocco, and their Algerian plants to be intermediate between var. *africana* and var. *thurifera*. Due to the potential that the Moroccan plants contain some unique genes, var. *africana* is recognized in this treatment.

Key to varieties:

1a. Seed cones 8 – 12 mm when mature, with 2-4 seeds...................................*J. thurifera* var. *thurifera*

1b. Seed cones 7 8 mm when mature, with 1 (2, rarely 3) seeds..................*J. thurifera* var. *africana*

Juniperus thurifera var. *africana* Maire, Bull. Soc. Hist. Nat. Afrique N. 17: 125 (1926). African thurifera. Type: uncertain, North Africa, probably Algeria or Morocco. Farjon (2005) lists: syntypes J. A. Battandier?, R. Maire? (MPU?) n. v.

J. africana (Maire) Villar, Types Sols Afrique N. 1:91 (1947)

J. thurifera subsp. *africana* (Maire) Gauquelin, Hassani et LeBreton, Ecol. Medit. 14 (3): 39 (1988) invalid (see Romo and Boratynski, Ann. Bot. Fenn. 44(1) 73, 2007).

J. thurifera subsp. *africana* (Maire) Romo and Boratynski, Ann. Bot. Fenn. 44(1) 73, 2007

Dioecious. Trees pyramidal up to 20 m. **Trunk bark** tan-brown, exfoliating in interlaced strips or long plates. **Branches** twigs 1 mm wide, quadrangular, scaly, distichously arranged. **Leaves** decurrent (whip) and scale-like, all decussate, scale leaves 1.5-2 mm, appressed but free at their incurved acuminate or acute apices, with an oblong furrowed gland on the back, and an entire or slightly toothed, but not scarious, border. **Seed cones** 7-8 mm, greenish purple when immature, dark purple when fully ripe, maturing in 2 yrs. **Seeds** 1(2-, rarely 3) seeds per cone, free. 2n=44. **Pollen shed** winter. **Habitat** red sandstone, high mountains, semi-arid highland Mediterranean climate. **Uses** none known. **Dist.:** Only known from the high Atlas Mtns, Morocco and a stand in Algeria. **Status**: populations are declining. Part of this is due to goats grazing and also due to increased aridity in the region. This variety is endangered.

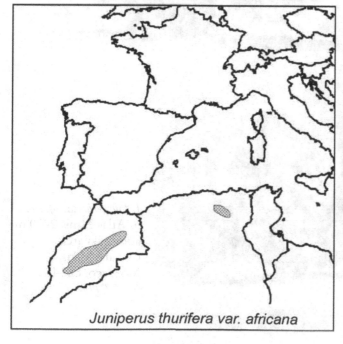

Juniperus thurifera var. africana

J. thurifera var. *africana* bark exfoliating in thick, narrow strips.

J. thurifera var. *africana*
leaves and seed cones.

J. thurifera var. *africana*,
with Jessica D. (Adams) Black (left),
Atlas Mtns. 2400 m, above
ski area, Oukaimeden,
Morocco. cf. *Adams 9415-9417*

J. thurifera var. *africana*,
in Atlas Mtns. 2400 m,
above ski area,
Oukaimeden,
Morocco. cf. *Adams
9415-9417*

Juniperus thurifera L. var. *thurifera*. Sp. Pl. 2:1039 (1753). Incense bearing juniper, Spanish juniper. Type: Spain, Aragon, Teruel, ("Mansana, Camarena"), *E. Reverchon 788* (neotype BM, designated by Farjon (p. 389, 2005).

    J. thurifera Hort. Par. ex Endl. Syn. Conif. 25. (1847)
    J. thurifera Bonpl. ex Endl. Syn. Conif. 63. (1847)
    J. thurifera Hort. ex Carriere Trait Gen. Conif., ed. 2, 50, 1867
    J. hispanica Mill., Gard. Dict., ed. 8: *Juniperus* No. 13 (1768)
    J. sabinoides Endl. Syn. Conif.: 24 (1847), *non* Griseb. (1846)
    J. bonatiana Vis., Mem. Ist. Venetia. 6:245 (1856)
    Sabina foetidissima (Willd.) Antoine var. *bonatiana* (Vis.) Antoine, Cupress.-Gatt.: 49 (1857)
    S. thurifera (L.) Antoine, Cupress.-Gatt.: 51 (1857)
    S. pseudothurifera Antoine, Cupress.-Gatt.: 60 (1857)
    J. cinerea Carriere, Traite Gen. Conif., ed. 2, 1:35 (1867)
    J. thurifera L. var. *gallica* Coincy, Bull. Soc. Bot. France 45:430 (1898)
    J. gallica (Coincy) Rouy, Fl. France 14:374 (1913)
    J. thurifera L. subsp. *gallica* (Coincy) A. E. Murray, Kalmia 12:21 (1982)

Dioecious. Trees pyramidal up to 20 m. **Trunk bark** tan-brown, exfoliating in interlaced strips or long plates. **Branches** twigs 1 mm wide, quadrangular, scaly, distichously arranged. **Leaves** decurrent (whip) and scale-like, all decussate, scale leaves 1.5-2 mm, appressed but free at their incurved acuminate or acute apices, with an oblong furrowed gland on the back, and an entire or slightly toothed, but not scarious, border. **Seed cones** 8 – 12 mm, greenish purple when immature, dark purple when fully ripe, maturing in 2 yrs. **Seeds** (1-) 2 – 4 (-5) per cone, free. 2n=44. **Pollen shed** Feb.- March. **Habitat** mountains of S, C and E Spain; French Alps. **Uses** none known. **Dist.**: Iberian Peninsula, French Alps. **Status**: common in habitats, not threatened except by goats grazing on the young plants.

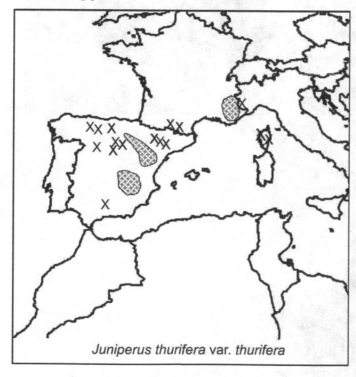

Juniperus thurifera var. thurifera

J. thurifera, bark exfoliating in narrow, interlaced strips.

J. thurifera distribution map, including data from Terrab et al. (2008).

J. thurifera var. *thurifera*
leaves and seed cones

1 cm

J. thurifera var. *thurifera*,
4 km n of Montdauphin, France
cf. *Adams 9036-9038*

J. thurifera var. *thurifera*, 1 km s of
Consuegra, Spain
cf. *Adams 9451-9453*

Juniperus tibetica Kom., Bot. Mater. Gerb. Glavn. Bot. Sada RSFSR 5: 27 (1924). Tibetan juniper, Da guo yuan bai. Type: China, Xizang (Tibet), Jinsha River, in forest near temple, *V. F. Ladygin* 25 (holotype SBT, isotype PE).

    J. potaninii Kom. Bot. Mater. Gerb. Glavn. Bot. Sada RSFSR 5:28 (1924)

    J. zaidamensis Kom., Bot. Mater. Gerb. Glavn. Bot. Sada RSFSR 5:29 (1924)

    J. zaidamensis Kom., f. *squarrosa* Kom., Bot. Mater. Gerb. Glavn. Bot. Sada RSFSR 5:29 (1924)

    J. distans Florin; Acta Hort. Gothob. 3: 6 (1927)

    Sabina tibetica (Kom.) W. C. Cheng & L.K. Fu, Fl. Reipubl. Pop. Sin. 7: 370 (1978)

Dioecious or monoecious. Trees to 30 m, rarely shrubs. **Trunk bark** brown, exfoliating in thin, elongated strips. **Branches** branchlets densely or loosely arranged, mostly straight, terete or slightly 4-angled, 1-2 mm in diam. **Leaves** both scale-like and decurrent (whip), decurrent leaves usually only present on seedlings and young plants, in whorls of 3, 4-8 mm, scale-like leaves decussate, sometimes in whorls of 3, ovate-rhombic, obtuse, 1-3 mm, abaxial glad central, conspicuous, slightly depressed, linear-elliptic or linear. **Seed cones** erect, brown when young, black, or purplish black when mature in 2 yrs., turbinate, 0.9-1.6 x 0.7-1.3 cm. **Seeds** 1 seed per cone, ovoid, rarely obovoid or globose, 7-11 x 6-8 mm, with deep resin pits. **Pollen shed** March-April. **Habitat** Forests on mountain slopes or in valleys. 2700-4800 m. **Uses** incense, construction. **Dist**.: s. Gansu, s. Qinghai, Sichuan, e. and s. Xizang, China. **Status**: vulnerable due to use for incense and construction and goats grazing.

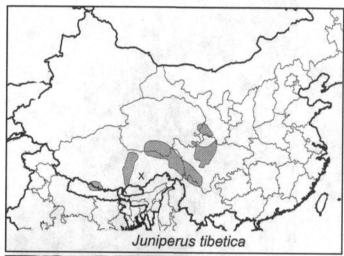

Juniperus tibetica

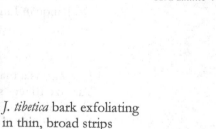

J. tibetica trees up to 12m tall on the bank of the Yangtze River, sw Qinghai, China. N 32° 59.606', E 97° 14.981', cf. *Adams 10345-10346*

J. tibetica bark exfoliating in thin, broad strips

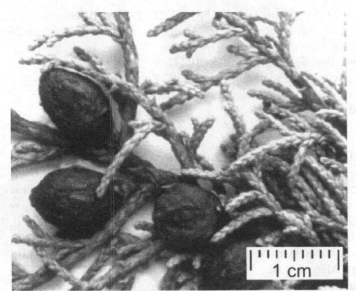

J. tibetica
leaves and seed cones

J. tibetica, as a pyramidal young shrub,
N 33° 01.934', E 97° 16.202' ,
 sw Qinghai, China. cf. *Adams 10342-10344*
The 2004 field to Tibet was arranged by
Dr. Jianquan Liu, Lanzhou, China.

J. tibetica, as a mature, 12m tree, on the bank of the
Yangtze River, sw Qinghai, China.
N 32° 59.606', E 97° 14.981',
cf. *Adams 10345-10346*

Juniperus tsukusiensis

Recent DNA analysis (Adams et al., 2011; Adams and Schwarzbach, 2013c, Figs. 1.3, 1.16.4, Chpt. 1) show *J. tsukusiensis* and *J. t.* var. *taiwanensis* to be a distinct species from *J. chinensis*.

Key to varieties:

1. Scale leaves 0.5-0.7 mm, overlapping by about 25%, endemic to Taiwan..... var. ***taiwanensis***

1. Scale leaves 0.8-1 mm, overlapping by less than 20%, endemic to Japan..... var. ***tsukusiensis***

Juniperus tsukusiensis var. ***taiwanensis*** (R.P. Adams and C-F Hsieh) R. P. Adams, Phytologia 93(1) 127, (2011). Taiwan juniper. Type: Taiwan, Mt. Chingshui, 2200 m, *Sheng-you Lu 14498.* (holotype TAIF!).

Juniperus chinensis var. *taiwanensis* R.P. Adams and C-F Hsieh, Biochem. Syst. Ecol. 30: 236 (2002).

Dioecious. Procumbent shrubs terminal branchlets ascending, leaves scale-like, apices obtuse, about 1 mm long, 1 mm wide. **Trunk bark** brown, exfoliating in plates and strips. **Branches** erect. **Leaves** decurrent (whip or juvenile) 6 mm long x 1 mm wide, and scale leaves 1.5 mm long x 1 mm wide. **Seed cones** about 5 mm in diam., bluish-black. **Seeds** 3, erect, triangular-elliptical. **Pollen shed** spring? **Habitat** talus rocks near the summit of Mt. Chingshui. **Uses** none known. **Dist.:** near the summit of Mt. Chingshui, Taiwan. **Status**: *J. chinensis* var. *taiwanensis* is known from only one location: Mt. Chingshui. This area is a protected area, but due to the very small population, the taxon should be considered as threatened, but not endangered. Endemic to Mt. Chingshui, Taiwan.

This variety differs from the typical variety by being a prostrate shrub and having scale leaves that are very short and wide (appearing as a string of beads) and with glands that are raised (as opposed to sunken in the typical variety). It differs from *J. chinensis* var. *tsukusiensis* in having scale leaves that are very short and wide and having the ultimate branchlets shorter (1 1.5 cm) than in *J. chinensis* var. *tsukusiensis* (1.5-2 cm).

J. tsukusiensis var. *taiwanensis* leaves and seed cones

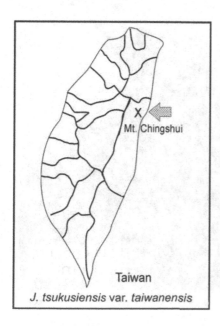

J. tsukusiensis var. *taiwanensis*

J. tsukusiensis var. *taiwanensis*, near summit of Mt. Chingshui, Taiwan, 2260 m. on limestone. cf. *Adams 9061-9063*

Juniperus tsukusiensis Masam. , Bot Mag. Tokyo 44:50 (1930). Type: Japan, Yakushima. G. Masamune s. n. (syntype IT)

 J. chinensis var. *tsukusiensis* (Masam.) Masamune, J. Soc. Trop. Agric. Formosa 2: 152. (1930).

Dioecious. Procumbent shrubs terminal branchlets ascending, leaves scale-like, apices obtuse, about 1 mm long, 1 mm wide. **Trunk bark** brown, exfoliating in plates and strips. **Branches** erect, foliage compact. **Leaves** decurrent (whip or juvenile) 6 mm long x 1 mm wide, and scale leaves 1.5 mm long x 1 mm wide, appear as a string of beads. **Seed cones** about 5 mm in diam., blue-black. **Seeds** 3, erect, triangular-elliptical. **Pollen shed** spring? **Habitat** mountains of Yaku shima (Island). **Uses** none known. **Dist.:** Yaku shima (Island), Japan) **Status**: occupies a very limited area, probably threatened.

J. tsukusiensis var. *tsukusiensis*

J. tsukusiensis var. *tsukusiensis* on Yaku shima. Photo from Jin Murata, cf. *Adams 8805-8808*

J. tsukusiensis var. *tsukusiensis* leaves

1 cm

Juniperus turbinata

See discussion under *J. phoenicea* for the justification changing the status of *J. phoenicea* var. *turbinata* to *J. turbinata*. and recent DNA sequencing analysis (Fig. 1.3, 1.16.1, 1.16.2, Chpt. 1).

Juniperus turbinata Guss., Fl. Sicula Syn. 2: 634 (1844), Turbinate phoenicean juniper. Type: Sicily, Montallegro, Secciara, nw of Agrigento, *G. Gasparrini* (FI?. PAV)

    J. oophora Kunze, Flora 29:637 (1846)

    J. phoenicea L. var. *turbinata* (Guss.) Parl., in Candolle, Prodr. 16(2): 487 (1868).

    Sabina turbinata (Guss.) Antoine, Cupress.-Gatt.: 41 (1857)

    J. phoenicea L. subsp. *turbinata* (Guss.) Nyman, Consp. Fl. Europ. 3: 676 (1881)

    J. canariensis Guyot & Mathou, Trav. Lab. Forest. Toulouse T. 1 (3,20): 7 (1942)

    J. phoenicea subsp. *eu-mediterranea* P. LeBreton & S. Thirend, Naturalis monspeliensia, ser. Bot. 47:8 (1981)

    J. turbinata Guss. subsp. *canariensis* (Guyot & Mathou) Rivas-Martinez & al., Itinera Geobot. 7: 511 (1993)

Monoecious or dioecious. Trees small up to 8 m, or shrub. **Trunk bark** brown, exfoliating in strips. **Branches** twigs 1 mm in diameter, terete, scaly. **Leaves** decurrent (whip) and scale-like, decurrent (juvenile) leaves 5-14 x 0.5-1 mm, patent, acute, mucronate, with two stomatiferous bands both above and beneath; adult leaves 0.7-1 mm, scale-like, ovate-rhombic, closely appressed, obtuse to somewhat acute, with an oblong furrowed gland on the back and a distinct scarious border, sometimes with minute serrations near the base (40X) but not serrated as the serrate junipers of the western hemisphere. **Seed cones** 7-11 mm, ripening in the second year, sub-globose (turbinate), blackish when very young, later green or yellowish and not or slightly pruinose, dark red when ripe. **Seeds** 3 (4-7)10 seeds per cone, free. 2n = 22. **Pollen shed** Oct-Nov (Arista et al., 1997). **Habitat** coastal sand dunes, Cambrian limestone (in mountains) and rocky seashores. **Uses** none known. **Dist.:** common in throughout the Mediterranean region, Algeria, Morocco, and Portugal, Canary Islands: Gomera, Gran Canaria, Hierro, La Palma and Tenerife, Madeira Island. **Status**: common and reproducing, not threatened.

Arista el al. (1997) and Arista and Ortiz (1995) have made robust studies of the reproductive isolation between *J. phoenicea* and *J. p.* subsp. *turbinata* (*J. turbinata*) in the Sierra de Grazalema, Spain. They have clearly established the pollen shedding times in *J. phoenicea* (late winter-early spring) and *J. p.* subsp. *turbinata* (*J. turbinata*) (Oct-Nov) as well as reporting exact morphological differences.

J. turbinata, note the turbinate cone.

J. turbinata, bark.

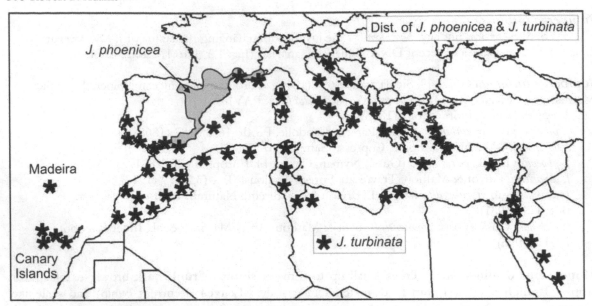

J. *turbinata*, at Tarifa sand dunes, Spain.
cf. *Adams 7202*.
This population appears to
be the most typical of
J. *turbinata* seen by the author.

J. *turbinata*, near Delphi, Greece
cf. *Adams 8793-8794*

Juniperus uncinata

Adams et al. (2009b) described a new variety (*J. recurva* var. *uncinata*, the hooked branchlet juniper) from Nepal. Analysis based on nrDNA and petN-psbM revealed that *J. recurva* var. *uncinata* differs from *J. recurva* by a number of SNPs.

The leaf oils of var. *uncinata* are very high in δ-3-carene (27.0 - 37.0%) and high in terpinolene, trans-muurola-4(14),5-diene, cubebol, δ-cadinene, with moderate amounts of α-pinene, sabinene, limonene, β-phellandrene, cis-muurola-3,5-diene, and trans-cadina-1(6),4-diene (see Adams et al., 2009b, Table 1). Several compounds were present in var. *recurva*, but absent in var. *uncinata*: α-thujene, cis-sabinene hydrate, trans-sabinene hydrate, hexenyl isovalerate, linalyl acetate, germacrene B, humulene epoxide II, and γ-eudesmol. Two compounds were found in all var. *uncinata* oils but were absent in var. *recurva* oils: cis-limonene oxide and β-oplopenone (Table 1, Adams et al., 2009b).

Recently, in publishing the phylogeny of *Juniperus*, Adams and Schwarzbach (2013e) have found that *J. recurva* var. *uncinata* in a well supported clade with *J. coxii* (Fig. 1.3, Chpt. 1, Fig. 5.32.) and *J. recurva* was in a separate clade with *J. indica* and *J. rushforthiana* (Fig. 5.26). *Juniperus recurva* var. *uncinata* was shown to be equally different from *J. coxii* (13 MEs) and *J. convallium* (13 MEs, Fig. 5.33), but 17 MEs different from *J. recurva*. Based on the DNA sequence and volatile oils data, Adams and Schwarzbach (2013b) recognized *J. uncinata*.

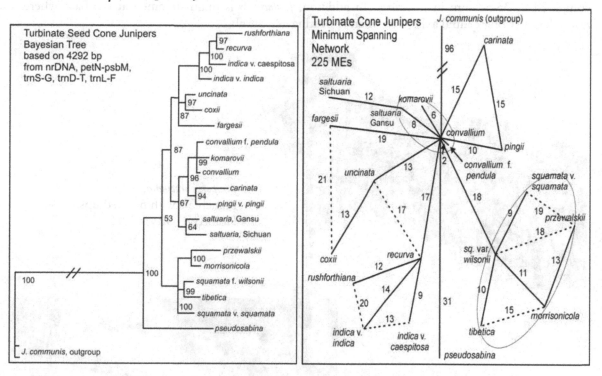

Figure 5.32. Bayesian tree of the turbinate junipers. Numbers at the branch points are the posterior probabilities (as percent).

Fig. 5.33. Minimum spanning network of turbinate junipers. The numbers next to lines are MEs (mutational events = base substitutions + indels). Dashed lines are secondary links for putative related taxa.

Juniperus uncinata (R. P. Adams) R. P. Adams, Phytologia 94(3) 399 (2012). TYPE: Nepal, 100 m s of Lauri Binayak, N 28 05.537', E 85 22.879', 3900 m, on south facing slope, 0.5 m x 0.5 m shrub, 1 Nov 1993, *Adams 7212* (HOLOTYPE: BAYLU, TOPOTYPES: *Adams 7213, 7214*, BAYLU), Fig. 2).
    Juniperus recurva var. *uncinata* R. P. Adams, Phytologia 91(3) 365 (2009).

Monoecious or rarely dioecious. **Shrubs**, multi-branched to 1 m. **Trunk bark** light grayish brown or brown; crown conical or broadly pyramidal. **Branches** ascending in apical part of plant and spreading toward base, branchlets hooked at tip. **Leaves** all decurrent (juvenile), mucronate tips, in whorls or 3, loosely appressed, greenish white or slightly glaucous adaxially, longer, whip leaves 20-28 mm. **Seed cones** axillary, slightly glaucous when young, maturing purplish black and not glaucous, ovoid, 6-12 x 5-9 mm. **Seeds** 1 per seed cone, ovoid or conical-ovoid, 5-9 x 3-6 mm. 2n=22 (Merha, 1988). **Pollen shed** summer. **Habitat** rocky areas, high mountains. **Uses** wood is burned as incense. **Dist.**: presently known only from the type locality, Lauri Binayak, Nepal, 3900m, but expected in the Himalayas (3000-4000 m). **Status**: common in areas but may be cut for incense, may become threatened.

      The terminal branchlet tips of *J. uncinata* are striking as they form a hook at their tips (below), in contrast to *J. recurva* that has lax tips that are not hooked. The leaves of *uncinata* are 20 - 28 mm long, compared to 35-50 mm in *J. recurva*. In addition, *J. uncinata* is multi-stemmed at the base whereas *J. recurva* has a strong central axis, even when appearing as a shrub.

Juniperus uncinata,
branchlets with hooked tips.
Holotype, *Adams 7212*.

J. uncinata
seed cones and leaves.

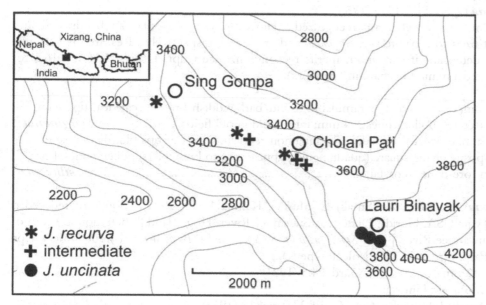

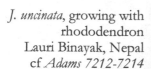

Distribution of *J. uncinata*, known only from type locality Lauri Binayak, Nepal, but expected to occur in Nepal and Xizang.

J. uncinata, holotype
Adams 7212
Lauri Binayak, Nepal
Notice hooked branchlets tips.

J. uncinata, growing with
rhododendron
Lauri Binayak, Nepal
cf *Adams 7212-7214*

Juniperus virginiana L., Sp. Pl. 2:1039 (1753).

In this treatment, two varieties are recognized. However, var. *virginiana* can be divided into pyramidal and strict trees to separate var. *virginiana* and var. *crebra*, respectively. Research is currently being conducted to determine if var. *crebra* merits recognition. See Chpt. 1 (Figs. 1.16.1, 1.16.5 for DNA Bayesian tree and a minimum spanning network).

Key to varieties:

1a. Seed cones 4-6 mm; crowns strict, pyramidal to round; bark reddish-brown; scale leaf tips acute, leaves in threes (terete); pollen cones 3-4 mm; inland and in old fields...............................var. ***virginiana***

1b. Seed cones 3-4 mm; crowns flattened; bark cinnamon-reddish; scale leaf tips bluntly obtuse to acute, leaves opposite giving square (quadrangular) appearance to branchlets; pollen cones 4-5 mm; on sand on fore-dunes (coastal)...var. ***silicicola***

Juniperus virginiana var. ***silicicola*** (Small) E. Murray, Kalmia 13: 8 (1983). Southern red cedar, coastal red cedar. Type: USA, Florida, Dixie Co., Suwannee River, Hog Island,["shell mound south of the mouth of the Suwannee River"] *J. K. Small (with G. K. Small and J. B. de Winkeler) 10030* (lectotype NY, designated by Farjon (p. 397, 2005, isolectotype K).

Sabina silicicola J. K. Small, N. Y. Bot. Gard. 24:5 (1923)

J. barbadensis C. Mohr *non* Linnaeus

J. silicicola (J. K. Small) L.H. Bailey, Cult. Conif. N Amer.18 (1933)

J. virginiana L. subsp. *silicicola* (Small) E. Murray, Kalmia 13: 8 (1983)

J. virginiana L. var. *silicicola* (Small) J. Silba, Phytologia Mem. 7: 37 (1984)

Dioecious. Trees small tree to 10 m, with a flattened crown, pyramidal when young and protected or crowded. **Trunk bark** cinnamon-reddish, exfoliating in narrow strips. **Branches** spreading to pendulous, ultimate twigs terete or 4-angled. **Leaves** both decurrent (whip) and scale. Scale leaves bluntly obtuse to acute. Whip and scale leaf margins entire (20 X and 40 X). Pollen cones 4-5 mm. **Seed cones** maturing in 1 year, blue, glaucous, resinous, ovoid 4-5 mm in diam. **Seeds** tan to chestnut brown, 1.5-3 mm long. **Pollen shed** late winter - early spring. **Habitat** coastal fore-dunes, coastal river sand banks, sea level - 15 m. **Uses** none at present. **Dist.**: along the coast from N.C., S.C., Ga., to western FL and AL. **Status**: This southern variety of *J. virginiana* appears to be restricted to coastal fore-dunes and differs little in morphology or leaf terpenoids (Adams, 1986) from the upland *J. virginiana*. Both of these taxa are distinct from the Caribbean junipers (*J. lucayana* Britt., Bahamas, Jamaica, Cuba; *J. bermudiana* L., Bermuda) see Adams, Zanoni and Hogge, 1984. There appears to be some intergradations of characters between *J. virginiana* and this variety in Georgia (Adams, 1986).

J. virginiana var. *silicicola*

J. virginiana var. *silicicola* bark exfoliating in thin, narrow strips.

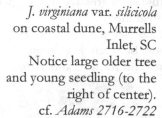

J. virginiana var. *silicicola* leaves and seed cones.

J. virginiana var. *silicicola* on coastal dune, Murrells Inlet, SC Notice large older tree and young seedling (to the right of center). cf. *Adams 2716-2722*

J. virginiana var. *silicicola*, on old coastal sand dune, Mullet Key, (St. Petersburg), Florida, USA. cf. *Adams 9186-9188*

Juniperus virginiana L. var. ***virginiana***, Sp. Pl. 2:1039 (1753). Red cedar, Virginia cedar, eastern red cedar. Type: USA, Location unknown, *leg. ign. LINN 1198.7* (lectotype LINN!, see Jarvis et al. 1993).

    J. caroliana Mill., Gard. Dict., ed. 8: *Juniperus* No. 4 (1768)

    J. arborescens Moench, Methodus: 699 (1794)

    J. caroliniana Du Roi, Harbk. Baumz., ed 2, 1:497 (1795)

    J. hermannii Spreng., Syst. Veg. 3:908 (1826)

    J. foetida Spach var. *virginiana* (L.) Spach, Ann. Sci. Nat. Bot., ser. 2, 16:298 (1841)

    J. virginiana L. var. *vulgaris* Endl., Syn. Conif.: 28 (1847)

    Sabina virginiana (L.) Antoine, Cupress.-Gatt.: 61 (1857)

    J. virginiana L. var. *crebra* Fernald & Griscom, Rhodora 37:133, t. 332 (1935)

    J. virginiana L. var. *ambigens* Fassett; (=X Ambigens, *virginiana* x *horizontalis*) Bull. Torrey
        Bot. Club 72: 380 (1945)

    J. virginiana L. subsp. *crebra* (Fernald & Griscom) E. Murray, Kalmia 12:21 (1982)

Dioecious. Trees single stemmed to 30 m, pyramidal to strict. **Trunk bark** brown, exfoliating in thin strips. **Branches** foliage erect or occasionally lax, green but turning reddish-brown in the winter, twigs (3-5 mm diam.) with persistent dead scale leaves, bark on twigs (6-15 mm diam.) not exfoliating in plates, if so brownish beneath. **Leaves** both decurrent (whip) and scale. Whip leaves growing only at branchlet tips (on mature trees), with an elliptical or elongated gland. Scale-like leaves overlapping (more than 1/4 length). Scale-leaf margins entire (20 X and 40 X). **Seed cones** blue-black to blue brownish, maturing in 1 year, borne terminally, 3-6(7) mm in diam., 1-2(3) seeded. **Seeds** tan to brown, 2-4 mm long. $2\underline{n} = 22$, $3\underline{n} = 33$ (Hall, Mukherjee and Crowley, 1979). **Pollen shed** March-April. **Habitat** upland or low woods, old fields, glades, fence rows and river swamps, near sea level to 1400 m. **Uses** production of eastern red cedar wood oil, furniture, fence posts, widely cultivated for landscaping. **Dist.**: Canada: Ont., Que.; Mexico, United States: all states <u>except</u>: Alaska, Ariz., Calif., Colo., Idaho, Mont., Nev., N. Mex., Ore., Utah, Wash., Wyo.. **Status**: an aggressive weedy species that invades abandoned fields and roadsides. Considered a pest. Not threatened.

    Juniperus virginiana hybridizes with the sibling species, *J. horizontalis* (see *J. horizontalis*, above) and *J. scopulorum* (see *J. scopulorum*, above). Earlier reports of hybridization between *J. ashei* and *J. virginiana* (Hall, 1952) were negated in subsequent studies (Adams, 1977; Flake, von Rudloff and Turner, 1969).

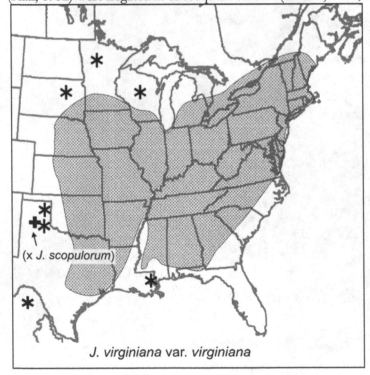

J. virginiana var. *virginiana*

J. virginiana var. *virginiana* bark exfoliating in thin, narrow strips.

J. virginiana var. *virginiana* leaves and seed cones

J. virginiana var. *virginiana* young trees seeded by birds under Oak trees in Tennessee. This is a common sight in many juniper species in the western hemisphere, but not so common in the eastern hemisphere, where goat and sheep grazing limits the growth of seedlings.

J. virginiana var. *virginiana*, pyramidal form, this form was denoted the 'Ozark Race' by M. T. Hall. Austin, Texas, USA.

Cedar apple (rust) on *J. virginiana*. Cedar apple rust, is caused by the fungus *Gymnosporangium*. The fungus needs two hosts to complete its like cycle: *Juniperus* and various Rose family plants (such as Saskatoon berry, hawthorns, and in some cases, apples). Cedar apples are found on many *Juniperus* species.

J. virginiana var. *virginiana*. Strict or columnar forms are associated with var. *crebra*, but the columnar form seems to arise in several areas. Muskogee, Oklahoma, USA. cf. *Adams 10210-10214*

Eastern Red Cedar is an aggressive, weedy species. *J. virginiana* var. *virginiana* (and most junipers) are disseminated by birds and a typical pattern in the USA is the 'fence row junipers' (left) where birds have dropped the seeds while sitting on the fence wire. It also invades disturbed sites (right) as well as old fields. *J. virginiana* var. *virginiana* is the most weedy juniper known, in that it can invade tall (0.5 m tall) grass. The control of *Juniperus* is a major problem in the United States. Interestingly, the junipers of the eastern hemisphere are seldom weeds. Of course, much of the juniper habitat in the eastern hemisphere has been grazed by goats for centuries, whereas goat grazing is a modern phenomenon in the western hemisphere and little practiced in the United States.

Juniperus zanonii R. P. Adams. Phytologia 92: 112 (2010). Zanoni juniper, Cerro Potosi juniper. Type: Mexico, Nuevo Leon, Cerro Potosi, 3550 m, *J. A. McDonald 1820*, 26 Jul 1985 (HOLOTYPE: TEX).

Dioecious. Decumbent shrubs less than 1m, trunk branching at the base. **Trunk bark** brown, longitudinally divided into narrow strips. **Branches** procumbent. **Leaves** decurrent (whip) and scale-like, foliage very densely compacted, margins denticulate with resin gland. **Seed cones** female cones with soft, fleshy pulp, globose, dark blue, with bloom. 5-8 mm diameter. **Seeds** 2-5 per cone **Pollen shed** winter? **Habitat** limestone, under *Pinus culminicola* and *P. hartwegii* near timberline. **Uses** none known. **Dist.**: At present, *J. zanonii* is known from Sierra La Viga, Sierra La Marta, Cerro Potosi and Sierra Pena Nevada of the seven alpine-subalpine areas of McDonald (1990, 1993). It seems likely that *J. zanonii* occurs on the three other alpine-subalpine areas (S. Potrero de Abrego, S. Coahuilon, and S. Borrado, see distribution map below). **Status**: This form is found in very scattered, sub-alpine areas. Because it is a small shrub of no economic importance and these areas are not suited for pasture, it is likely not threatened.

In the previous edition (Adams, 2008), I treated *J. monticola* f. *compacta* from the central Mexico trans-volcanic belt as conspecific with the sub-alpine juniper from Cerro Potosi (now *J. zanonii*) as *J. compacta* (Adams et al., 2007). But a recent comparison of the Cerro Potosi juniper with *J. monticola* f. *compacta* from volcano Popocatepetl revealed that the junipers are not conspecific. Figure 5.34 shows that *J. zanonii* is most closely related to *J. saltillensis* and equally distant from *J. monticola* f. *compacta* and *J. jaliscana* based on SNPs from nrDNA and petN-psbM (adapted from Adams et al. 2010b).

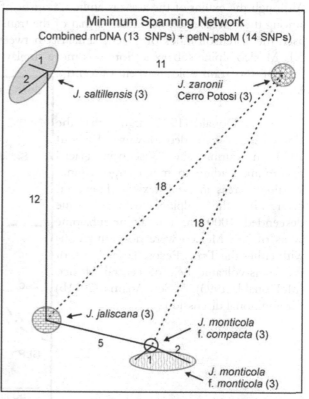

Using additional cpDNA regions, Adams and Schwarzbach (2013e) show that *J. zanonii* is in a clade with *J. saltillensis* but quite distinct (Figs. 1.3, 1.14, Chpt. 1). In fact using nrDNA and 4 cp DNAs (~4300 bp) revealed (Fig. 1.14.1) that *J. zanonii* is separated by 16 MEs (mutational events = base substitutions + indels), which is a bit more than seen in the nrDNA + petN data (Fig. 5.27).

In addition, the leaf oil of *J. zanonii* was found to be distinct from *J. monticola* and *J. saltillensis* (Adams and Zanoni, 2010).

The table below shows some characteristic differences between *J. zanonii* and *J. monticola* f. *compacta*. Although the DNA shows *J. zanonii* to be more closely related to *J. saltillensis* than to *J. monticola*, the leaves and habit of *J. zanonii* are similar to *J. monticola*. Of course, it may be that the shorter, colder growing season in the alpine-subalpine areas leads to similar, compact foliage in both taxa.

Figure 5.34. Minimum Spanning network based on 27 SNPs. Note the distinct differentiation of *J. zanonii* from *J. saltillensis*, *J. monticola* f. *compacta* and *J. jaliscana*. (adapted from Adams et al., 2010b).

Morphological differences between *J. zanonii* and *J. monticola* f. *compacta*.

	J. zanonii	*J. m.* f. *compacta*
Branchlets (3-5 mm diam.)	rough, with persistent dead leaves	smooth
Scale leaves	often with ridge on beak	domed beak, no ridge
Scale leaf glands (when visible)	oval, flat or raised	elongated, sunken groove on leaf
Whip leaf glands	oval to elongated raised or flat	ill-defined elongated, sunken to flat
Seeds/ cone	2-5	(2-) 3-7 (-9)
Habitat	limestone, under *Pinus culminicola* and *P. hartwegii* near timberline	volcanic lava and rocks above timberline

Juniperus zanonii was named after Thomas A. Zanoni (1949-, presently at NYBG). Tom was my first Ph. D. student and spent much time in Mexico collecting *Juniperus*. Tom and I visited (again) the top of Cerro Potosi to collect junipers in 1991 and, fortunately, I put samples in silica gel for DNA analysis, as we did not do that in his work in the 1970s. Tom led some great field trips in Mexico and Guatemala.

McDonald (1990, 1993) examined and mapped the alpine-subalpine vegetation of NE Mexico. Although the extent of the areas is limited, floristic affinities were shown (McDonald, 1990) to be strong among these areas and not similar to that of the trans-volcanic belt of central Mexico. McDonald (1993) used the Sorenson index of genetic similarity between alpine-subalpine vegetations and showed that the NE Mexico alpine-subalpine flora was more similar to that of the White Mtns., NM, U. S. A., than to that of the trans-volcanic belt of central Mexico.

McDonald (1993) argues that the alpine zones extended downward about 1000 m during the Wisconsin glacial maximum, leading to much larger alpine-subalpine areas in NE Mexico. However, even if the alpine-subalpine zone descended 1000 m, the alpine-subalpine areas of NE Mexico were not continuous with either the Trans-Pecos, Texas Mtns, or the trans-volcanic belt of central Mexico (McDonald, 1993). See Adams (2011b) for additional discussion.

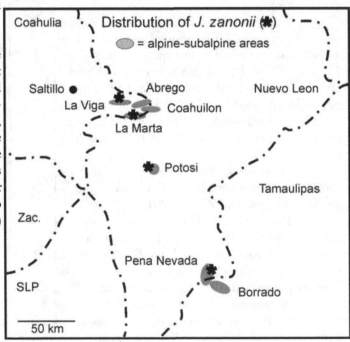

Distribution of *J. zanonii* overlaid onto a map of alpine-subalpine areas in NE Mexico based on McDonald, 1990, 1993.

J. zanonii
leaves and seed cones

Walter Kelly with *Juniperus zanonii* (lower
right) under *Pinus culminicola* (upper right)
at Cerro Potosi.

Tom A. Zanoni,
collecting samples of *J. zanonii* near the
summit of Cerro Potosi.

Notice the tilted limestone bed
(bottom) and the *Pinus culminicola*
overstory as well as the fog bank.

Chapter 6: Hybridization

During the past 70 years, the genus *Juniperus* has been the subject of numerous studies concerning gene flow between species. The early work by Fassett (1944, 1945a, 1945b) indicated hybridization between the closely related species *Juniperus horizontalis*, *Juniperus scopulorum*, and *Juniperus virginiana*. Portions of that work have been re-examined using terpenoids (von Rudloff, 1975a, 1975b; Flake et al., 1978; Comer et al., 1982; Adams, 1983), numerical analysis of morphological characters (Van Haverbeke, 1968a,b; Schurtz, 1971), and combined morphological, isozyme, and terpenoid data (Palma-Otal et al., 1983). These studies were all in general agreement that gene flow either has occurred or is now occurring between the species studied.

On the other hand, the putative allopatric introgression between two very different species, *Juniperus ashei* Buch. and *Juniperus virginiana* (species in different subsections of the genus), has fared quite differently. The case was initially based (Hall, 1952a, 1952b) on the use of a few selected morphological characters and the scatter diagrams of the type devised by Anderson (1949). Later research utilized chemical characters that are not visible to the collector in the field and, thus, not subject to a non-random collection bias. These studies showed no evidence of gene flow between these rather different species (terpenoids, von Rudloff et al., 1967, and Flake et al., 1969, 1973; terpenoids and morphology, Adams and Turner, 1970, and Adams, 1975a, 1977).

Gene flow has also been suggested to occur between *J. ashei* and *Juniperus pinchotii* (Hall et al., 1962), between *J. pinchotii* and *Juniperus deppeana.* (Hall et al., 1962) and between *J. pinchotii* and *J. monosperma* (Hall and Carr, 1968; Correll and Johnston, 1970). Correll and Johnston (1970) included *Juniperus erythrocarpa* (now *J. coahuilensis*) as part of *J. pinchotii*. They noted that the rose-fruited individuals (*J. coahuilensis*) in trans- Pecos Texas were intermediate (which they are) between *J. monosperma* and *J. pinchotii* and concluded that hybridization was occurring. Hybridization between *J. pinchotii* and *J. monosperma* was not substantiated in studies using both morphology and terpenoids (Adams, 1972, 1975b; Zanoni and Adams, 1975, 1976), although one tree was found in Palo Duro Canyon that might show evidence of gene flow (Adams, 1972). In general, the studies of the 22 taxa of junipers in Mexico and Guatemala failed to show any evidence of hybridization except between infraspecific taxa or the most closely related species (Zanoni and Adams, 1975, 1976).

It is interesting that several species that have areas of sympatry have different seasons of pollen shedding, that insures genetic isolation (*J. pinchotii - monosperma, pinchotii - coahuilensis, ashei - virginiana, ashei - pinchotii*, etc.).

However, there are some cases where hybridization seems to be occurring. Fassett (1944, 1945a, b, c) analyzed possible hybridization between *J. scopulorum* and *J. virginiana* (Fassett, 1944); *J. horizontalis* and *J. scopulorum* (1945a); *J. horizontalis* and *J. virginiana* (1945b); *J. horizontalis-scopulorum-virginiana* (Fassett, 1945c). In each of these cases he concluded that hybridization was occurring. Fassett named the following hybrids:

J. virginiana var. *ambigens* Fassett (Bull. Torrey Bot. Club, 72:380, 1945)
  for *J. horizontalis* x *J. virginiana* (in Wisconsin and Maine).
J. scopulorum var. *patens* Fassett (Bull. Torrey Bot. Club, 72: 482, 1945)
  for *J. horizontalis* x *J. scopulorum* from the Big Horn Mtns., and Banff, Alberta.

Van Haverbeke (1968a,b) examined *J. scopulorum* and *J. virginiana* across the Missouri basin and concluded that these junipers may have arisen by divergence of *J. scopulorum* into the Missouri basin, rather than introgression from *J. virginiana*. However, he noted a zone of putative hybridization in western Kansas, western Nebraska, sw South Dakota and central North Dakota.

Comer et al. (1982) used leaf terpenoids to examine seedlings of *J. scopulorum* and *J. virginiana* from throughout the great plains but grown in a common nursery. They found some evidence of hybridization, but mostly the *J. virginiana* plants tended to be either show introgression from *J. scopulorum* or differentiation towards *J. scopulorum*. *Juniperus scopulorum* was strongly differentiated from *J. virginiana*.

Subsequent work on the Wisconsin Driftless Area junipers (Palma-Otal et al. 1983) using morphology, leaf terpenes, and isozymes seems to confirm hybridization between *J. horizontalis* and *J. virginiana* in that area.

Adams (1983a), using terpenoid data showed hybridization between *J. horizontalis* and *J. scopulorum* (Fig. 6.1). Notice the U shaped ordination that has been shown to be obtained from hybridization (Adams, 1982). This study confirmed earlier studies (Fassett, 1945b,c) that hybridization is occurring between *J. horizontalis* and *J. scopulorum*.

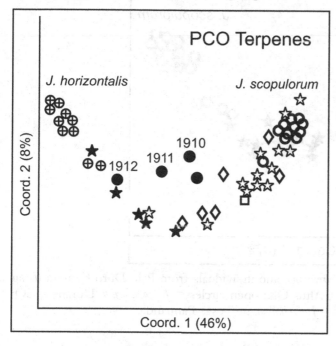

Fig. 6.1. PCO of *J. horizontalis*, *J. scopulorum* and other individuals:
crossed circles = *J. horizontalis*, Saskatoon, Canada
open circles = *J. scopulorum*, Durango, CO
open stars = putative hybrids, Banff, Alberta, Canada
open box = Missouri River, MT
open diamonds = Bridger, MT
solid circles = Amidon, ND Dakota
solid stars = Grass Range, MT
(based on Adams, 1983a).

Adams (1983a) also investigated the status of a '*J. virginiana*' population in Palo Duro Canyon, Texas. This population appears to be a relict from the Pleistocene. Figure 6.2 shows that *J. virginiana* is tightly clustered, except for one individual. The *J. scopulorum* individuals cluster together except for 3 individuals that are intermediate towards the Palo Duro plants (Fig. 6.2).

In general, the Palo Duro plants cluster as one might expect hybrids (cf. Figs. 6.1, 6.2 and Adams 1982), except for one individual that is clearly associated with *J. virginiana*.

Based on these data, it appears that the Palo Duro plants are relict, stabilized hybrids from the Pleistocene when *J. scopulorum* and *J. virginiana* were in contact in this region of the Great Plains. This population is currently being re-examined using DNA data to try to ascertain this more clearly.

Vasek (1966) reported that in northwest Nevada, *J. occidentalis* becomes more like *J. osteosperma* as one travels eastward in Nevada, suggesting introgression from *J. occidentalis* into *J. osteosperma*. Vasek (1966) noted that the three species differed in whether the glands ruptured with an exudate (*J. occidentalis*), did not rupture (*J. osteosperma*) or rarely ruptured (*J. californica*). This is a key character separating *J. occidentalis* and *J. osteosperma* (and other species). Figure 6.3 (left) shows that the glands in the scale leaves (and whip leaves) are near the surface in *J. occidentalis* (and *J. deppeana*, *J. pinchotii*, etc.) and rupture. In many junipers, the terpenoids form a white, crystalline exudate (from camphor and bornyl acetate). The leaves of *J. osteosperma* (Fig. 6.3, center) are imbedded and not visible nor do they rupture (common in many junipers). The leaves of *J. californica* (Fig. 6.3, right) are near the surface and are sometimes ruptured (also a common condition in many juniper species).

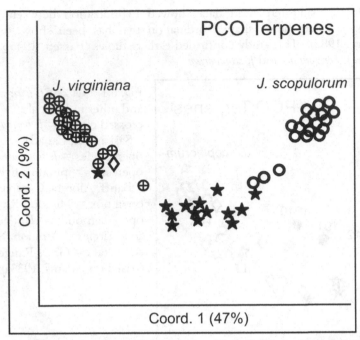

Figure 6.2. Ordination of *J. scopulorum* and *J. virginiana*, and individuals from Palo Duro Canyon using leaf terpenoids. crossed circles = *J. virginiana*, Altus OK; open circles = *J. scopulorum*, Durango, CO; stars = Palo Duro Canyon, TX; open diamond = *J. virginiana*, ne Texas Panhandle.

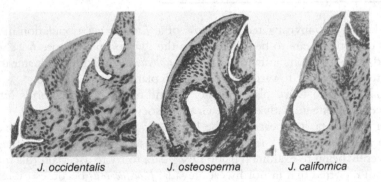

Fig. 6.3 Cross sections of scale leaves showing the proximity of the glands to the leaf surface. Photos from Frank C. Vasek, pers. comm. (see also Vasek, 1966).

Terry, Nowak and Tausch (2000) and Terry (2010) examined cpDNA and nrDNA (rDNA) and based on cpDNA haplotypes, concluded that Vasek (1966) was correct that introgression of *J. occidentalis* was most likely occurring into *J. osteosperma*.

In the southeastern most part of the range of *J. coahuilensis* var. *coahuilensis* there is extraordinary variation in populations near Dr. Arroyo, Galeana, and Alvaro Obregon. Adams (1994) examined the terpenoids of plants from this area as well as throughout the range of *J. coahuilensis*, along with *J. angosturana*. Figure 6.4 shows an ordination based on terpenoids. Notice that the U ordination is shown, that suggests hybridization and possible introgression (as shown in artificial hybridization, Adams, 1982).

Both *J. angosturana* and *J. coahuilensis* (from Chihuahua) are well defined (Fig. 6.4), but one Dr. Arroyo individual appears to be a backcross to *J. angosturana*. The Alvaro Obregon and Dr. Arroyo individuals form a pretty tight cluster that would be typical of stabilized hybrids (or a third taxon).

Clearly, the Galeana population has individuals that are intermediate as well as those typical of *J. coahuilensis*. Gene flow appears to be occurring into *J. coahuilensis* in this area. The soils in these areas are heavily disturbed creating new, open habitats for unusual individuals. As such they present an ideal place for hybrids to prosper.

Adams and Kistler (1991) examined putative hybridization between *J. coahuilensis* and *J. pinchotii* in the trans-Pecos Texas region. *Juniperus coahuilensis* (= *J. erythrocarpa*) and *J. pinchotii* occupy rather separate habitats, with *J. coahuilensis* found in the *Bouteloua* grasslands at elevations from 1,500 to - 2,000 m in trans-Pecos Texas, southern New Mexico, and southern Arizona and southward into Mexico. *Juniperus pinchotii* is found at lower elevations on eroded soils in west Texas and northern Coahuila (Fig. 6.5).

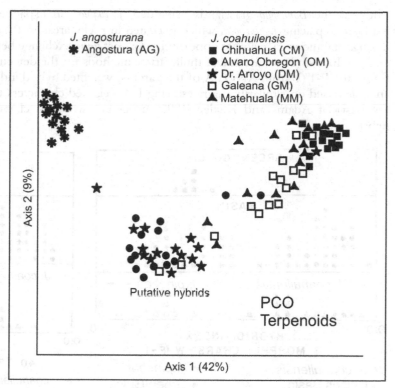

Figure 6.4. PCO based on leaf terpenoids; see text for discussion (from Adams, 1994).

Aside from a zone of sympatry in eastern Coahuila near Saltillo, the Basin area of the Chisos Mountains (Fig. 1) represents the most distinct zone of sympatry between *J. coahuilensis* (= *J. erythrocarpa*) and *J. pinchotii*.

The zone of sympatry between *J. coahuilensis* and *J. pinchotii* changed repeatedly during the Pleistocene pluvial periods. The last extensive zone of sympatry was about 10,000 to 15,000 ybp when the life zones around the Chisos Mountains descended 400 to 800 m (Wells, 1966). This led to an expansion of the range of *J. coahuilensis* into the margins of the Chihuahuan desert. Although *J. coahuilensis* undoubtedly expanded westward, a discussion of its expansion eastward towards the range of *J. pinchotii* is relevant. The area between the Chisos Mountains and Del Rio, Texas was a Pinyon-juniper woodland (Wells, 1966; Bryant, 1969; Adams, 1977) that probably contained *J. coahuilensis*, *J. pinchotii*, *J. ashei*, and possibly *J. monosperma* (entering from the northwest). Elements of *J. pinchotii* still persist as far south as Saltillo, Coahuila, where they are sympatric with *J. coahuilensis* (Adams, 1975b).

Figure 6.5. Populations of *J. coahuilensis and J. pinchotii* sampled (from Adams and Kistler, 991).

Thus, as the *Bouteloua* grasslands expanded, *J. coahuilensis* expanded its range into these new sites, gradually replacing *J. pinchotii*, which persisted in rocky areas in the drier sites. This continual mixing during migrations gave ample opportunity for past gene exchange between these taxa.

In a study on the use of multivariate methods for the detection of hybridization, Adams (1982) found the F-1 (F from ANOVA of the parents) weighted hybrid index (WHI) and principal coordinate analysis based on similarity matrices using F-1 weighted characters to be most useful. That work was the basis of Adams and Kistler (1991) using F-1 weighted characters in multivariate detection of hybridization.

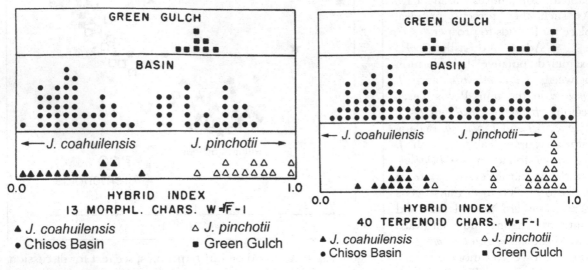

Figure 6.6. Morphological hybrid index. (adapted from Adams and Kistler, 1991).

Figure 6.7. Terpenoid hybrid index. (adapted from Adams and Kistler, 1991).

The morphological WHI shows (Fig. 6.6) *J. pinchotii* and *J. coahuilensis* to be well resolved except for one intermediate *J. coahuilensis* individual (1401) from Alpine. There is also one outlier of *J. pinchotii* (1382). Not all the individuals sampled in the study could be used in the morphological analyses because female cones were not present on some plants. This is a limitation of morphological data that does not apply to chemical data.

The Basin individuals (Fig. 6.6) have a bimodal distribution ranging from typical *J. pinchotii* to typical *J. coahuilensis*. Several of the Basin individuals appear to be intermediate. The Green Gulch individuals were ordinated about as expected from our field examination, fairly typical of *J. pinchotii*.

The WHI based on the terpenoids (Fig. 6.7) revealed that the Marathon population of *J. pinchotii* has one individual (1393) that clusters strongly with *J. coahuilensis* and three somewhat divergent (introgressed?) individuals. This probably accounts for the slight intermediacy found previously (population 14 in Adams, 1975b). The Basin individuals again show signs of bimodality, but the distribution is continuous except for a few individuals similar to typical *J. pinchotii*. The Green Gulch individuals show a definite bimodality: some individuals are typical *J. pinchotii*, while others have a greater similarity to *J. coahuilensis* but are not typical of that taxon.

The WHI based on terpenoids and based on morphology agreed in general but disagreed in many specific cases. The correlation between these indices was 0.80 (95% confidence interval = 0.72, 6). Several Basin individuals showed intermediacy in both indices. For example, if the WHI for chemical data = C and WHI for morphology = M; then C and M for *J. pinchotii* = 1.0; C and M for *J. coahuilensis* = 0.0, then C = 0.56, M = 0.62 for sample 1313, C = 5, M = 0.55 for sample 1314, and C = 0.61, M = 0.52 for sample 1346.

Principal coordinate analysis (PCO) has been shown to be superior to other multivariate methods for the detection of hybridization (Kistler, 1976; Adams, 1982). PCO using morphological data reveals (Fig. 6.8) the Basin (and Green Gulch) individuals cluster in a U or V shape with the first coordinate ordinating between *J. pinchotii* and *J. coahuilensis* explaining 48.5% of the variation. The second axis ordinates between parental types and first filial generation (F1s) and second filial generation (F2s) (see Kistler, 1976; Adams, 1982, 1983). Thus, the hybrids should be found midway on axis 1 and low (-0.35) on axis 2 (the dashed line, Fig. 6.8); F2s tend to scatter about the F1s and backcrosses toward the recurrent parent (see Kistler, 1976, for computer simulated data). Thus, samples 1346 and 1341 might be backcrosses (BC) to *J. pinchotii*, and samples 1313 and 1323 might be BC2 or BC3 (see Figs. 6.8 and 6.9).

PCO using terpenoids shows a similar overall pattern (Fig. 6.9) to PCO with morphology in that a U- or V-shaped ordination is achieved with possible F1s, F2s, and backcrosses. As with the WHI, terpenoids show more of a continuous distribution than does morphology. Placement of specific individuals differs considerably in some cases. Individual 1313 looks like *J. pinchotii* or an advanced generation backcross to *J. pinchotii* morphologically, but appears as an F2 or BC1 (to *J. pinchotii*) using terpenes. The morphology of 1323 placed it close to *J. pinchotii* or an advanced generation backcross (Fig. 6.8) whereas terpenoids show it to be intermediate (Fig. 6.9). On the other hand, placement of 1346 is similar using either morphology (Fig. 6.8) or terpenoids (Fig. 6.9).

A detailed examination of individuals in the dashed line region of Figs. 6.8 and 6.9 revealed that very few were intermediate in both morphology and chemistry. This lack of intermediacy in both suites of characters lends support to the idea that few first generation hybrids are present. However, the lack of correspondence between morphological and chemical data when identifying hybrids has been noted by Levin (1968). In such cases, this probably indicates that the chemical characters segregate independently of the morphological characters.

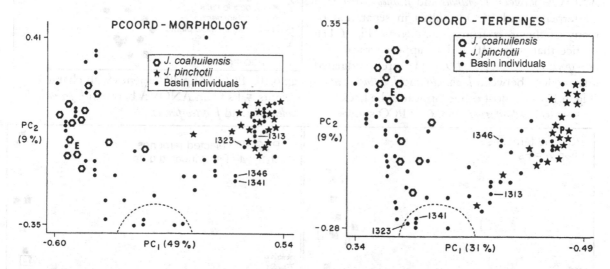

Figure 6.8. PCO based on morphology.
(adapted from Adams and Kistler, 1991).

Figure 6.9. PCO based on leaf terpenoids.
(adapted from Adams and Kistler, 1991).

In addition to the use of principal coordinate analysis, multiple step-wise discriminate function analysis (canonical variate analysis, CVA) of the terpenoids gave almost identical results to PCO (Kistler, 1976), except the groupings were not as well ordinated in two dimensions. This is similar to studies with simulated data (Kistler, 1976) and sunfish hybrids (Adams, 1982). The CVA with morphological data also gave similar results as the PCO (Kistler, 1976), except no groups were apparent and the ordination was essentially only 1-dimensional.

To have hybridization, several factors must occur: the taxa must be in close enough proximity to exchange pollen; their pollination times must overlap; there must be a suitable (intermediate) habitat (Anderson, 1949) for the F1s to grow (of course, one could list numerous other factors involving genetic compatibility). The Chisos Basin population satisfies the proximity problem and provides considerable intermediate habit. *J. pinchotii* pollinates in September and October in the area and *J. coahuilensis* pollinates in November and December (R. P. Adams, pers. obs.; Kruse, pers. obs.)

However, variation in moisture and temperature can move these dates by several weeks, and, thus, considerable overlap in pollination times is to be expected.

Both morphological and terpenoid data indicated that hybridization is occurring between *J. coahuilensis* and *J. pinchotii* in the Basin of the Chisos Mountains. The variation forms a continuum between the species suggesting considerable backcrossing. The two reference populations of *J. coahuilensis* (west of Alpine) and *J. pinchotii* (northeast of Marathon) appear to contain a few introgressed individuals.

Recently, I examined hybridization between *J. grandis* and *J. osteosperma* in northwest Nevada at the Leviathan Mine area (Adams, 2013a). This is part of the area where Vasek (1966) reported hybridization between *J. osteosperma* and *J. occidentalis*. Adams (2013a), using 49 leaf terpenes, F (from ANOVA) weighted, found the population at Leviathan Mine Road to be composed of one typical *J. grandis*, 13 trees with oils similar to *J. osteosperma* and one tree (#9, Fig. 6.9) intermediate between *J. grandis* and *J. osteosperma*. No trees of *J. occidentalis* were found (Fig. 6.10).

Re-analysis using F weights derived from ANOVA between *J. grandis* and *J. osteosperma* gave 42 terpenes and PCO resulted in separation of many of the trees from *J. osteosperma* (Fig. 6.11). Notice that only #10 and 13 appear typical of *J. osteosperma*. Trees #9, 11 are ordinated intermediate between *J. grandis* and *J. osteosperma* (Fig. 6.11) and most others appear to be backcrosses to *J. osteosperma*. Fig. 6.12 PCO shows

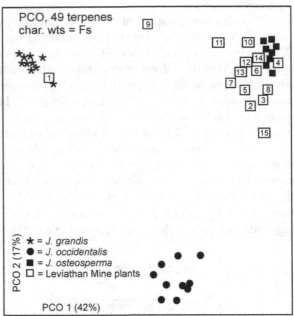

Fig. 6.10. PCO using 49 terpenes with character weights = Fs from ANOVA between *J. grandis*, *J. occidentalis*, and *J. osteosperma*.

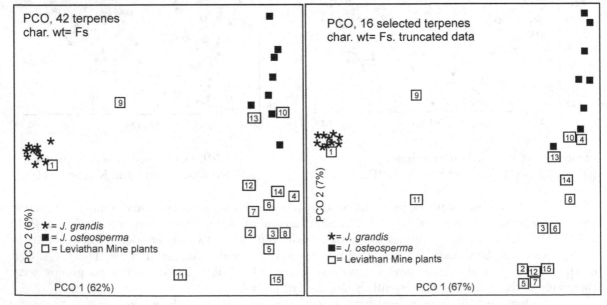

Fig. 6.11. PCO, 42 terpenes, wts= Fs, between *J. grandis* and *J. osteosperma*.

Fig. 6.12. PCO, 16 selected terpenes, wts=Fs, truncated to values between putative parents.

the results from selecting 16 of the highest Fs terpenes, and truncating transgressive values back to the range encountered in the parents (*J. grandis* and *J. osteosperma*). Notice the additional separation of some of the putative backcrosses (2, 5, 7, 12, 15).

It appears that the Leviathan mine population samples contain one typical *J. grandis*, 2 hybrids, 5-10 backcrossed (to *J. osteosperma*) individuals and 3 plants whose oils are fairly typical of *J. osteosperma*. The terpene data support the haplotype data (Fig 6.13) of Terry (2010). It is interesting that Terry (2010) and Fig. 6.13 show 5 haplotypes in the Leviathan mine population, of which only 2 of the 5 haplotypes appear in *J. osteosperma* populations (none of the 5 haplotypes appears in *J. occidentalis* populations). It seems likely that haplotypes 6, 7, and 8 are from *J. grandis* germplasm.

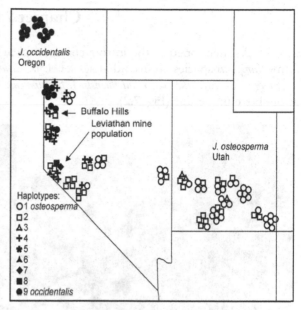

Fig. 6.13. Dist. of haplotypes (trnL-trnF and trnS-trnG) in *J. occidentalis* and *J. osteosperma* (based on Terry, 2010).

In a second study, Adams (2013b) examined the leaf terpenoids in a putative hybrid population involving *J. occidentalis* and *J. osteosperma*. in northwestern Nevada at Buffalo Hills (Fig. 6.13). This is an area that Vasek (1966) showed to be completely introgressed between *J. occidentalis* and *J. osteosperma*. and the haplotype data do indicate that something unusual is occurring in this area (Fig. 6.13).

PCO using 28, F-1 weighted terpenes revealed that the trees at Buffalo Hills are all intermediate between *J. occidentalis* and *J. osteosperma* (Fig. 6.14). No trees with oils typical of *J. occidentalis* or *J. osteosperma* were found (Fig. 6.14). Two trees (#12, 14) may be backcrosses to *J. osteosperma* and a few trees may be backcrosses to *J. occidentalis* (#1, 7, 10), but most trees appear to be hybrids.

It is interesting that these two populations are so different in composition with the Leviathan Mine being closely allied to *J. osteosperma* and the Buffalo Hills being quite intermediate. However, it should be noted that one of the putative parents at Leviathan Mine (*J. grandis*) is very rare in that area.

using 28 terpenoids, F-1 weighted.

In summary, there are well documented cases of hybridization in *Juniperus* (particularly in the western hemisphere where the ranges seem to overlap more than in the eastern hemisphere). But widespread hybridization does not appear to be the general rule in *Juniperus*.

Chapter 7: Ecology

As mentioned in the introduction, *Juniperus* is one of the most diverse genera of the conifers. Some *Juniperus* species are found at sea level: *J. lutchuensis* is found in the rock on seashores (Fig. 7.1) as is *J. virginiana* var. *silicicola*, *J. lutchuensis*, *J. procumbens*. Other species, such as *J. monticola* f. *compacta* occur at or above timberline (Fig. 7.2).

Figure 7.1. *Juniperus taxifolia* var. *lutchuensis* on the seashore, Oshima Island, Japan.

Figure 7.2. *Juniperus monticola* f. *compacta*, at timberline, Nevada de Toluca, Mexico

Limestone is the preferred substrate for many *Juniperus* species such as *J. ashei* (Fig. 7.3) and *J. thurifera* in Spain (Fig. 7.4).

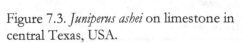

Figure 7.3. *Juniperus ashei* on limestone in central Texas, USA.

Figure 7.4. *Juniperus thurifera* on limestone near Sepulveda, Spain.

However, some junipers grow on sand dunes, such as *J. communis* var. *megistocarpa*, *J. macrocarpa* (Fig. 7.5), *J. turbinata* (Fig. 7.6), *J. davurica* var. *arenaria* (Fig. 7.7), *J. rigida* var. *conferta* and *J. virginiana* var. *silicicola* (Fig. 7.8).

Figure 7.5. *Juniperus macrocarpa* on sand dunes, Tarifa Sand Dunes, Spain.

Figure 7.6. *J. turbinata*, on sand dunes, Tarifa Sand Dunes, Spain.

Figure 7. 7. *Juniperus davurica* var. *arenaria*, on high sand dunes at Lake Qinghai, China.

Figure 7.8. *Juniperus virginiana* var. *silicicola* growing on stabilized sand dune, Mullet Key, FL, USA.

Perhaps the most atypical habitat is that of *J. communis* var. *charlottensis* on Queen Charlotte Island, British Columbia, Canada, where it grows in sphagnum moss swamps (Fig. 7.9), *Arctostaphylos* and stunted spruce.

Figure 7.9. *Juniperus communis* var. *charlottensis* growing in a sphagnum bog on Queen Charlotte Island, Canada.

Juniperus communis var. *depressa* grows on glaciated granite in Canada (Fig. 7.10).

Figure 7.10. *Juniperus communis* var. *depressa* on granite exposed by Pleistocene glaciers. Nemeiben Lake, Saskatchewan, Canada.

In North America, several species grow on the margins of deserts. These species include *J. californica*, *J. coahuilensis*, *J. monosperma*, *J. osteosperma*, and *J. pinchotii*. Figure 7.11 shows *J. californica*, chemotype A, in the Mojave Desert near Amboy, CA, USA. Figure 7.12 shows *J. californica*, chemotype B, in the Mojave Desert near Yucca, AZ, USA. The junipers in the background (hourglass figures) show the effects of browsing by goats and/or deer.

Figure 7.11. *J. californica* with Yucca in the Mojave desert, near Amboy, CA.

Figure 7.12. *Juniperus californica* in grassland at higher elevation at the margin of the Mojave desert, Yucca, AZ. The 'hourglass' plants in the background showing effects of browsing by goats and/ or deer. Adams et al. (2011) have recently shown that goats browse on *J. ashei* with lower total leaf oils.

Figure 7.13. *J. coahuilensis* in the Chihuahuan desert south of Marfa, TX, USA.

Figure 7.14. *J. osteosperma* with *Artemisia* and *Chrysothamnus* in the high desert of Utah, USA.

Juniperus species are very opportunistic colonizers. Open habitat is colonized by a specialized taxa as millions of juniper seeds are dispersed into all kinds of habitats.

A key element in the evolution of *Juniperus* (and their success in colonizing various habitats) was the evolution of a fleshy seed (female) cone. *Juniperus* species have evolved a fleshy female cone in which the cone scales are fused. These are often called "fruits" or "berries". This reproductive structure is especially consumed by birds and small mammals (Phillips, 1910; McAtee, 1947; Holthuizen and Sharik, 1985; Santos et al., 1999). It is very common to see junipers 'planted' by birds along fence rows (Fig. 7.15) and under trees (Fig. 7.16). In fact, the long distance dispersal by birds has resulted in *Juniperus* being established on Atlantic islands such as the Azores, Bermuda, the Caribbean and Canary Islands.

Figure 7.15. *J. virginiana* along a fence row in Tennessee, USA where birds have sit and excreted "planted" seeds.

Figure 7.16. *J. virginiana* under 'nurse tree' canopies where birds sit digesting the seed cones, Tennessee, USA.

In North America, *Juniperus* species have become weedy and invaded millions of acres of rangeland and abandoned farms (Adams et al., 1998). The invasion of range grasslands by *Juniperus* in the United States has now expanded to affect millions of acres. In fact, Juniper species affect over 21.5 million acres in Texas alone (Smith and Rechenthin, 1964). The principal species involved are *Juniperus ashei* (rock cedar or mountain cedar), *J. deppeana* (alligator bark juniper), *J. monosperma* (one-seeded juniper), *J. pinchotii* (copper fruited juniper) and *Juniperus virginiana* (eastern red cedar). The common substrate inhabited by *Juniperus* (as a genus) is limestone (Adams, 1993). However, in the eastern United States one can find old trees of *J. virginiana* along fence rows in tall grasses and deep soils. This may be due to openings in the habitat due to fires or erosion or may be due to the sheer numbers of seeds "planted" by birds sitting on fence rows and the chance favorable germination and survival of a seedling in these conditions. In Texas, *J. ashei*, was thought to have been confined to rocky outcrops and steep slopes until historical times (Owens, 1996).

Juniperus ashei occurs on limestone outcrops from northern Mexico to Oklahoma and Arkansas, with the bulk of the distribution found in the limestone hill country of Texas (Adams, 1977). Analyses based on volatile leaf terpenoids and morphology revealed the major geographical trend divided the populations into a recently derived group (Pleistocene) that was very uniform and the ancestral populations in northern Mexico, and far west Texas (Adams, 1977).

Juniperus ashei is now invading deep, blackland soils in central Texas. Adams et al. (1998) discovered young (ca. 5 - 7 years old) *J. ashei* plants growing on the edge of an old field in deep, blackland soil in the Leon River flood plain of central Texas and a population of old (50 - several hundred years) *J. ashei* tree growing on limestone about 30 - 50 meters away. They hypothesize that some novel mutation(s) or gene combinations are involved in the success of *J. ashei* in invading this tall grass, deep blackland soil area.

The invasion of *J. ashei* into sites not previously considered typical is shown in Figure 7.17. The limestone hill site is typical for millions of acres of *J. ashei* habitat. It is in the limestone hills that one finds pure stands of *J. ashei* with no other terrestrial plant species present. The deep, blackland site (Fig. 7.17) is in the flood plain of the Leon River and is the corner of a field that has apparently not been cultivated in several years (the oldest junipers found at this site were 6 years old).

Note the near proximity of the limestone hill plants to those across the road in the deep blackland soil (Fig. 7.17). The deep soil site is not typical of the habitat for *J. ashei*. The site is filled with competitive grasses, herbs and other species.

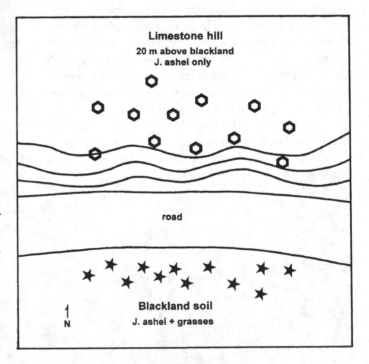

Figure 7.17. Site map showing the locations of *J. ashei* individuals sampled near the Leon River in central Texas

Clearly, *J. ashei*, under these conditions is at a disadvantage during the first few years growth. This seemed a suitable site to examine the theory that the invading junipers have a new mutation enabling them to compete in this environment.

After screening over one hundred RAPD markers, Adams et al. (1998) did not find a clear marker that has a 100% fidelity in separating the limestone based and blackland soil junipers. Primer 347 did however reveal some differences that generally, but not completely discriminated between the populations (Fig. 7.18). Note (Fig. 7.18) the bright fast running band in three of the five limestone individuals and the lack of slow running bands in the blackland individuals. It seems likely that additional screening may uncover some bands that do completely separate the blackland and limestone plants.

However, if one subjects the data matrix to principal coordinate analysis to factor the matrix, one can examine the trends (if present) in the data. Although eleven eigenroots were larger than the average diagonal element, the eigenroots appear to asymptote after the first two roots and no biological patterns could be discerned in eigenroots 3 - 11. The first three eigenroots accounted for 11.4,

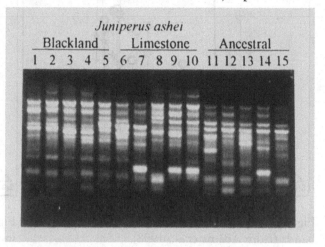

Figure 7.18. Comparison of RAPDs for *J. ashei* individuals invading blackland soil (lanes 1-5), limestone hill nearby (lanes 6-10), and the ancestral population at Ozona (lanes 11-15).

7.1, and 6.4% of the variance among the 29 individuals. This seems to indicate that there is not strong groupings among the individuals (i.e., subsets within the set). However, if we ordinate the 29 individuals onto the first three coordinates, we do see biological information (Fig. 7.19). Notice that the ancestral (Ozona) population is clearly separated from the other individuals on axis 1 (11.4% of the variance). This trend is congruent with the previous work (Adams, 1977), which showed the Ozona (and northern Mexico and far west Texas) population to be different from the more recent (Pleistocene) populations in central Texas. There is also some separation between the limestone (open hexagons) and the blackland soil (stars) (Fig. 7.19).

In order to further examine the separation between the limestone and blackland growing junipers, the five Ozona junipers were removed from the data set and the remaining 24 junipers (12 Limestone, 12 blackland) were subjected to PCO. Ordination shows the limestone and blackland junipers are nearly separated (Fig. 7.20). Notice the lone blackland juniper that clusters with the limestone junipers and one of the limestone junipers clusters with the blackland junipers. Clearly these two populations are almost genetically identical. It is likely that they differ only in a few genes - but a very important difference, in that the junipers on the blackland have the unusual (for *J. ashei*) ability to compete with tall grasses while very young.

One should remember that the 3-D ordination in figure 7.20 only accounts for about 24% of the variation among these 24 trees and 76% of the variation is not shown. Much of this variation is due to individual polymorphisms. A second factor to consider is that we are only examining length and sequence polymorphisms for 175 sites in the genome. Although Williams et al. (1990) have shown RAPD bands to be randomly distributed across the chromosomes, we still have only a small portion of the total genome represented.

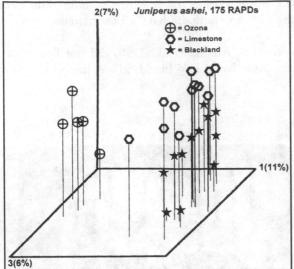

Figure 7.19. PCO based on 175 RAPD bands (from Adams et al., 1998).

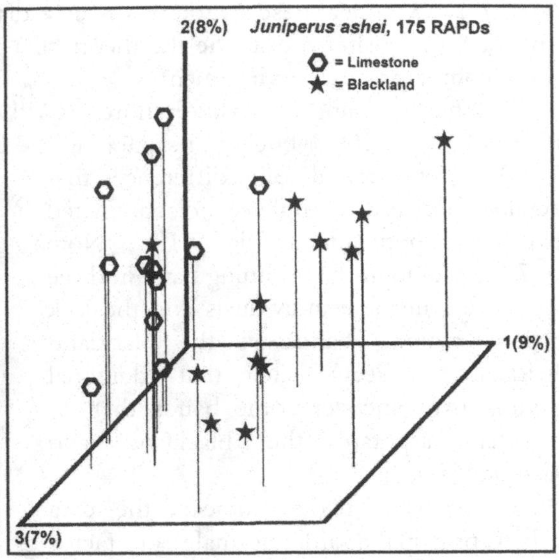

Figure 7.20. PCO of limestone and Blackland individuals only (from Adams et al., 1998).

Is there a gene for "weediness" enabling these *J. ashei* to invade these deep soils with tall grasses? Although we have yet to find a single band that clearly separated all of the "blackland" plants from the "limestone" plants, the ecological evidence is supportive of intense selection. In spite of the fact that these two sites are separated by only about 30 - 50 m and this juniper species is dioecious and wind pollinated, Adams et al. (1998) have evidence of differentiation across this short distance.

Figure 7.21. *Juniperus ashei* invading tall grasses in an old field near Austin, TX, USA.

Three of the largest juniper plants in the Blackland site were cut and found to have 6 growth rings, indicating that the population is very new. It is likely that the field edge was abandoned about 8 - 10 years ago and the junipers became established.

Adams et al. (1988) concluded that it is seems likely that the major hindrance in the establishment of juniper in grasslands is the relatively slow growth of juniper seedlings during the first few years. In tall grasses, (Figs. 7.21) the young seedling faces stiff competition for sunlight, moisture and nutrients, but a mutation that provides faster growth and/or drought tolerance would be a selective advantage in these situations.

One of the major factors in the distribution and abundance of *Juniperus* in the eastern hemisphere is the millennia of goats browsing. The vegetation of the Mediterranean and middle-east are particularly degraded due to pasturing of goats. *Juniperus* has evolved a protection system in which awn shaped leaves are present on seedlings in all species. These sharp awn shaped leaves (Fig. 7.22) do deter most large herbivores such as cattle, but apparently, goats can tolerate these sharp, needle-like leaves. As a consequence, in the Mediterranean and middle-east one sees many junipers that have an hour-glass form (Figs. 7.23, 7.24) where the goats have eaten the lower branches. What one does not see is the loss of seedlings in areas (where *Juniperus* has open habitat), that has led to an almost denuding of the landscape. In these areas, one only finds rocks and eroded soils, devoid of vegetation that could protect the soils from erosion.

Figure 7.22. Awn-shaped leaves of *J. communis*.

Figure 7.23. *Juniperus deltoides*, browsed by goats, near Mt. Parnassos, Greece.

Figure 7.24. *Juniperus polycarpos* var. *turcomanica*, browsed by goats, Kopet Mtns., Turkmenistan.

The genus is hardy and with the bird and small mammal modes of seed dispersal, plants can quickly occupy open habitat. Although several species are critically endangered (mostly on islands), the future of the genus appears competitive on the continents of the northern hemispheres.

Chapter 8: Seed Dispersal in *Juniperus*

Recently Adams and Thornburg (2010) reviewed seed dispersal in *Juniperus*. In an effort to present a more complete coverage of *Juniperus*, their paper is presented below (slightly modified).

The seminal paper in juniper seed dispersal was by Phillips (1910). His oft-cited paper was based on six years of field observations and interviews with naturalists concerning the dispersal of seeds of several *Juniperus* species. Phillips noted that because many juniper species' berries (female cones) mature in the fall and stay on the tree until spring, these present birds with a high energy food source when other seeds may be in short supply. In addition, the berries are often blue with a white bloom that makes them discernible even at considerable distances (Phillips 1910, Adams 2008). Phillips (1910) notes that E. A. Mears fed caged Bohemian waxwings (*Ampelis garrulous*) berries of *J. scopulorum* Sarg. and observed that over 900 berries passed through the birds in 5 hours. Clearly, these birds can consume and rapidly spread juniper seed. Phillips noted that a flock of 50 robins consumed all of the fruits on a juniper tree in a single day! He also noted the occurrence of 'fence row' junipers that were presumably 'planted' by birds resting on fences.

Phillips (1910) compiled a list of birds that had been found to consume juniper berries (Table 8.1). In addition, he reported that in Texas, the feces of raccoons, foxes, wildcats and chipmunks contained large amounts of juniper seeds from November to March. But he concluded that mammals are of rather minor importance in juniper-seed dispersal, compared to birds. Phillips also gave an interesting account of a Texas cattle herd that was driven to Kansas and grazed a few days on the treeless prairie, that later gave rise to a small juniper population.

Table 8.1. Birds that have been found to consume juniper berries (Phillips, 1910).

Juniperus species
Canachites canadensis - Canada Grouse
Corvus brachyrhynchos - Common Crow
Empidonax trailli - Traill Flycatcher
Oreortyx p. plumiferus - Plumed Quail
Pedioecetes phasianellus - Sharp-tailed Grouse

Juniperus communis
Merula migratoria - Robin
Parus atricapillus - Black-capped Chickadee

Juniperus sabina
Tyrannus tyrannus - Eastern Kingbird

Juniperus scopulorum
Ampelis garrulous - Bohemian Waxwing

Juniperus utahensis (J. osteosperma)
Meleagris gallopavo - Wild Turkey

Juniperus virginiana
Ampelis cedrorum - Cedar Waxwing
Carpodacus purpureus - Purple Finch
Corvus brachyrhynchos - Common Crow
Colaptes auratus - Yellow-shafter Flicker
Dendroica coronata - Myrtle Warbler
Dryobates pubescens - Downy Woodpecker
Hesperiphona vespertina - Evening Grosbeak
Hylocichla guttala -Hermit Thrush
Lagopus leucurus - White-tailed Ptarmigan
Merula migratoria - Robin
Mimus polyglottos - Mockingbird
Passerella iliaca - Fox Sparrow
Pinicola enucleator - Pine Grosbeak
Sayornis phoebe - Say's Phoebe
Sialia sialis - Bluebird
Sphyrapicus varius - Yellow-bellied Sapsucker
Tyrannus tyrannus - Kingbird

McAtee (1947) published an exhaustive annotated bibliography of papers dealing with the distribution of seeds by birds and noted that shade-tolerant cedars often develop under "perch trees."

Abbott and Belig (1961) found that red squirrels fed on *J. communis* berries during the winter, but avoided the prickly foliage of common juniper.

Poddar and Lederer (1982) reported that Townsend's solitaires feed exclusively on berries of *J. occidentalis* during the winter. They analyzed the nutritional content and found 4% protein, 16% lipid and 46% carbohydrate and concluded that *J. occidentalis* berries provide sufficient nutrients and energy to sustain solitaires during the winter.

One of the most detailed studies of the fate of juniper berries was conducted by Holthuijzen and colleagues (overview and references in Holthuijzen, Sharik and Fraser, 1987). Figure 8.1 shows a summary of their results. Noteworthy is that 65.5% of the berries were

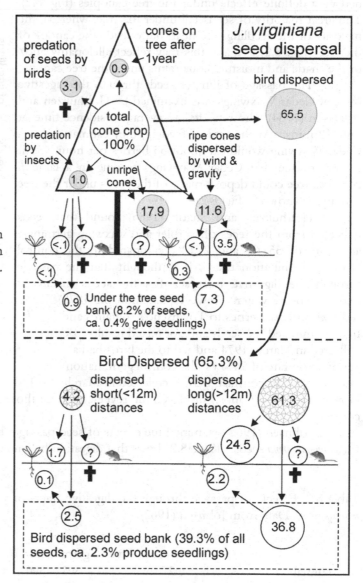

Figure 8.1. Dispersal of seeds in *Juniperus virginiana* (adapted from Holthuijzen et al. 1987).

dispersed by birds (Fig. 8.1). In addition, a large proportion (17.9%) of the cone-crop fell as unripe cones. This appears to be a common situation as I (RPA) have observed many unripe berries under various juniper species. However, very few of the seeds deposited under the tree canopy produce seedlings (ca. 0.4% of the seed crop, Holthuijzen et al. 1987). This seems to imply that the allelopathic nature of juniper foliage may inhibit seed germination under the canopy. Of course, shading and competition for moisture are additional factors.

Of the 65.5% berries dispersed by birds, about 93.6% were dispersed greater than 12 m from the tree. Of the 2.7% that produce seedlings, about 85% of these arose from long distance dispersal of seeds. Of course, a considerable number of seeds are deposited on the soil surface (1.7%, 24.5%) and these may germinate in favorable years.

Holthuijzen and Sharik (1985) showed that seed dispersal and seedling patterns in *J. monosperma* had very definite effects under the tree canopies (Fig. 8.2). The great majority of seeds fall under the tree, whereas, the majority of the seedlings are near, but outside the canopy of the tree (Fig. 8.2). Again, this may reflect allelopathy and/or the shade and moisture constraints under the tree canopy.

The passage of juniper seeds through the digestive tract of Cedar Waxwings was examined by Holthuijzen and Adkisson (1984) who reported an average residence time of only 12 mins. At the rate of feeding, they estimated that a Cedar Waxwing would eat about 53 berries per hour. So it is easy to see that Cedar Waxwings feeding in a large *J. virginiana* tree could deposit many of the seeds under the tree canopy, as shown in Fig. 8.2.

Holthuijzen and Sharik (1985) found that seeds collected from the feces of warbler and waxwing germinate at a rate of 55.0% and 27.6%, compared to the control (16.1%) germination tests. It is thought that the passage through the digestive tract scarifies the seeds making it easier to absorb water. However, Salomonson (1978) fed *J. monosperma* berries to Townsend's solitaires and then germinated the seeds and got mixed results. Berries collected in March, 1974 and fed to the birds had a germination rate of 24% vs. the control germination rate of 18%. In contrast, berries collected in March, 1974, then stored in dry, dark boxes at 21°C until November, 1974 (6 mos.) displayed the reverse pattern: those fed germinated at 35% vs. 45% for the control.

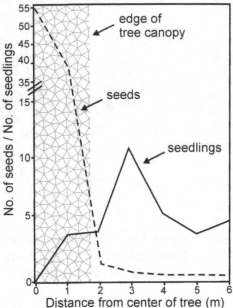

Figure 8.2. Number of seeds (per 0.1 m²) vs. seedlings (adapted from Holthuijzen and Sharik 1985).

Johnsen (1962) examined the effects of seed passage through various animals and germination for *J. monosperma* seeds. Table 8.2 shows that digestion by animals speeds up the

Table 8.2. Effects of passage through the digestive tracts of various animals on the germination of *J. monosperma*. Data from Johnsen (1962).

animal	\multicolumn{10}{c}{cumulative % germination by weeks (w1..w10)}									
	w1	w2	w3	w4	w5	w6	w7	w8	w9	w10
none (control)	0%	1	6	16	29	29	36	41	44	44%
bird	11	15	20	39	43	43	44	44	44	44
coyote or fox	2	10	28	37	40	44	45	45	45	45
packrat	8	14	22	40	44	44	45	45	45	45
jackrabbit	7	13	25	45	50	51	51	51	51	51
sheep	3	12	31	42	45	46	46	46	46	46

germination, but did not appear to increase the percentage of germination. It is assumed that the control germination seeds were kept moist for the entire 10 weeks. In northern Arizona, it would be very unlikely that soils might be kept moist for 10 weeks, so rapid germination following a rain would appear to be favored (i.e., favoring animal excreted seeds).

In the Canary Islands, *Juniperus cedrus* is dispersed by thrushes, ravens and lizards (B. Rumeu, pers. comm.). Control germination reached about 21% (after 200 days), whereas seeds passed by ravens and thrush attained about 33% germination. In contrast, seeds passed through the digestion tract of lizards showed reduced germination of about 10% (B. Rumeu, pers. comm.). Rumeu et al. (2009) reported that during the winter, they obtain about 98% of their diet from *J. cedrus* berries.

Schupp et al. (1997) examined feces of Nuttall's cottontail rabbits, mule deer, elk and coyotes for the presence of *J. occidentalis* seeds. Table 8.3 shows that, in general, these mammals do not consume vast quantities of juniper berries, except for coyotes that have a significant amount of seeds in some scats. We have observed *J. osteosperma* seeds in coyote scat in Arizona (Fig. 8.3), where it makes up the major portion of the scat.

Table 8.3. Numbers and frequency of *J. occidentalis* seeds in defecations of 4 mammals.

	# pellets or scats	# with seeds	total # seeds
cottontail	2046 pellets	8	8
mule deer	19,414 pellets	6	6
elk	562 pellets	0	0
coyote	29 scats	4	437

Figure 8.3. *Juniperus osteosperma* seeds in coyote scat (near Cottonwood, AZ)

Horncastle and Hellgren (2004) reported opossums and deer mice removed *J. virginiana* berries from beneath trees. They remarked that rodents appear to be seed predators and are not effective in dispersal.

Chavez-Ramirez and Slack (1993) analyzed scat from four carnivores (raccoon, ringtail, and brown and gray foxes. They found that during the winter months when *J. ashei* fruits are plentiful, all of these carnivores consumed considerable amounts of the berries. Because these carnivores have large home ranges (3.2 km to 10.4 km) they might be effective in long range dispersal. However, Chavez-Ramirez and Slack (1993) noted that, in at least 15 cases, rodents had consumed as much as 50% of the *J. ashei* seeds in mammal feces.

Harvester ants (*Pogonomyrmex* sp.) are well known to collect seeds and plant materials (Rissing, 1988). In a study at the Desert Botanica Garden, Phoenix, the latter workers found that the diet of *P. rugosus* diet consisted of 87% seeds from three annual plant species. The harvester ant's range varied from 12.7 to 22.1 m (avg. +/- 4 m, 1 SD) over a 3 year period. The larger range (22.1 m) occurred in 1984 when a drought restricted the growth of the favored annual species.

Recently we observed harvester ants carrying berries of *J. arizonica* from a tree about 15 m from their nest/ mound (Fig. 8.4). A few cleaned seeds were found next to the mound and one clean seed was found in the nest (fig. 8.5).

In our situation, the ants generally become active in the morning, forming only a single marching column about eight inches wide to the juniper tree. They harvested the berries from the ground only, and returned at a rate of one berry every two minutes. This continued all day (~8 hr) for a week. At such a rate as many as 1600 berries may have been carried to their nest. During this period, the column from the juniper tree to the nest carried only juniper seeds. A few ants, not in this column, cut up leaves from a *Penstemon* near the hole. On another occasion, an ant column was harvesting only bird seed in a bird feeding area. When juniper berries were place along the 'bird seed' column, the ants refused to pick up any berries. It appears that the ants just harvest one kind of material at a time.

Figure 8.4. Harvester ant carrying *J. arizonica* fruit (seed cone) to its nest/ mound.

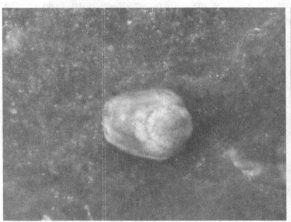

Figure 8.5. Cleaned seed of *J. arizonica* found inside the ant nest. Cleaned seeds were also found above ground, outside the ant mound.

Figure 8.6. Fungal farm inside the ant nest.

In summary, it appears that harvester ants bring juniper berries into their nest, eat off the pulp and then carry the seeds out to the edge of their mound.

This is not the first report of harvester ants collecting juniper berries. Willard and Crowell (1965) reported that *Pogonomyrmex owyheei* carried juniper (*J. occidentalis*) twigs and berries to their nests.

Although harvester ants may play a role in the dispersal of *J. arizonica* seeds, there is little doubt that birds are the main dispersal agents of seeds of that taxon as well as nearly all *Juniperus* species (Holthuijzen et al. 1987). In fact, Santos et al. (1999) found that birds were about 30 times as effective as mammals in the dispersal of *J. thurifera* seeds in Spain.

Chapter 9: Sex Expression in *Juniperus*

Vasek (1966), in a seminal study of the junipers of the western United States, tagged individual trees of *J. occidentalis* subsp. *australis* Vasek (now *J. grandis* R. P. Adams) and *J. osteosperma* (Torr.) Little and recorded their sexual expression for up to five years. Table 1 shows the results of his study. Although *J. o.* subsp. *australis* was reported as 90-95% dioecious (Vasek, table 4, 1966), some of the trees changed from male cones to producing both male and female cones (MBMBM, OMMB, MMMB, table 1) and some trees changed from female to none (FFFO, FOOO, table 1), but none changed from male to female or from female to male.

Juniperus osteosperma is about 85-90% monecious (Vasek, 1966, Table 4). A few trees changed from female to both (FBBBB, FBBB, FBB, FFBB, table 9.1) and some changed from male to both (MBBB, MMBB, Table 9.1) and 2 trees changed in all combinations (FFMB, OMFB, Table 9.1). Thus, of the 84 trees tagged and followed, 3/43 *australis* and 7/41 *osteosperma* trees showed sex changes.

Table 9.1. Sexual expression of tagged plants in the San Bernardino Mtns. (*J. occidentalis* subsp. *australis*) and White Mtns. (*J. osteosperma*). M = male cones only, F = female cones only, B = both male and female cones, O = no cones produced. Each letter represents an observation for one year. For example, OMFB means that the tree was observed 4 years with no cones (O) in the first year, male cones (M) in year 2, female cones (F) in year 3 and both male and female cones (B) in year 4. (data from Vasek, 1966)

J. occidentalis subsp. *australis* (*J. grandis*)				*J. osteosperma*			
# trees	pattern	# trees	pattern	# trees	pattern	# trees	pattern
1	MMMMM	4	FFF	11	BBBB	1	FBBBB
5	MMMM	2	FF	15	BBB	3	FBBB
2	MMMO	1	BBB	7	BB	1	FBB
5	MMM	1	OOOO	2	FFFF	2	FFB
3	MM	1	MBMBM	1	FFF	2	MBBB
10	FFFF	1	OMMB	1	FFFO	1	MMBB
5	FFFO	1	MMMB	1	MM	1	FFMB
1	FOOO			1	OOO	1	OMFB

It is also interesting that Vasek (1966) observed a female *J. occidentalis* subsp. *australis* tree with a broken, forked branch. The lower (cambium intact) portion produced female cones and the upper (presumably stressed) portion produced male cones. This seems to imply that environmental conditions may play a role in sex expression.

Jordano (1991) reported on sex expression in *J. phoenicea* L., a species that is largely monecious. He found that strongly male trees did not convert to females (Table 9.2) and that the strongly female trees did occasionally convert to strong male trees (Table 9.2). One of the inconstant (inconsistent) male trees did have some female cones the next year. Jordano (1991) reported that strong males

Table 9.2. Gender expression in consecutive years of *J. phoenicea* (data from Jordano, 1991).

	Gender expression in the next year				
Gender in current year	Male	Inconstant male	Monecious	Inconstant female	Female
Male	2	8	0	0	0
Inconstant male	6	17	1	1	0
Monecious	0	4	1	0	0
Inconstant female	0	0	1	0	1
Female	3	1	0	0	21

produced fewer than 10 female cones, whereas inconstant males rarely exceeded 200 females cones. Strong female trees produced more than 100 cones, except in years of crop failures. In addition, he reported that male trees produced smaller female cones with fewer seeds and these female cones tend to be aborted before maturity (Jordano, 1991). He speculated that self-fertilized might be the cause of self-abortion.

In March, 2009, we (Adams and Thornburg, 2011) found 18 male, *J. arizonica,* trees that had mature female cones (Table 9.3). In June, 2010, 3 additional male trees were discovered with female cones (Table 9.3, trees 19-21). Interestingly, none of the trees that bore new female cones in 2008 produced female cones (YF) in the winter of 2009 (Table 9.3). The mature fruit (MF) were from the 2008 pollination season (winter). Since none of the 21 trees had young fruit in March, 2009, then, of course, they would not have any mature fruit (MF) in June, 2010. Many of the trees that bore a few female cones in 2008 (YF, Table 9.3) did not bear any female cones in either 2009 or 2010. Two trees (#2 and 10) bore more, or about the same, number of female cones from 2008 to 2010 (but none in 2009).

Table 9.3. Sex expression in tagged male *J. arizonica* trees on successive years (2008, 2009, 2010). MF = Mature Fruit (1 yr old), YF = Young current year's Fruit. No data is available for MF in 2008. YF in 2008 is based on MF found on the same tree in 2009.

Tree	2008 YF	2009 YF	2009 MF	2010 YF	2010 MF
1	50+	0	50	12	0
2	1+	0	1	9	0
3	3+	0	3	2	0
4	1000+	0	1000+	8	0
5	2+	0	2	0	0
6	11+	0	11	0	0
7	6+	0	6	0	0
8	1+	0	1	1	0
9	2+	0	2	0	0
10	10+	0	10	14	0
11	5+	0	5	0	0
12	5+	0	5	0	0
13	2+	0	2	0	0
14	13+	0	13	0	0
15	2+	0	2	0	0
16	9+	0	9	0	0
17	100+	0	100+	13	0
18	3+	0	3	0	0
19	?	0	?	5	0
20	?	0	?	0	0
21	?	0	?	1	0

Most of the male trees had only a few fruits (female cones), but trees #4 and 17 had at least 100 female cones (Table 9.3). Trees #1 and 4 that bore larger numbers of female cones (50, 100) in 2008 (YF, Table 9.3), had fewer fruits (12 and 8, respectively) in 2010 (Figure 9.1).

It might be noted that no strongly female trees with abundant female cones were found with male cones. However, seeing a few of the small male cones amongst the female cones is very difficult.

In addition to counting the cones, in April, 2009, forked limbs were cut about 1/3 through the top portion (to mimic the broken branch noted by Vasek, 1966) on 25 male trees. In the spring of 2010, these cut branches were observed. All of these produced male cones on both the upper and lower forked branchlets. It seems reasonable that stressed branchlet (Vasek, 1966) might produce male cones if a species has that facultative ability. I (RPA) have observed that often one finds very few female cones on junipers in areas of severe drought. However, floral sex ratios in monecious plants change in response to hormones: both auxins and gibberellins (Heslop-Harrision, 1972; Friedlander et al. 1977)

Figure 9.1. Single female cone surrounded by male cones on *J. arizonica.* from Adams and Thornburg, 2011.

Freeman et al. (1981) reported that *J. osteosperma*, a monecious species, had a significantly higher frequency of male cones (54.85%) than female cones (17.50%) on trees in a xeric population (Table 9.4) but slightly more female cones (34.25%) than male cones (30.60%) in a mesic population site.   It seems likely that the additional water and other resources needed to produce the larger seed cones also curtails seed production in dry years. The author has noted that in drought years, few seed cones are produced in arid land *Juniperus* species and favorable years produce bumper corps of seed cones.

Table 9.4. Comparison of the frequencies of male and female cones on terminal limbs (2 per tree), 25 trees per site of *J. osteosperma*, a monecious species.

Xeric site			Mesic site			
male	female	nothing	male	female	nothing	F value for site-sex interaction
54.85	17.50	27.15%	30.60	34.25	31.15%	30.57**

Freeman et al. (1981) found a similar pattern with Gambel oak (*Quercus gambelii*) and black greasewood (*Sarcobatus vermiculatus*); the plants in mesic sites had a higher ratio of female flowers vs. male flowers. Freeman et al. (1981) concluded that there is a tendency for male flowers (cones) to be more prevalent in xeric sites and female flowers (cones) to be more prevalent in mesic sites.

Vasiliauskas and Aarssen (1992) examined the growth and special distribution of male and female *J. virginiana* trees. They reported that sex ratios were not related to age structure, stand density, or local competition intensity. However, they did find that male trees were taller than female trees and concluded that female trees pay a slightly greater cost for reproduction in terms of reduced vegetative growth.

However, Marion and Houle (1996) found no differences in radial growth patterns, annual elongation of the main axis, or size between male and female plants of *J. communis* var. *depressa* in a north-south transect on the eastern coast of Hudson Bay. But they did report that the northern-most populations had a male-biased sex ratio in contrast to the southern-most populations that had a female-biased sex ratio. If the northern-most populations are under more environmental stress, then there appears to be an increase in males with more stressful environments.

Gehring and Whitham (1992) reported that, for *J. monosperma* Engelm., plants highly infested with mistletoe (*Phoradendron juniperinum* Engelm.) growing on a stressful (volcanic cinder, ash and lava) site, female plants were more highly infested than male plants. But on a less stressful site (sandy-loam), there was no significant difference between the infestation rates for females or males. Again, there does seem (at least in *J. monosperma*) to be some costs associated with berry (female cone) production under stressful conditions.

The Adams and Thornburg (2011) study merely focused on the production of a few female cones on otherwise male trees in the dioecious species *J. arizonica*. Is *J. arizonica* truly dioecious? It has been my (RPA) experience that one can find a few monecious individuals among thousands of trees examined for all the dioecious species of *Juniperus* (and nearly all species are dioecious). The presence of a few monecious individuals would not invalidate one saying that a given species is dioecious. In the present case of *J. arizonica*, it may be more correct to describe the species as dioecious, but rarely monecious.

The apparent ease with which male *J. arizonica* plants appear to produce a few female cones seems to indicate the dioecious/ monecious mode is somewhat porous and may be easy to bridge. Could *J. arizonica* have the facultative ability to produce viable seed from 'male' trees to aid in colonization by long distance dispersal? If only a few male tree seeds are dispersed (by chance), then it could be advantageous to produce some seed by a 'partially' monoecious plant(s) to start a new population. Of course, we do not yet know if the seeds produced on the 'male' plants in this study are viable (Adams and Thornburg, in progress).

Sex changes and conversion from dioecious to monecious *Juniperus* plants (Vasek, 1966; Jordano, 1991, this study) raise some evolutionary questions that deserve a closer look in the future.

Chapter 10: Cultivated Junipers

There has been a proliferation of cultivars of *Juniperus*. Krussmann (1991) lists over 220 cultivars of *Juniperus*, including 18 cultivars in the Pfitzer group. It is very difficult to identify many cultivars by their morphology, much less to associate these cultivars with their wild (native) *Juniperus* species. Yet, it is critically important to associate a cultivar with its wild relative. The wild relative tells us about the ecology of the species, its range of adaptability, as well as its bioactivity (toxicity, allergenicity, etc.). The city of Albuquerque, New Mexico has banned the planting of male juniper cultivars due to production of allergenic pollen from most of our cultivated junipers (ex. *Juniperus x pfitzeriana*, Fig. 10.1).

A detailed examination of juniper cultivars is beyond the scope of this treatment and the reader is referred several treatments (Grace, 1983; Gelderen and Smith, 1989; Krussmann, G. 1991; Rushforth, 1987; Welch, 1966) as well as several web sites.

Le Duc et al. (1999) used RAPDs to examine the origin of the Pfitzer group. Historically, cultivars of the Pfitzer Group have been ascribed to *Juniperus chinensis* (*Cupressaceae* Bartling). Van Melle (1947) studied an extensive number of both preserved and living specimens of both cultivated and non-cultivated plants. He concluded that *J. chinensis* "Pfitzeriana" was of hybrid origin (*J. chinensis* x *J. sabina*), and proposed the name, *Juniperus x media* for the purported hybrid. He recognized var. *pfitzeriana* (male), var. *globosa* (male), var. *arbuscula* (female) and var. *plumosa* (female). The probable origin of the Pfitzer group was seed sent back to France in the 1860's by Armand David from Ho Lan Shan of "Inner Mongolia". Plants from the seed were extensively grown by French and Belgian nurserymen by the 1870's (Van Melle, 1947). "Pfitzeriana" was selected by Spath Nursery in the 1890's and has become one of the most commonly planted cultivars of juniper (Krussmann, 1991).

Welch (1966) was one of the first proponents of the new nomenclature. Some universities were incorporating the new nomenclature into woody plant courses by the 1970's, but the horticultural community has been slow to adopt Van Melle's classification. Currently, some horticultural authors (Dirr, 1998; Fling, 1998) await proof before adopting *J. x media* as the correct name for the Pfitzer

Figure 10.1 *Juniperus x pfitzeriana*, cultivated, male, Gruver, TX, USA, about 40 years old.

group. However, at least fourteen important horticultural references have adopted Van Melle's treatment (Lewis, 1995). Krussmann (1991), although adopting the new classification, qualified his position by stating that "Detailed cytological investigation could determine if it (*Pfitzeriana*) is a hybrid or a form of *J. sabina*...". He recognized 28 cultivars in the Pfitzer group.

Van Melle's name, *J. x media*, is rendered illegitimate under Article 64.1 of the International Code of Botanical Nomenclature (Greuter, 1994). It is a homonym of *J. media* V. D. Dmitriev that has priority (Czerepanov, 1973). Lewis (1995) argued for the conservation of the name *J. x. media* because of its historical use, dating from 1947. The request was rejected and the name proposed by Schmidt (1983) of *J. x pfitzeriana* has been accepted. The original Pfitzer plant, still alive at the Spath Arboretum, Berlin, Germany, has been assigned the denomination *Juniperus x pfitzeriana* (Spath) Schmidt "Wilhelm Pfitzer".   A comparison of the volatile leaf essential oils by Fournier et al. (1991) showed that *Juniperus* "Pfitzeriana" and several of its cultivars contained significant percentages of both bornyl acetate and sabinyl acetate, whereas *J. chinensis* contained only sabinyl acetate and *J. sabina* only bornyl acetate. Their chemical evidence supports the argument for the putative hybrid origin of *J.* "Pfitzeriana".

Le Duc et al. (1999) found fourteen RAPD primers that resulted in a total of 122 useable DNA bands. Computation and subsequent PCO ordination using the first three coordinate axes revealed several patterns (Fig. 10.2). The first three principal coordinates extracted 23, 20 and 10% of the variance among the fifteen OTUs. All samples of *Juniperus chinensis* cluster tightly together (Fig. 10.2). A similar grouping appears among *J. sabina* samples, with the *J. sabina* "Tamariscifolia" nesting within the cluster (Fig. 8.2). The purported *J. x media* cultivars exhibited no affinity to either *J. sabina* or *J. chinensis*, but stand apart as a distinct cluster, with the exception of "Fruitlandii" (FR). Principal coordinate 3 accounted for 10% of the variation and chiefly separates "Fruitlandii" (FR) from the other junipers (Fig. 8.2).

Coordinate 4 (7.64%) (not shown) separated "Kalley's Compact" (KC) and "Gold Coast" (GC) from the other junipers. No pattern was evident in any of the other coordinates.

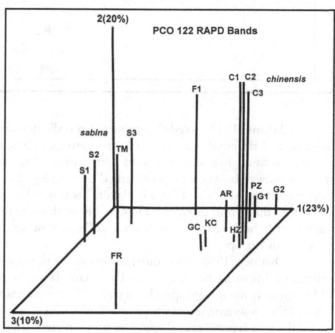

Figure 10.2. PCO ordination of *J. chinensis* (C1, C2, C3), *J. sabina* (S1, S2, S3), and cultivars of the Pfitzer group. (from Le Duc et al. 1999).

Adams (1982) has shown that the most effective way to visualize both artificial and natural hybridization was to plot the first two axes that separate the putative parents. However, samples of *J. chinensis* and *J. sabina* from the Ho Lan Shan area were not available for comparison, and the *J. chinensis* and *J. sabina* samples were utilized as proxies. Therefore, a synthetic F1 hybrid was created for the purpose of the analysis (Le Duc et al, 1999) by adding together all the bands that were present in either *J. chinensis* or *J. sabina*, because RAPD markers are inherited as simple dominants (Tingey and Tufo, 1993). Figure 10.3 shows this ordination and one can see that the synthetic F1 appears midway between *J. chinensis* and *J. sabina*. The ordination is similar to that obtained by Adams (1982) for both artificial and natural hybridization.

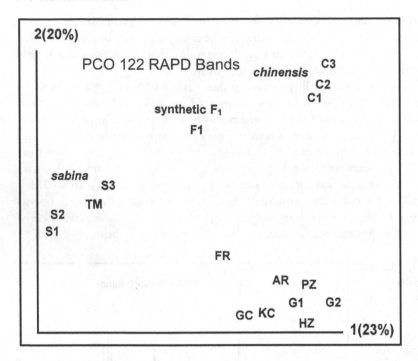

Figure 10.3. PCO based on 122 RAPD bands of axes 1 and 2 (from Le Duc et al. 1999).

Adams (1982) noted that principal coordinates separate groups, and a group of hybrids arising from several different parents can form a group. Thus, it is difficult to prove whether intermediate plants constitute hybrids or a third "intermediate" taxon. The computer generated F_1 (Fig. 10.3) appears midway between clusters formed by samples of *J. sabina* and *J. chinensis*, but separated from the Pfitzer cluster. Although FR ("Fruitlandii") is somewhat removed from the other Pfitzer cultivars, it is still within the hybrid distance. These cultivars show a strong cluster group, but recall that the original purported hybrid was derived from a collection of seeds grown out and dispersed throughout the nurseries of Europe.

Ownbey (1950) listed three criteria in the classical treatment of the taxonomic status for hybrids of *Tragopogon* L. (*Asteraceae*). These criteria can be rephrased as questions:

1. Are the taxa natural groups, characterized by a combination of distinctive morphological features (*and or DNA differences, my addition*)?
2. Are the taxa reproducing themselves under natural conditions?
3. Is there free gene exchange between taxa?

In the present case, criteria 1 is fulfilled because the DNA and terpene data supports Van Melle's premise that the Pfitzer group of junipers is distinct from *J. chinensis* (and *J. sabina*). Unfortunately, we lack field knowledge that would enable us to evaluate criteria 2 and 3.

The present data provides additional evidence which supports Van Melle's conclusion that the Pfitzer group is separate from *J. chinensis* and indicates hybrid origin, the parents being *J. chinensis* and *J. sabina*.

Figure 10.4. *Juniperus sabina* cv. *Tamariscifolia*, Kew Gardens, London.

Based on the decision of the International Commission for the Nomenclature of Cultivated Plants, the correct name is *Juniperus* × *pfitzeriana* (Spath) Schmidt [Pfitzer group] and includes "Fruitlandii", "Gold Coast", "Hetzii", "Kalley's Compact", and the "Wilhelm Pfitzer", "Pfitzeriana Aurea", and "Pfitzeriana Glauca". Le Duc et al.(1999) showed that "Tamariscifolia" (Fig. 10.4) is a part of *J. sabina* and recognized it as *Juniperus sabina* cv "Tamariscifolia".

It seems likely that other taxonomic problems in referring cultivars to their parental species will be resolved by DNA technology in the future.

Chapter 11: Commercial Use of Junipers

Cedarwood oil

The major commercial product from *Juniperus* is cedarwood oil. Cedarwood oil is an important natural product for components used directly in fragrance compounding or as a source of raw components in the production of additional fragrance compounds. The oil is used to scent soaps, technical preparations, room sprays, disinfectants, and similar products, as a clearing agent for microscope sections and with immersion lenses (Guenther 1952).

The price varies but has generally been about $4.50/lb. for Virginiana cedarwood oil, $3.50/lb. for Texas cedarwood oil and $1.50-$1.75/lb for Chinese cedarwood oil. The Chinese cedarwood oil, although almost identical in composition, is lesser valued because its fragrance is very different from the Texas and Virginia cedarwood oils. Chinese cedarwood oil is reported to come from *Cupressus funebris*, but due to the similarity of its oil to *J. ashei* and *J. virginiana* oils, it may come from *J. chinensis*. The commercial cedarwood oils are obtained from 3 genera of Cupressaceae: *Juniperus* (Texas and Virginia oils); *Cupressus* (China) and *Cedrus* (Morocco, India) according to Bauer and Garbe (1985). The heartwood oils of the *Cupressaceae* are well known for having the same components across the family (i.e., evolutionarily conserved), so the occurrence of similar oils in different genera should not be surprising.

The world production (1984) has been reviewed by Lawrence (1985), who reported the following (source and metric tons): Texas (*J. ashei* Buch.) – 1400; Virginia (*J. virginiana* L., S.E. United States) – 240; China (*C. funebris* Endl.) – 450; India (Himalaya, *Cedrus atlantica* Manetti)- 20; Morocco (Atlas Mtns., *Cedrus deodora* Loud.)- 7; Kenya (*J. procera* Endl.) – no production at present.

Cedarwood oils have not been examined thoroughly or systematically. Many of the analyses of cedarwood oil were done by Runeberg (1960a-e, 1961) and associates (Pettersson and Runeberg 1961; Pilo and Runeberg 1960) (Table 11.1). For many years *J. ashei* was reported to contain only α-cedrene and cedrol (Guenther 1952; erroneously referred to as *J. mexicana* in Erdmann and Norin 1966 and Walker 1968, see Zanoni and Adams 1979 for nomenclatural discussion). However, more recently, Kitchens et al. (1971) reported β-cedrene, thujopsene, widdrol, pseudocedrol, β-chamigrene, prim cedrol, widdrene (= thujopsene?), isowiddrene, α-chamigrene, and cuparene (three isomers). Unfortunately, as is typical of most of the reports on identifications, not enough data were given by Kitchens et al. (1971) to allow confirmatory studies of their work.

Juniperus californica was reported (Pettersson and Runeberg 1961) to have cedrol as the major component (52%) of the heartwood volatile oils with a considerable amount of thujopsene (26%). A more recent study (Adams 1987) reported low yields (Table 11.1) of cedrol from two chemical races (A & B, Vasek and Scora 1967) of *J. californica*. It is presumed that Pettersson and Runeberg (1961) may have analyzed the wood of *J. occidentalis* instead of *J. californica*.

Juniperus communis was reported to have mostly thujopsene (37%); however, it is not clear how the author (Bredenberg 1961) arrived at these percentages. *Juniperus horizontalis* is a prostrate plant that forms mats. Due to its low wood biomass, its oil composition is primarily of academic interest. Narasimhachari and von Rudloff (1961) reported that *J. horizontalis* contained α-cedrene, thujopsene, cuparene, cedrol, and widdrol but relative concentrations were not reported (Table 11.1). *Juniperus occidentalis* was examined by Kurth and Lackey (1948) who merely reported that the wood contained α-cedrene and cedrol. A more recent analysis of both varieties (Table 11.1, Adams 1987) showed the varieties to be high in cedrol and thujopsene.

Juniperus osteosperma (referred to as *J. utahensis* by Runeberg 1960b) had 47.8% thujopsene, with about equal amounts of α-cedrene, cuparene, and widdrol. Adams (1987) found that the taxon was high in thujopsene but reported that cedrol was also a major component (Table 11.1).

Table 11.1. Literature reports on the heartwood oils of *Juniperus* species. Approximate percent concentrations of key components were obtained when possible from the original literature cited. ACD = α-cedrene; BCD = β-cedrene; THJ= cis-thujopsene; CUP = cuparene; CED = cedrol; WID = widdrol .

Section *Juniperus* (awn-like leaves only)

Species	ACD	BCD	THJ	CUP	CED	WID	Reference
J. cedrus	-	-	82.5	3.7	2.2	2.6	Runeberg, 1960a
J. communis	-	-	37.0	3.0	2.0	1.0	Bredenberg, 1961
J. deltoides (reported as J. oxycedrus from Turkey, but surely J. deltoides)	1.2	-	21.5	-	2.0	19.9	Ucar and Balaban, 2002
J. rigida var. conferta	-	-	+	+	+	-	Doi and Shibuya, 1972

Section *Sabina* (scale-like leaves)

Species	ACD	BCD	THJ	CUP	CED	WID	Reference
J. angosturana	1.4	0.7	17.5	0.3	51.0	t	Adams, 2009e
J. arizonica	1.9	0.6	14.5	1.7	3.1	7.6	Adams, 2009e
J. ashei	0.6	0.4	12.0	1.2	36.7	12.0	Adams, 2009e
J. ashei (=J. mexicana)	1.8	1.6	60.4	2.8	19.0	1.1	Adams, 1987
J. californica 'A'	4.9	2.7	19.7	6.4	8.0	8.0	Adams, 1987
J. californica 'B'	3.9	1.9	18.7	4.7	9.3	9.2	Adams, 1987
J. californica	1.2	0.6	2.2	4.8	11.7	11.7	Adams, 2009e
J. chinensis	-	-	11.6	4.3	72.9	6.0	Pilo and Runeberg, 1960
J. coahuilensis	1.1	0.5	18.4	1.2	17.4	-	Adams, 2009e
J. coahuilensis	1.9	1.6	67.9	3.0	8.5	0.5	Adams, 1987
J. comitana	0.5	0.2	2.9	0.3	43.2	-	Adams, 2009e
J. deppeana	16.9	3.9	14.9	3.9	26.4	1.0	Adams, 1987
J. deppeana	7.6	2.0	9.1	1.6	25.7	12.0	Adams, 2009e
patoniana	1.5	0.8	7.2	1.3	30.4	10.0	Adams, 2009e
zacatacensis	4.1	1.4	10.3	2.1	1.3	16.0	Adams, 2009e
robusta	3.9	0.9	6.3	1.2	2.0	15.6	Adams, 2009e
gamboana	2.5	1.0	5.5	0.8	55.2	0.1	Adams, 2009e
J. durangensis	1.2	0.9	7.9	0.0	21.9	5.0	Adams, 2009e
J. excelsa (sapwood)	0.4	-	2.7	-	9.0	22.5	Ucar and Balaban, 2002
J. flaccida	3.5	1.0	4.8	1.0	15.9	32.0	Adams, 2009e
J. foetidissima	58.3	-	-	-	8.3	5.0	Runeberg, 1961
	8.2	2.0	10.5	1.9	13.0	15.3	Tunalier et al, 2004
	2.3	-	23.8	-	10.0	15.8	Ucar and Balaban, 2002
J. grandis	1.4	0.9	15.4	1.4	41.8	-	Adams, 2009
J. grandis	3.3	1.3	20.1	1.5	38.2	1.6	Adams, 1987
J. horizontalis	+	-	+	+	+	+	Narasimhachari & von Rudloff, 1961
J. jaliscana	2.6	0.7	7.6	1.0	1.1	10.0	Adams, 2009e
J. martinezii	2.0	0.9	6.8	0.2	34.0	3.7	Adams, 2009e
J. monosperma	0.7	0.5	9.0	1.3	14.1	7.1	Adams, 2009e
J. monosperma	2.7	1.8	61.0	3.8	4.1	1.7	Adams, 1987
J. monticola	1.9	0.9	14.3	1.0	10.4	7.0	Adams, 2009e
J. occidentalis	8.8	2.6	18.9	1.5	38.9	1.6	Adams, 1987
J. occidentalis	0.9	0.8	6.4	0.9	43.3	t	Adams, 2009e
J. osteosperma	1.3	0.7	13.7	1.1	36.3	t	Adams, 2009e

Section *Sabina* (scale-like leaves) continued

Species	ACD	BCD	THJ	CUP	CED	WID	Reference
J. osteosperma (=*J. utahensis*)	4.0	1.8	40.0	2.6	13.2	1.5	Adams, 1987
J. phoenicea	-	-	79.3	2.9	7.2	0.1	Runeberg, 1960c
J. pinchotii	2.8	1.2	4.8	0.1	4.4	-	Adams, 1987
J. pinchotii	3.6	1.6	12.0	2.4	38.6	4.2	Adams, 2009e
J. poblana	3.3	0.9	6.6	1.1	21.0	20.6	Adams, 2009d
J. procera	41.8	-	-	2.5	41.8	-	Pettersson & Runeberg, 1961
J. recurva	3.5	0.9	5.1	1.8	49.0	16.7	Oda et al., 1977
J. saltillensis	1.2	0.7	9.0	1.2	46.9	t	Adams, 2009d
J. semiglobosa	-	+	-	-	+	-	Goryaev et al., 1962
J. standleyi	2.6	1.2	10.3	0.6	2.8	25.2	Adams, 2009d
J. thurifera	23.3	-	15.5	3.9	27.1	-	Runeberg, 1960d
J. scopulorum	4.3	2.4	57.9	6.1	6.1	3.0	Adams, 1987
J. virginiana	27.2	7.7	27.6	6.3	15.8	1.0	Adams, 1987
J. zanonii	0.4	0.4	2.5	0.8	54.2	-	Adams, 2009d

Juniperus virginiana wood was not directly analyzed by Runeberg (1960e). Using a commercial sample of cedarwood oil said to be from *J. virginiana*, he found mostly α-cedrene and thujopsene with a very small amount of cedrol (4%). However, the commercial cedarwood oil may have been precipitated or fractionally distilled to remove cedrol because Adams (1987) noted that *J. virginiana* wood (collected from native trees in Texas) contained about 16% cedrol (Table 11.1). Wenninger et al (1967) analyzed the sesquiterpene hydrocarbons of American cedarwood oil (*J. virginiana* ?) and reported the oil contained 55-65% sesquiterpene hydrocarbons with α-cedrene and thujopsene as the major components. Runeberg (1960e) stated that the highest yield of oil, about 3.5% of the wood (dry wt.?). was obtained from sawmill waste from older trees (i.e., trees with a greater ratio of heartwood to sapwood). Guenther (1952) obtained only a 0.2% yields by distilling sapwood of *J. virginiana*; he noted that young trees (commonly called "sap cedars") yielded less than 1% oil, compared with older trees (commonly called "virgin cedars"), which yielded 3.5%.

Sample Collection

It is assumed that the reader is not only interested in the analysis of commercial cedarwood oil, but also investigating the cedarwood oil from trees. In general, at least 5 trees should be sampled (obviously 10 is preferred). Unfortunately, little is known about seasonal variation of wood oils. Depending on the local customs and laws, one may be faced with only a few options in collecting wood samples. I have even resorted to visiting a local wood working shop in Ethiopia to obtain wood samples of *J. procera*. The ideal situation is to visit a site where trees are being cut for firewood, posts, lumber, etc. and obtain wood blocks directly. Failing this, one may look for broken limbs and cut off the stump section near the stem. I have also used a drill with a large wood bit (2 cm) to obtain wood from the trunk. One must be very careful to keep track of both sap- and heart-wood shavings if biomass and yields are to be determined.

If a wood block is available, a wood planer is useful to produce fresh wood shavings for steam distillation. A power drill can also be used to produce wood chips for distillation, but care must be taken so the bit does not get hot and cause a loss of the oil. Cedarwood oil appears to be very stable in intact wood blocks as cedarwood cut in the 1930s is still being collected and distilled near Junction, Texas with the oil apparently acceptable.

Oil Extraction

Commercially, cedarwood oil is extracted in a variety of manners ranging from home-built stills to a newly designed process by Texarome Inc., Leakey, Texas (Figs. 11.1, 11.2). Distillation times vary from 8 hrs or more in traditional stills, but only 30 - 60 seconds in the Texarome process (Gueric Boucard, pers. comm.). One should bear in mind that changing the distillation time, sizes of wood chips extracted, or temperature and pressure of the steam can make vast differences in the cedarwood oil composition, so that comparisons of various commercial oils may vary as much by distillery as by region or species utilized.

Figure 11.1. Texarome cedarwood oil extraction near Leakey, TX, USA. *Juniperus ashei* is on the hillside in the background (Photo courtesy, Texarome, Inc.).

Figure 11.2. Close-up of modern oil extraction equipment at Texarome, Inc. Residence time is 30 to 60 seconds. (Photo courtesy, Texarome, Inc.).

Figure 11.3. Cedar wood (*J. ashei*) being taken to the PAX extraction plant in Junction, TX, USA.

Figure 11.4. Cedar wood being unloaded at the PAX Corp. extraction plant near Junction, TX, USA This cedar wood was cut and stacked in fields in the 1930s as part of a government program to improve rangelands. The sapwood has rotted away, but the heartwood still contains the typical cedarwood oil after about 50 years in the weather.

Figure 11.5. Tanks for steam distillation, PAX Corp., Junction, TX, USA.

Figure 11.6. Extracted cedar wood chips, PAX Corp., Junction, TX, USA.

Juniper (cedar) is cut from ranches in central Texas and hauled to oil extraction plants (Fig. 11.3), unloaded (Fig. 11.4) and chipped into small pieces (<1cm) for steam distillation. Traditional steam distillation at the PAX plant in Junction, TX takes about 8 hours (Fig. 11.5). The extracted (spent) chips are then carried to a large pile (Fig. 11.6). Juniper wood fiber, in 1996 (Garriga, et al. 1997), was worth about $35 per ton and used for fiber-board, drilling mud for oil well drilling, animal bedding, mulch for nurseries and tar paper. In 1996, the Advanced Environmental Recycling Technologies, Inc. (AERT) company began using the spent cedar chips in the Junction area to manufacture fiber-board for construction.

Unextracted cedarwood chips are sold by Cedarcide Inc. for use around houses to repel insects (Figs. 11.7, 11.8, 11.9).

Figure 11.7. *J. virginiana* cedarwood ready for chipping, Cedarcide Inc., Houston, TX.

Figure 11.8. Cedar wood chipping, Cedarcide, Inc.

Figure 11.9 (right). Cedar wood chips for insect repelling insects by Cedarcide, Inc.

Supercritical carbon dioxide extraction of cedarwood has been accomplished with good results (Eller and King, 2000). This method has advantages over steam distillation in that by varying extraction parameters, one can obtain oils that have more or less of the cedrenes or cedrol/widdrol, etc. However, this technology is generally rather expensive and it is not known when or if small cedarwood distillers will adopt this technology.

For laboratory use, one should be most careful about placing the wood into water and boiling out the oil. Several studies (Fischer et al 1987; Koedam and Looman 1980; Koedam et al., 1981; Schmaus and Kubeczka, 1985) have shown that plants produce acidic conditions when boiled and this leads to terpene rearrangements and decompositions. This is shown for cedarwood oil in Fig. 11.10.

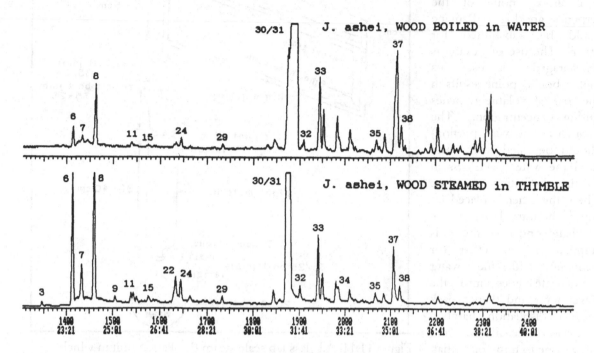

Figure 11.10. Ion chromatograms of cedarwood oils obtained by boiling wood chips in water (top), versus oil obtained by steam distillation in which the wood is suspended above the steam source (bottom) (from Adams, 1991). Boiling wood (or leaves) often results in decomposition.

The initial pH of the water in the boiling flask was 7.12. After only 2 hrs boiling the wood chips directly in the flask, the pH was 6.17 and the composition was quite changed (Fig. 11.10, upper). Steam distillation using an apparatus with the plant material suspended above the steam generator flask (Fig. 11.11), resulted in the chromatogram in Figure 11.10 (lower). In this case, the pH of the water in the steam generator flask was 8.62 after the steam distillation. The shift in the base line (Fig. 11.10, upper) is indicative of decomposition. Note particularly, the low yield of α- and β- cedrenes (peaks 6, 7). There is a large increase in the oxygenated sesquiterpenoids (peaks 30 and upward).

Fischer et al. (1987) discuss the fact that the original (*in situ*) flavor components of Marjoram may be quite different than those of the commercial oils. However, if one is to work within the legal and market framework that has already been established for cedarwood oil, it seems that practical work may be forced to use steam distillation extraction. Von Rudloff (1967) examined the use of direct distillation (plant material in boiling water), a Markham type device and a modified Clevenger-type circulatory apparatus. He preferred the modified Clevenger-type circulatory apparatus and that is essentially what I recommend (Fig. 11.11). I have added ball joints so the apparatus is easier to align and the ether trap can be adjusted. Notice that the plant material is placed in the cylindrical part so that only steam comes in contact with the plant material. An external heating jacket can be added to the cylindrical part to increase the distillation efficiency if desired. The condenser has also been modified so that the water jacket completely covers the ether trap area. This has resulted in much less loss of ether during distillation.

I prefer ether as the terpene trap because the ether can be evaporated by a stream of nitrogen in a hood and almost none of the terpenes are lost. Pentane could be substituted for ether. The use of hexane is discouraged because its higher boiling point results in the loss of volatile terpenes during concentration. The condenser (lower portion) should be filled with water until the water overflows in to the distillation chamber. Then, the ether is placed on top of the water later. As the distillate condenses, the oil is trapped in the ether (or pentane) and the water condensate goes into the lower layer and thence back into the distillation chamber. The low density of ether allows one to trap oils that have a density greater than water. The apparatus can be run without attendance and any

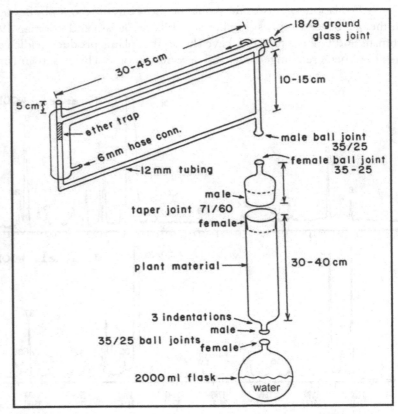

Figure 11.11. All glass lab scale steam distillation unit in which the plant material is suspended above the steam generator, a 2000 ml flask (from Adams, 1991).

terpenes that carry over in the water are automatically volatilized as the condensate flows back into the distillation chamber.

When using the apparatus with finely ground wood or small wood chips, I have found it useful to place the ground wood into a sandwich of nylon screen (as used for window screens) and then place the elongated sandwich into the cylindrical chamber. If loose, finely ground material is placed directly into the cylindrical chamber, it will pack down and block the steam, channels will form and the distillation will not proceed regularly. In addition, the distillate water, returning from the condenser will accumulate on top of the plug and one faces the danger of running the steam generator flask to dryness. Care should be taken when handling the ether and the entire apparatus should be placed in a well ventilated hood when used.

The oil samples can be dried over anhydrous sodium sulfate to remove water in the ether if desired. We routinely pre-weigh our vials (with either compression or screw caps but in either case, using Teflon coated caps), and evaporate the ether in a hood with nitrogen. A GC run is then used to determine the percent ether remaining in the sample and the final weight of the oil is then calculated. The samples should be stored at -20° C or colder for long term storage. Sealing the samples under nitrogen is also advisable for very long storage. Although decomposition of various oil samples has been mentioned to me by many colleagues, we have not experienced a problem over the past 25 years. Either our cedar and juniper oils are very stable or the aforementioned procedures mitigate decomposition. I expect that those who distill directly in water, obtain oils that are quite acidic and this may be the reason that oil decomposition is a problem. In any case, one can not assume that there will be no decomposition during long-term storage (months to years).

Chemical Analysis

Traditionally, cedarwood oils are defined (Walker 1968) on the basis of several physical properties: specific gravity at 10° C (or 20° C) = 0.94 -0.99; optical rotation = -16 to -60: refractive index at 20° C = 1.48 – 1.51; and solubility (at 20° C) in 90 or 95% ethanol (varies with source). Although this treatment will focus on the individual chemical components, one should be aware of the practical use of the aforementioned physical properties.

Gas Chromatography

Gas Chromatography has become an integral part of any essential oil analysis today. All of our primary analyses are on a J & W fused silica capillary columns, DB-5, 30m, 0.26mm i.d., 0.25 micron coating thickness.

Identification

Early work on the identifications of terpenoids used component trapping from preparative GC, with subsequent liquid infrared (IR) spectral analysis for identification. The introduction of capillary columns have reduced the samples to the point that those techniques are no longer practical. The more recent development of vapor phase IR with on-the-fly analysis offers considerable promise as libraries are being compiled. However, the most practical method of identification is generally combined GC/MS or GC/MS/computer searches.

GC/MS

A large library of mass spectra is readily available from sources such as the U. S. NBS (National Bureau of Standards, formerly the EPA/NIH data base) with thousands of spectra. Unfortunately, searches from these large data bases, with the current technology (i.e., simple matching coefficients and no retention data) do not yield reliable identifications (see Adams et al. 1979 for discussion).

Analyses of the three major cedarwood oils are shown in Fig. 11.12. Notice that the three oils share the major components (α-cedrene, β-cedrene, thujopsene, cedrol and widdrol). Although the minor components vary quantitatively among the oils, there is a remarkable uniformity. The off-flavor of the Chinese cedarwood oil (cf. *C. funebris*) is apparently due to minor components from mixing various species during extraction (see Adams and Lu, 2008 for discussion on the origin of Chinese cedarwood oil).

Adams (1991) reported ion trap mass spectra (ITS) for the major components. Although the ITS spectra are generally quite similar to quadrupole mass spectra (Adams 1989; Adams, 2001), there is a large reduction in ion 151 in both cedrol and widdrol on the ion trap. It might be noted that cedrol is very sensitive to space charging effects (overloading) and tuning on the ion trap. We use cedrol as a tuning standard for quadrupole mass spectroscopy due to its sensitivity.

The reader is invited to compare spectra from Adams (1989) versus Adams (2006). The current (4th edition) of my book on identification of terpenoids contains 2205 compounds and comes with libraries in all commercially utilize mass spectrometers (Adams, 2007, **see www.allured.com**).

Activities

Clark et al. (1990) made hexane and methanol extracts of heartwood, bark/sapwood and leaves of twelve taxa of *Juniperus* from the United States and these were assayed for anti-fungal and anti-bacterial activities. The hexane extracts of the heartwood (which contains the cedarwood oil) of several junipers appear comparable in anti-bacterial activity to streptomycin. No anti-fungal activities comparable to amphotericin B were found in either the hexane or methanol extracts of heartwood. Additional research is needed to isolate and determine the anti-bacterial components.

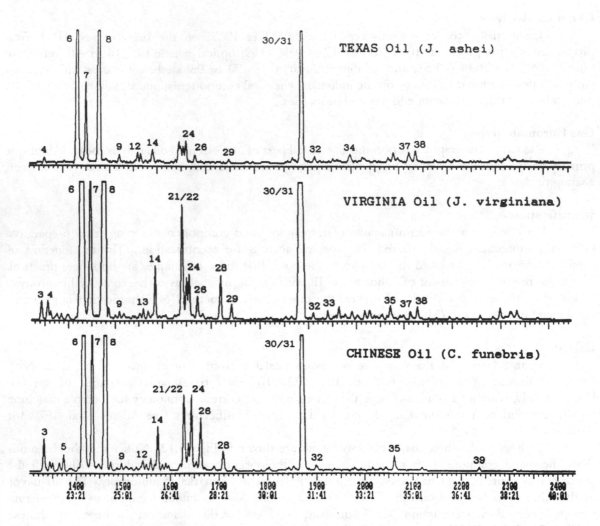

Figure 11.12. Ion chromatograms of commercial cedarwood oils (Texas - *J. ashei*, Virginiana - *J. virginiana*, and Chinese - cf. *Cupressus funebris*) (from Adams, 1991).

Antimicrobial Insecticidal and Insect Repellency Properties

Oda et al. (1977) examined the insecticidal activities of several fractions from the heartwood of *Juniperus recurva* from Nepal. The insecticidal activities were found in the steam volatile fraction (i.e., cedarwood oil). A detailed examination of individual components revealed the following LD50 μg/mosquito: **α-cedrene- 33.5**; β-cedrene- not active; **thujopsene- 4.5**; acoradiene- not active; β-chamigrene- not active; curparene - not active; **8, 14- cedranoxide-10.7**; 8-cedrene-13-al- not active; **cedrol- 21.2**; widdrol- not active; 8-cedren-13-ol acetate- not active; **8-cedren-13-ol- 6.6**; 8S, 13 & 14-cedrane-diols- not active. Clearly the most insecticidal components were thujopsene and 8-cedren-13-ol. Again, additional research is warranted on both *Juniperus* and *Cupressus* (and other *Cupressaceae* species) for natural insecticidal compounds to replace chlorinated pesticides of current usage. Cedarcide Industries, Houston, TX (Dave Glassel, pers. comm.) markets Virginia cedarwood oil for use in foggers to repel mosquitoes (Fig. 11.13).

Figure 11.13. Cedarcide products, based on cedarwood oil, used for mosquito control.

Appel and Mack (1989) reviewed the literature to date and reported that cedarwood products (chips) had been shown to be repellent or toxic to several fabric pests, the webbing clothes moth, black carpet beetle and furniture carpet beetle.

Appel and Mack (1989) found that eastern red cedar wood (*J. virginiana*) cedar flake board (used for lining closets and trunks), was repellent to the German cockroach, but not the American or smoky-brown cockroaches. Nevertheless, cedarwood planks are widely used in the USA for lining clothes closets to repel clothes moths.

Meissner and Silverman (2001) reported that eastern cedar wood chips repelled the Argentine ant and the odorous house ant. In addition, when placed in closed container with cedarwood chips, both ant species were killed.

Termiticidal Activities

The control of termites is a world wide problem. Current preservatives use arsenic, chlorinated and copper based products, all of which are toxic to humans and/or carcinogenic. Carter (1976) found that termites (*Reticulitermes flavipes*) could not survive on sawdust from *Juniperus virginiana* nor on filter paper treated with a pentane extract (cedarwood oil) of *J. virginiana* sawdust.

Subsequently, Adams et al (1988) found extremely high termiticidal activities in the heartwood sawdust from all 12 of the United States junipers examined. Hexane extracts of the heartwoods revealed that treated paper showed considerable termiticidal activities in seven of these species.

Zhu et al. (2001) reported that cedarwood oil was effective in repelling and killing the Formosan termite. Because the junipers are used for posts in the United States, it is obvious that wood preservatives are in the wood. These same kind of observations about wood rotting should be used to select promising species for additional termiticidal (and wood rotting) tests around the world (particularly in the *Cupressaceae*).

Cedarcide Inc. (Houston, TX) seems to be the only commercial manufacturer of a wood preservation based on cedar wood oil (and polysiloxane) in the USA (Agent Gold, Fig. 11.14, and CedarShield). The search for a "green" wood preservation is important in the United States, so it is surprising that it has taken so long for cedarwood oil to be utilized. Because cedarwood oil is on the GRAS (Generally Regarded As Safe) list (USA), lots of difficult regulatory problems can be avoided.

Figure 11.14. "Agent Gold" wood preservative from Cedarcide, Inc.

Cade oil (Oleum Cadinum, B. P., Huile de Cade)

Cade oil (Oil of Cade) is an empyreumatic wood oil obtained from the destructive distillation of the wood of *Juniperus oxycedrus* (and some other junipers such as *J. communis, J. sabina, J. thurifera*) in the south of France. It is used as a stimulant antiseptic in chronic skin diseases such as psoriasis and eczema (The British Pharmaceutical Codex, online). Cade oil from *J. oxycedrus* is high in δ-cadinene and calamenene (Chalchat et al., 1988), but apparently lacking cedrol and widdrol. Chalchat et al. (1998) reported that Cade Oil from *J. phoenicea* was very low in cadinenes, but high in α-cedrene and also lacked cedrol and widdrol.

Wood Products

Cedarwood (primarily *J. virginiana* and *J. v.* var. *silicicola*) has been used for making pencils for many decades in southeastern United States. However, about 1940, producers switched to incense cedar in California, due to the overharvest and closed the pencil factory in Cedar Key, FL.

The use of *Juniperus* wood for furniture is not widespread, presumably because the wood is subject to cracking and splitting. However, it is still used to make 'cedar chests', as a wood lining for closets and for various craft items (Figs. 11. 14, 11.15, 11.16).

Figure 11.14. Thermometer craft made from *J. virginiana* wood.

15 cm (6 inches)

Figure 11.15. Small jewelry box made from *J. virginiana* wood.

Fig. 11.16. Carved head made from *J. barbadensis* var. *lucayana*. From Nassau, Bahama Islands, 1972. Unfortunately, cedarwood is highly prized for making red, aromatic carvings in the Caribbean.

Appendix I: Literature Cited
(you can go to www.juniperus.org and download PDF files of Dr. Adams' papers).

Abbott, H. G. and W. H. Belig. 1961. Juniper seed: A winter food for red squirrels in Massachusetts. J. Mammalogy 42: 240-244.

Achak, N. A. Romane, M. Alifriqui and R. P. Adams. 2008. Effect of leaf drying and geographic sources on the essential oil composition of *Juniperus thurifera* L. var. *africana* Maire from the Tensift - Al Haouz, Marrakech region. J. Ess. Oil. Res. 20: 200-204.

Adams, R. P. 1972. Chemosystematic and numerical studies of natural populations of *Juniperus pinchotii* Sudw. Taxon 21: 407-427.

Adams, R. P. 1975a. Gene flow versus selection pressure and ancestral differentiation in the composition of species: analysis of populational variation in *Juniperus ashei* Buch. using terpenoid data. J. Molecular Evol. 5: 177-185.

Adams, R. P. 1975c. Statistical character weighting and similarity stability. Brittonia 27: 305-316.

Adams, R. P. 1975c. Numerical-chemosystematic studies of infraspecific variation in *Juniperus pinchotii* Sudw. Biochem. Syst. Ecol. 3: 71-74.

Adams, R. P. 1977. Chemosystematics - Analyses of populational differentiation and variability of ancestral and recent populations of *Juniperus ashei*. Ann. Mo. Bot. Gard. 64: 184-209.

Adams, R. P. 1982. A comparison of multivariate methods for the detection of hybridization. Taxon 31:646-661.

Adams, R. P. 1983a. Infraspecific terpenoid variation in *Juniperus scopulorum*: Evidence for Pleistocene refugia and recolonization in western North America. Taxon 32: 30-46.

Adams, R. P. 1983b. The junipers (*Juniperus: Cupressaceae*) of Hispaniola: Comparison with other Caribbean species and among collections from Hispaniola. Moscosa 2:77-89.

Adams, R. P. 1986. Geographic variation in *Juniperus silicicola* and *J. virginiana* of the southeastern United States: Multivariate analysis of morphology and terpenoids. Taxon 35: 61-75.

Adams, R. P. 1987. Investigation of *Juniperus* species of the United States for new sources of cedarwood oil. Econ. Bot. 41: 48-54.

Adams, R. P. 1989a. Biogeography and evolution of the junipers of the West Indies. pp. 167-190 In Biogeography of the West Indies, ed. C. A. Woods. Gainesville, FL, Sand Hill Crane Press.

Adams, R. P. 1989b. Identification of essential oils by ion trap mass spectroscopy. Academic Press, New York.

Adams, R. P. 1991. Cedarwood oil - Analyses and properties. In: Modern methods of plant analysis - Oils and Waxes. H.-F. Linskens and J. F. Jackson, eds., Springer-Verlag Publ., New York.

Adams, R. P. 1993. *Juniperus*, pp. 412-420. In: N. R. Morin (Ed.), Flora of North America. Vol. 2. Pteridophytes and Gymnosperms., Oxford University Press, NY.

Adams, R. P. 1994. Geographic variation and systematics monospermous *Juniperus* (*Cupressaceae*) from the Chihuahuan desert based on RAPDs and terpenes. Biochem. Syst. Ecol. 22: 699-710.

Adams, R. P. 1995. Revisionary study of Caribbean species of *Juniperus* (*Cupressaceae*) Phytologia 78: 134-150.

Adams, R. P. 1998. The leaf essential oils and chemotaxonomy of *Juniperus* sect. *Juniperus*. Biochem. Syst. Ecol. 26: 637-645.

Adams, R. P. 1999. Systematics of multi-seeded eastern hemisphere *Juniperus* based on leaf essential oils and RAPD DNA fingerprinting. Biochem. Syst. Ecol. 27: 709-725.

Adams, R. P. 2000a. Systematics of smooth leaf margin *Juniperus* of the western hemisphere based on leaf essential oils and RAPD DNA fingerprinting. Biochem. Syst. Ecol. 28: 149-162.

Adams, R. P. 2000b. Systematics of *Juniperus* section *Juniperus* based on leaf essential oils and RAPD DNA fingerprinting. Biochem. Syst. Ecol. 28: 515-528.

Adams, R. P. 2000c. Systematics of the one seeded *Juniperus* of the eastern hemisphere based on leaf essential oils and random amplified polymorphic DNAs (RAPDs). Biochem. Syst. Ecol. 28: 529-543.

Adams, R. P. 2000d. The serrate leaf margined *Juniperus* (Section *Sabina*) of the western hemisphere: Systematics and evolution based on leaf essential oils and Random Amplified Polymorphic DNAs RAPDs). Biochem. Syst. Ecol. 28: 975-989.

Adams, R. P. 2001a. Geographic variation in leaf essential oils and RAPDs of *J. polycarpos* K. Koch in Central Asia. Biochem. Syst. Ecol. 29: 609-619.

Adams, R. P. 2001b. Identification of essential oils by gas chromatography/quadrupole mass spectroscopy. Allured Publ., Carol Stream, IL.

Adams, R. P. 2004a. The junipers of the world: The genus *Juniperus*. Trafford Publ., Victoria, BC, Canada.

Adams, R. P. 2004b. *Juniperus deltoides*, a new species and nomenclatural notes on *Juniperus polycarpos* and *J. turcomanica* (*Cupressaceae*). Phytologia 86: 49-53.

Adams, R. P. 2004c. The co-occurrence and systematic significance of pregeijerene B and 8-α-acetoxyelemol in *Juniperus*. BSE 32: 559-563.

Adams, R. P. 2007. Identification of essential oil components by gas chromatography/mass spectrometry, 4th Edition. Allured Publ., Carol Stream, IL.

Adams, R. P. 2007. *Juniperus maritima*, the seaside juniper, a new species from Puget Sound, North America. Phytologia 89(3): 263-283.

Adams, R. P. 2008a. Distribution of *Juniperus ashei* var. ashei and var. *ovata* around New Braunfels, Texas. Phytologia 90(1): 97-102.

Adams, R. P. 2008b. Point of View: *Juniperus bermudiana*: a species in crisis, should it be rescued from introduced Junipers? Phytologia 90(2): 134-136.

Adams, R. P. 2008c. Taxonomy of *Juniperus communis* in North America: Insight from variation in nrDNA SNPs. Phytologia 90(2): 176-192.

Adams, R. P. 2008d. *Juniperus* of Canada and the United States: Taxonomy, key and distribution. Phytologia 90(3): 237-296.

Adams, R. P. 2009a. Variation among the smooth leaf margin *Juniperus* of Mexico: Analyses of nrDNA, 4CL, ABI3 and petN-psbM SNPs. Phytologia 91(3): 571-580.

Adams, R. P. 2009b. Variation among the smooth leaf margin *Juniperus* of Mexico: Analyses of nrDNA, 4CL, ABI3 and petN-psbM SNPs. Phytologia 91(3): 571-580.

Adams, R. P. 2009c. The leaf essential oil of *J. maritima* R. P. Adams compared with *J. horizontalis*, *J. scopulorum* and *J. virginiana* oils. Phytologia 91(1): 31-39.

Adams R. P. 2009d. Variation in *Juniperus durangensis* and related junipers (Cupressaceae): Analysis of nrDNA and petN SNPs. Phytologia 91(2):353-358.

Adams, R. P. 2009e. Analysis and taxonomic utility of the cedarwood oils of the serrate leaf junipers of the western hemisphere. Phytologia 91: 117-139.

Adams, R. P. 2011. The junipers of the world: The genus *Juniperus*. 3rd ed. Trafford Publ., Victoria, BC.

Adams, R. P. 2011a. Infraspecific terpenoid variation in *Juniperus scopulorum*: Pleistocene refugia and Post-Pleistocene recolonization. Phytologia 93(1): 3-12.

Adams, R. P. 2011b. The taxonomic affinity of a juniper population from Colonia Pacheco, Mexico. Phytologia 93(1): 51-64.

Adams, R. P. 2012. *Juniperus occidentalis* forma *corbetii* R. P. Adams, a new shrubby variant: Geographic variation in the leaf essential oils. Phytologia 94: 22-34.

Adams, R. P. 2013a. Hybridization between *Juniperus grandis*, *J. occidentalis* and *J. osteosperma* in northwest Nevada I: Terpenes, Leviathan mine, Nevada. Phytologia 95: 58-69.

Adams, R. P. 2013b. Hybridization between *Juniperus grandis*, *J. occidentalis* and *J. osteosperma* in northwest Nevada II: Terpenes, Buffalo Hills, Northwestern Nevada. Phytologia 95: 107-114.

Adams, R. P. 2013c. *Juniperus communis* var. *kelleyi*, a new variety from North America. Phytologia 95(3): 215-221.

Adams, R. P., A. L. Almirall and L. Hogge. 1987. Chemosystematics of the junipers of Cuba, *J. lucayana* and *J. saxicola* using volatile leaf oils. Flavor and Fragrance Journal 2: 33-36. 253.

Adams, R. P. , J. Altarejos, C. Fernandez, and A. Camacho 1999. The leaf essential oils and taxonomy of *Juniperus oxycedrus* L. subsp. *oxycedrus*, subsp. *badia* (H. Gay) Bedeaus, and subsp. *macrocarpa* (Sibth. & Sm.) Ball. J. Essent. Oil Res. 11: 167-172.

Adams, R. P. and L. Baker. 2007. Pleistocene infraspecific evolution in *Juniperus ashei* Buch. Phytologia 89(1): 8-23.

Adams, R. P., A. F. Barrero and A. Lara. 1996. Comparisons of the leaf essential oils of *Juniperus phoenicea, J. phoenicea* ssp. *eu-mediterranea* Lebr. & Thiv. and *J. phoenicea* var. *turbinata* (Guss.) Parl. J. Ess. Oil Res. 8: 367-371.

Adams, R. P., P. S. Beauchamp, V. Dev and S. M. Dutz. 2007. New Natural Products isolated from one-seeded *Juniperus* of the Southwestern United States: Isolation and occurrence of 2-ethenyl-3-methyl phenol and its derivatives. J. Ess. Oil Res. 19: 146-152.

Adams R. P., A. Boratynski, M. Arista, A. E. Schwarzbach, H. Leschner, Z. Liber, P. Minissale, T. Mataraci and A. Manolis. 2013. Analysis of *Juniperus phoenicea* from throughout its range in the Mediterranean using DNA sequence data from nrDNA and petN-psbM: The case for the recognition of *J. turbinata* Guss. Phytologia 95: 202-209.

Adams, R. P. and R. Chaudhary. 1996. Leaf essential oil of *Juniperus indica* Bertol. from Nepal. J. Essent. Oil Res. 8: 677-680.

Adams, R. P., R. P. Chaudary, R. N. Pandey and L. Singh. 2009. *Juniperus recurva* var. *uncinata*, the hooked branchlet juniper, a new variety from Nepal. Phytologia 91(3): 361-382.

Adams, R. P., G-L. Chu, and S-Z. Zhang. 1994. The volatile leaf oils of *Juniperus prezewalskii* Kom. and forma *pendula* (Cheng & L. K. Fu) R. P. Adams & Chu Ge-Lin from China. J. Essent. Oil Res. 6: 17-20.

Adams, R. P., G-L. Chu and S-Z. Zhang. 1994. Comparison of the volatile leaf oils of *Juniperus chinensis* L., *J. chinensis* var. *Kaizuka* Hor. and cv. *pyramidalis* from China. J. Essent. Oil Res. 6: 149-154.

Adams, R. P. and T. Demeke 1993. Systematic relationships in *Juniperus* based on random amplified polymorphic DNAs (RAPDs). Taxon 42: 553-571.

Adams, R. P., T. Demeke and H. A. Abulfatih. 1993. RAPD DNA fingerprints and terpenoids: Clues to past migrations of *Juniperus* in Arabia and east Africa. Theoret. Appl. Genetics 87: 22-26.

Adams, R. P., A. D. Dembitsky and S. Shatar. 1998. The leaf essential oils and taxonomy of *Juniperus centrasiatica* Kom., *J. jarkendensis* Kom., *J. pseudosabina* Fisch., Mey. & Ave.-Lall., *J. sabina* L. and *J. turkestanica* Kom. from central Asia. J. Essent. Oil Res. 10: 489-496.

Adams, R. P., M. S. G. Elizondo and T. A. Zanoni. 2012. Variation in the leaf essential oils, DNA sequences and morphology in *Juniperus durangensis*. Phytologia 94: 40-52.

Adams, R. P., L. E. Flournoy, R. L. Singh, H. Johnson and H. Mayeux. 1998. Invasion of grasslands by *Juniperus ashei*: A new theory based on random amplified polymorphic DNAs (RAPDs). Biochem. Syst. Ecol. 26: 371-377.

Adams, R. P., M. Granat, L. R. Hogge, E. von Rudloff E. 1979. Identification of lower terpenoids from gas-chromatography-mass spectral data by on-line computer method. J. Chromat. Sci. 17: 75-81.

Adams, R. P. and F. Hojjati. 2012. Taxonomy of *Juniperus* in Iran: Insight from DNA sequencing. Phytologia 94: 219-227.

Adams, R. P., F. Hojjati and A. E. Schwarzbach. 2014. Taxonomy of *Juniperus* in Iran: DNA sequences of nrDNA plus three cpDNAs reveal *Juniperus polycarpos* var. *turcomanica* and *J. seravschanica* in southern Iran. Phytologia 96: 19-25.

Adams, R. P., C. Hsieh, J. Murata, and R. N. Pandey. 2002a. Systematics of *Juniperus* from eastern Asia based on Random Amplified Polymorphic DNAs (RAPDs). Biochem. Syst. Ecol. 30: 231-241.

Adams, R. P., C. Hsieh, J. Murata and A. E. Schwarzbach. 2011. Systematics of *Juniperus chinensis* and *J. tsukusiensis* from Japan and Taiwan: DNA sequencing and terpenoids. Phytologia 93(1): 118-131.

Adams, R. P., G. Hunter, and T. A. Fairhall. 2010. Discovery and SNPs analyses of populations of *Juniperus maritima* on Mt. Olympus, a Pleistocene refugium? Phytologia 92(1): 68-81.

Adams, R. P., F. Jimenez Rodriquez, J. Hoppe and T. A. Zanoni. 2010e. Discovery of a new population of *Juniperus gracilior* var. *urbaniana* from the Dominican Republic: Analyses of leaf terpenoids and SNPs from nrDNA and trnC-trnD. Phytologia 92(3): 413-423.

Adams, R. P. and M. E. Kaufmann. 2010a. Geographic variation in the leaf essential oils of *Juniperus grandis* and comparison with *J. occidentalis* and *J. osteosperma.* Phytologia 92(2): 167-185.

Adams, R. P. and M. E. Kauffmann. 2010b. Geographic variation in nrDNA and cp DNA *Juniperus californica, J. grandis, J. occidentalis* and *J. osteosperma* (Cupressaceae). Phytologia 92(2): 266-276.

Adams, R. P. and J. R. Kistler. 1991. Hybridization between *Juniperus erythrocarpa* Cory and *Juniperus pinchotii* Sudworth in the Chisos Mountains, Texas. Southwest. Natl. 36: 295-301.

Adams, R. P. and S-F Lu. 2008. The botanical source of Chinese cedarwood oil: *Cupressus funebris* or Cupressaceae species? J. Ess. Oil. Res. 20: 235-242.

Adams, R. P., K-S. Mao and J-Q. Liu. 2013. The volatile leaf oil of *Juniperus microsperma* and its taxonomy. Phytologia 95: 87-93.

Adams, R. P., C. A. McDaniel, F. L. Carter FL. 1988. Termiticidal activities in the heartwood, bark/sapwood and leaves of *Juniperus* species from the United States. Biochem. Syst. and Ecol. 16: 453-456.

Adams, R. P., J. A. Morris, R. N. Pandey and A. E. Schwarzbach. 2005. Cryptic speciation between *Juniperus deltoides* and *J. oxycedrus* (Cupressaceae) in the Mediterranean. Biochem. Syst. Ecol. 33: 771-787.

Adams, R. P., J. A. Morris and A. E. Schwarzbach. 2008a. Taxonomic study of *Juniperus excelsa* and *J. polycarpos* using SNPs from nrDNA and cp trnC-trnD plus essential oils and RAPD data. Phytologia 90(2): 203-220.

Adams, R. P., J. A. Morris and A. E. Schwarzbach. 2008b. Taxonomic affinity of Rushforth's Bhutan juniper and *Juniperus indica* using SNPs from nr DNA and cp trnC-trnD, terpenoids and RAPD data. Phytologia 90(2): 233-245.

Adams, R. P., L. E. Mumba, S. A. James, R. N. Pandey, T. Gauquelin and W. Badri. 2003. Geographic variation in the leaf essential oils and DNA fingerprints (RAPDs) of *Juniperus thurifera* L. from Morocco and Europe. J. Essent. Oil Res. 15: 148-154.

Adams, R. P., J. P. Muir, C. A. Taylor and T. R. Whitney. 2012. Differences in chemical composition between browsed and non-browsed *Juniperus ashei* Buch. trees. Biochem. Syst. Ecol. 46: 73-87.

Adams, R. P. and S. Nguyen. 2005. Infra-specific variation in *Juniperus deppeana* and f. *sperryi* in the Davis Mountains, Texas: Variation in leaf essential oils and Random Amplified Polymorphic DNAs (RAPDs). Phytologia 87: 96-108.

Adams, R. P., S. Nguyen and N. Achak. 2006. Geographic variation in *Juniperus phoenicea* (Cupressaceae) from the Canary Islands, Morocco and Spain based on RAPDs analysis. Phytologia 88(3): 270-278.

Adams, R. P., S. Nguyen and J. Liu. 2006. Geographic variation in the leaf essential oils of *Juniperus sabina* and var. *arenaria.* J. Ess. Oil Res. 18: 497-502.

Adams, R. P., S. Nguyen, J. A. Morris and A. E. Schwarzbach. 2006. Re-examination of the taxonomy of the one-seeded, serrate leaf margined *Juniperus* of Southwestern United States and northern Mexico (Cupressaceae). Phytologia 88(3): 299-309.

Adams, R. P. and S. Nguyen. 2007. Post-Pleistocene geographic variation in *Juniperus communis* in North America. Phytologia 89(1): 43-57.

Adams, R. P., R. N. Pandey, J. W. Leverenz, N. Dignard, K. Hoegh and T. Thorfinnsson. 2003. Pan-Arctic variation in *Juniperus communis*: historical biogeography based on DNA fingerprinting. Biochem. Syst. Ecol. 31: 181-192.

Adams, R. P., R. N. Pandey, S. Rezzi and J. Casanova. 2002. Geographic variation in the Random Amplified Polymorphic DNAs (RAPDs) of *Juniperus phoenicea, J. p.* var. *canariensis, J. p.* subsp. *eu-mediterranea*, and *J. p.* var. *turbinata.* Biochem. Syst. Ecol. 30: 223-299.

Adams, R. P. and R. N. Pandey. 2003. Analysis of *Juniperus communis* and its varieties based on DNA fingerprinting. BSE 31: 1271-1278.

Adams, R. P., B. R. Ruiz, M. Nogales and S. S. Fontinha. 2009. Geographic variation and systematics of *Juniperus phoenicea* L. from Madeira and the Canary Islands: Analyses of volatile leaf oils. Phytologia 91: 40-53.

Adams, R. P., B. R. Rumeu, M. Nogales, and S. S. Fontinha. 2010d. Geographic variation and systematics of *Juniperus phoenicea* L. Madeira and the Canary Islands: Analyses of SNPs from nrDNA and petN-psbM DNA. Phytologia 92: 59-67.

Adams, R. P., B. Rumeu, S. S. Fontinha and M. Nogales. 2010a. Geographic variation in the leaf essential oils of *Juniperus cedrus* Webb. & Berthol. from Madeira and the Canary Islands. Phytologia 92(1): 31-43.

Adams, R. P., B. Rumeu, S. S. Fontinha and M. Nogales. 2010b. Speciation of *Juniperus* cedrus and *J. maderensis* in the Canary Islands and Madeira based on terpenoids and nrDNA and cp petN sequences. Phytologia 92(1): 44-55.

Adams, R. P. and A. E. Schwarzbach. 2006a. A new variety of *Juniperus sabina* from Mongolia: *J. sabina* var. *mongolensis*. Phytologia 88 (2): 179-185.

Adams, R. P. and A. E. Schwarzbach. 2006b Infraspecific adjustments in *Juniperus deppeana* (Cupressaceae). Phytologia 88: 227-232.

Adams, R. P. and A. E. Schwarzbach. 2011. DNA barcoding a juniper: the case of the south Texas Duval county juniper and serrate junipers of North America. Phytologia 93: 146-154.

Adams, R. P. and A. E. Schwarzbach. 2012a. Taxonomy of *Juniperus* section *Juniperus*: Sequence analysis of nrDNA and five cpDNA regions. Phytologia 94: 280-297.

Adams, R. P. and A. E. Schwarzbach. 2012b. Taxonomy of the turbinate seed cone taxa of *Juniperus* section *Sabina*: Sequence analysis of nrDNA and four cpDNA regions. Phytologia 94: 388-403.

Adams, R. P. and A. E. Schwarzbach. 2012c. Taxonomy of the multi-seeded, entire leaf taxa of *Juniperus* section *Sabina*: Sequence analysis of nrDNA and four cpDNA regions. Phytologia 94: 350-368.

Adams, R. P. and A. E. Schwarzbach. 2013a. Taxonomy of *Juniperus deppeana* varieties and formas based on nrDNA (ITS), petN-psbM, trnS-trnG, trnD-trnT, trnL-trnF sequences. Phytologia 95: 161-166.

Adams, R. P. and A. E. Schwarzbach. 2013b. Taxonomy of the serrate leaf *Juniperus* of North America: Phylogenetic analyses using nrDNA and four cpDNA regions. p. 172-178

Adams, R. P. and A. E. Schwarzbach. 2013c. Taxonomy of the turbinate shaped seed cone taxa of *Juniperus*, section *Sabina*: Revisited. Phytologia 95: 122-124.

Adams, R. P. and A. E. Schwarzbach. 2013d. The multi-seeded, entire leaf taxa of Juniperus section Sabina: inclusion of *Juniperus microsperma*. Phytologia 95: 118-121.

Adams, R. P. and A. E. Schwarzbach. 2013e. Phylogeny of Juniperus using nrDNA and four cpDNA regions. Phytologia 95: 179-187.

Adams, R. P., A. E. Schwarzbach, J. A. Morris, and S. Gonzalez Elizondo. 2007. *Juniperus compacta*, a new species from Mexico (Cupressaceae). Phytologia 89(3): 361-369.

Adams, R. P., A. E. Schwarzbach and J. A. Morris. 2008. The evolution of Caribbean *Juniperus* (Cupressaceae): Terpenoids, RAPDs, and DNA SNPs data. Phytologia 90(1): 103-119.

Adams, R. P., A. E. Schwarzbach and J. A. Morris. 2010. *Juniperus zanonii*, a new species from Cerro Potosi, Mexico. Phytologia 92(1): 105-117.

Adams, R. P., A. E. Schwarzbach and S. Nguyen. 2006 Re-examination of the taxonomy of *Juniperus flaccida* var. *martinezii*, and var. *poblana* (Cupressaceae). Phytologia 88(3): 233-241.

Adams, R. P., A. E. Schwarzbach, S. Nguyen, and J. A. Morris. 2007. Geographic variation in *Juniperus deppeana*. Phytologia 89(2): 132-150.

Adams, R. P. A. E. Schwarzbach, S. Nguyen, J. A. Morris and J-Q. Liu. 2007. Geographic variation in *Juniperus sabina* L., *J. sabina* var. *arenaria* (E. H .Wilson) Farjon, *J. sabina* var. *davurica* (Pall.) Farjon and *J. sabina* var. *mongolensis* R. P. Adams. Phytologia 89(2): 153-166.

Adams, R. P., A. Schwarzbach and R. N. Pandey. 2003. The Concordance of Terpenoid, ISSR and RAPD markers, and ITS sequence data sets among genotypes: An example from *Juniperus*. BSE 31: 375-387.

Adams, R. P., S. Shatar and A. D. Dembitsky. 1994. Comparison of the volatile leaf oils of *Juniperus davurica* Pall. from Mongolia, with plants cultivated in Kazakhstan, Russia and Scotland. J. Essent. Oil Res. 6: 217-221.

Adams, R. P., M. Socorro G. Elizondo, and M. G. Elizondo and E. Slinkman. 2006. A new subalpine variety of *Juniperus blancoi* Martinez (Cupressaceae) from Durango, Mexico. Biochem. Syst. Ecol. 34: 205-211.

Adams, R. P., A. N. Tashev, K. H. C. Baser and A. K. Christou. 2013. Geographic variation in the volatile leaf oils of *Juniperus excelsa* M.-Bieb. Phytologia 95: 279-285.

Adams, R. P., S. Terzioglu and T. Mataraci. 2010c. Taxonomy of *Juniperus oxycedrus* var. *spilinanus* in Turkey: leaf terpenoids and SNPs from nrDNA and petN. Phytologia 92: 156-166.

Adams, R. P., R. K. Thappa, S. G. Agrawal, B. K. Kapahi and Y. K. Sarin. 1992. The volatile leaf oils of *Juniperus semiglobosa* Regel from India compared with *J. excelsa* M.-Bieb. from Greece. J. Essent. Oil Res. 4: 143-149.

Adams, R. P., R. K. Thappa, S. G. Agrawal, B. K. Kapahi, T. N. Srivastava, and R. P. Chaudhary. 1998. The leaf essential oil of *Juniperus recurva* Buch.-Ham. ex D. Don from India and Nepal compared with *J. recurva* var. *squamata* (D. Don) Parl. J. Essent. Oil Res. 10: 21-24.

Adams, R. P. and D. Thornburg. 2011. Sexual change in *Juniperus arizonica*: facultative monecious? Phytologia 93(1): 43-50.

Adams, R. P. and D. Thornburg. 2010. A review of seed dispersal in *Juniperus*. Phytologia 92(3): 424-434.

Adams, R. P., and B. L. Turner. 1970. Chemosystematic and numerical studies of natural populations of *Juniperus ashei*. Taxon, 19: 728-751.

Adams, R. P. and Y. Turuspekov. 1998. Taxonomic reassessment of some Central Asian and Himalayan scale-leaved taxa of *Juniperus* (*Cupressaceae*) supported by random amplification of polymorphic DNA. Taxon 47: 75-84.

Adams, R. P., E. von Rudloff, T. A. Zanoni, and L. Hogge. 1980. The terpenoids of an ancestral/advanced species pair of *Juniperus*. Biochem. Syst. Ecol. 8: 35-37.

Adams, R. P., E. von Rudloff, L. Hogge and T. A. Zanoni. 1981. The volatile terpenoids of *Juniperus blancoi* and its affinities with other entire leaf margin junipers of North America. J. Nat. Prod. 44: 21-26.

Adams, R. P., E. von Rudloff, L. Hogge. 1983. Chemosystematic studies of the western North American Junipers based on their volatile oils. Biochem. Syst. Ecol. 11: 189-193.

Adams, R. P. and D. Wingate. 2008. Hybridization between *Juniperus bermudiana* and *J. virginiana* in Bermuda. Phytologia 90(2): 123-133.

Adams, R. P. and T. A. Zanoni. 1979. The distribution, synonymy, and taxonomy of three junipers of the southwestern United States and northern Mexico. Southwestern Natl. 24: 323-330.

Adams, R. P. and T. A. Zanoni. 1993. *Juniperus monticola* (Cupressaceae) revisited. Taxon 42: 85-86.

Adams, R. P., T. A. Zanoni, and L. Hogge. 1980. The volatile terpenoids of *J. monticola* f. *monticola*, f. *compacta* and f. *orizabensis*. J. Natural Prods. 43: 417-419.

Adams, R. P., T. A. Zanoni, and L. Hogge. 1984a. The volatile leaf oils of *Juniperus flaccida* var. *flaccida* and var. *poblana*. J. Natural Prods. 47: 1064-1065.

Adams, R. P., T. A. Zanoni, L. Hogge. 1984b. Analyses of the volatile leaf oils of *Juniperus deppeana* and its infraspecific taxa: Chemosystematic implications. Biochem. Syst. Ecol. 12: 23-27.

Adams, R. P., T. A. Zanoni, and L. Hogge. 1985a. The volatile leaf oils of two rare junipers from western Mexico: *Juniperus durangensis* and *J. jaliscana*. J. Natural Prods. 48: 673-675.

Adams, R. P., T. A. Zanoni, and L. Hogge. 1985b. The volatile leaf oils of *Juniperus comitana*, *J. gamboana*, and *J. standleyi*. J. Natural Prods. 48: 678-681.

Adams, R. P., T. A. Zanoni, E. von Rudloff, and L. Hogge. 1981. The south-western USA and northern Mexico one-seeded junipers: their volatile oils and evolution. Biochem. Syst. Ecol. 9: 93-96.

Adams, R. P., S-Z. Zhang, and G-L. Chu. 1993a. The volatile leaf oil of *Juniperus convallium* Rehd. & Wils. J. Essent. Oil Res. 5: 571-573.

Adams, R. P., S-Z. Zhang, and G-L. Chu. 1993b. The volatile leaf oil of *Juniperus saltuaria* Rehd. & Wils. J. Essent. Oil Res. 5: 675-677.

Adams, R. P., S-Z. Zhang, and G-L. Chu. 1996. Essential oil of *Juniperus squamata* D. Don var. *fargesii* Rehd. & Wils. leaves from China. J. Essent. Oil Res. 8: 53-56.

Akimov, A. P., S. I. Kuznetsov, G. I. Nilov, N. N. Chirkina, A. P. Krylova and R. M. Litvinenko. 1976. Essential oils of *Junipers* from ancient Mediterranean region. Composition, properties and prospectives of use. Tr. Nikitsk. Botan. Sad. 69: 79 - 93.

Anderson, E. 1949. Introgressive hybridization. Wiley and Sons, New York, 109 pp.

Anderson, E. 1953. Introgressive hybridization. Biological Review 28: 280-307.

Arista, M., P. L. Ortiz and S. Talavera. 1997. Reproductive isolation of two sympatric subspecies of *Juniperus phoenicea* (Cupressaceae) in southern Spain. Plant Syst. Evol. 208: 225-237.

Arista, M. and P. L. Ortiz. 1995. *Juniperus phoenicea* subsp. *turbinata* en la Sierra de Grazalema. Acta Botanica Malacitana 20: 303-304.

Ashworth, V. E. T. M., B. C. O'Brien and E. A. Friar. 1999. Fingerprinting *Juniperus communis* L. cultivars using RAPD markers. Madrono 46: 131-141.

Ashworth, V. E. T. M., B. C. O'Brien and E. A. Friar. 2001. Survey of *Juniperus communis* L. varieties from the western United States using RAPD fingerprints. Madrono 48: 172-176.

Axlerod, D. I. 1958. Evolution of the Madro-Tertiary geoflora. Bot. Rev. 24: 433-509.

Barbero, M., P. Lebreton and R. Quezel. 1994. Sur les affinites biosystematiques et phytoecologiques de *Juniperus thurifera* L. et du *Juniperus excelsa* Bieb. Ecologia Mediterranea 20: 21-37.

Bauer K, and D. Garbe. 1985. Common fragrance and flavor materials. CVH Verlagsgesellschaft, Weinheim, W. Germany.

Bennett, K.D., S. Boreham, M.J. Sharp and V.R. Switsur. 1992. Holocene history of environment, vegetation and human settlement on Catta Ness, Lunnasting, Shetland. Jour. of Ecol. 80: 241-273.

Bennett, F.D. and I.W. Hughes. 1959. Biological control of insect pests in Bermuda. Bulletin of Entomology Research 50: 423-436.

Boratynski, A., K. Browicz and J. Zielinski. 1992. Chorology of tree and shrubs in Greece. Poznan/ Kornik: Sorus.

Bradshaw, H. D., Jr., K. G. Otto, B. E. Frewen, J. K. McKay, and D. W. Schemske. 1998. Quantitative trail loci affecting differences in floral morphology between two species of monkeyflower (*Mimulus*). 149: 376-382.

Bredenberg, J. B. 1961. The chemistry of the order Cupressales. 36. The ethereal oil of the wood of *Juniperus communis* L. Acta Chem Scand 15: 961-966.

Britton, N.L. 1908. North American trees. Henry Holt & Co., N.Y.

Broecker, W.S. 1965. Isotype geochemistry and the Pleistocene climatic record. pp 737-754 In: The Quaternary of the United States. H.E. Wright and D.G. Frey. (eds.). Princeton Univ. Press, Princeton, N.J.

Browicz, K. 1982. Chorology of trees and shrubs in south-west Asia and adjacent regions. Vol. 1. Polish Scientific Publ., Warsaw.

Brown, T. A. 1986. Gene Cloning, an introduction. Van Norstrand Reinhold, Berkshire, UK.

Bryan, K. and R.C. Cady. 1934. The Pleistocene climate of Bermuda. American Journal of Science 27: 241-264.

Bryant, V. M., Jr. 1969. Late full-glacial and post-glacial pollen analysis of Texas sediments. Unpubl. Ph.D. dissert., Univ. of Texas, Austin, 168 pp.

Carabia, J.P. 1941. Contribuciones al estudio del flora Cubana. Caribbean Forrester 2: 83-92.

Carter, F. L. 1976. Responses of subterranean termites to wood extractives. Material u Organismen Beiheft 33: 357-364.

Chalchat, J. C., R. Ph. Garry and A. Michet. 1988. chemical composition of the hexane fraction of empyreumatic wood oils from *Juniperus oxycedrus* L. and *Juniperus phoenicea* L. Flavour and Frag. J. 3: 19-22.

368 Robert P. Adams

Charles, D. R. and Goodwin, R. H. 1953. An estimate of the minimum number of genes differentiating two species of golden-rod with respect to their morphological characters. Am. Natl. 77: 53-69.

Chavchanidze, V. Y. and L. G. Kharabava. 1989. Juniper essential oils. Subtrop. Kul't. 4: 131-143.

Chavez-Ramirez, F. and R. D. Slack. 1993. Carnivore fruit-use and seed dispersal of two selected plant species of the Edwards Plateau, Texas. Southwestern Natl. 38: 141-145.

Ciesla, W. M., G. Mohammed and A. H. Buzdar. 1998. Juniper dwarf mistletoe, *Arceuthobium oxycedri* (DC.) M. Bieb., in Balochistan Province, Pakistan. Forestry Chronicle 74: 549-553.

Clague, J. J. 1989. Inlandsis de la Cordillère. pp. 41-44. In: Le Quarternaire du Canada et du Groenland, sous la direction de R. J. Fulton. Commission géologique du Canada. Vol. 1, Ottawa.

Clark, A. M., J. D. McChesney, R. P. Adams. 1990. Antimicrobial properties of the heartwood, bark/sapwood and leaves of *Juniperus* species. Phytotherapy Res. 4: 15 - 19.

Cool, L. G. and R. P. Adams. 2003. A pregeijerene isomer from *Juniperus erectopatens* foliage. Phytochem. 63: 105-108.

Comer, C. W., R. P. Adams and D. F. van Haverbeke. 1982. Intra- and interspecific variation of *Juniperus virginiana* and *J. scopulorum* seedlings based on volatile oil composition. Biochem. Syst. Ecol. 10: 297-306.

Correll, D. S., and M. C. Johnston. 1970. Manual of the vascular plants of Texas. Texas Res. Foundation., Renner, Texas, 1881 pp.

Correll, D.S. and H.B. Correll. 1982. Flora of the Bahama Archipelago. Cramer Publ., Hirschberg, W. Germany.

Cory, V. L. 1936. Three junipers of western Texas. Rhodora 38: 182-187.

Cox, W.M. 1959. Bermuda's beginning. C. Tinling & Co., Ltd., Liverpool, England.

Curray, J.R. 1965. Late Quaternary history, continental shelves of the United States. pp 723-736. In: The Quaternary of the United States. H.E. Wright and D.G. Frey, (eds.). Princeton Univ. Press, Princeton, N.J.

Czerepanov, S. K. 1973. Addidamenta et corrigeanta ad Floram URSS. (Russ.) Leningrad, Nauka.

Dembitsky, A. D. 1969. Chemical content of steam oils of some wild growing plants from southern USSR. Uzbek Academy of Science, Tashkent, Uzbekistan.

Demeke, T. and R. P. Adams. 1994. The use of PCR-RAPD analysis in plant taxonomy and evolution. In: PCR Technology: Current Innovations. Griffin, H. G. and Griffin, A. M. (Eds.), CRC Press, Boca Raton, FL.

Dirr, M. 1998. Manual of Woody Plants. Stipes, Champaign, IL.

Douaihy B, Vendramin GG, Boratyn´ski A, Machon N, Bou Dagher-Kharrat M. 2011. High genetic diversity with moderate differentiation in *Juniperus excelsa* from Lebanon and the eastern Mediterranean region. AoB PLANTS 2011 plr003, doi:10.1093/aobpla/plr003

Dyke, A. S., J.-S. Vincent, J. T. Andrews, L. A. Dredge and W. R. Cowan. 1989. L'Inlandsis laurentidien: Introduction à la géologie quarternaire du Bouclier canadien. In: Le Quarternaire du Canada et du Groenland, chap. 3, sous la direction de R. J. Fulton. Commission géologique du Canada. Vol. 1, Ottawa. pp. 188-202.

Eller, F. J. and J. W. King. 2000. Supercritical carbon dioxide extraction of cedarwood oil: a study of extraction parameters and oil characteristics. Phytochemical Analysis 11: 226-231.

Endlicher, S. L. 1847. *Synopsis confierarum*. Scheitlin & Zollikofer, Sankt-Gollen.

Erdmann, H., Norin, T. 1966. The chemistry of the Cupressales. In: L. Zechmeister, (ed) Fortschritte der Chemis organischer Naturstoffe. Springer-Verlag, New York p. 207-287.

Exell, A.W. and H. Wild. 1960. *Flora Zambesiaca*, (Exell AW, Wild H, eds) Vol. 1, Part 1, Crown Agents for Overseas Govt., London.

Farjon, A. 1992. The taxonomy of the multiseed junipers (*Juniperus* sect. *Sabina*) in southwest Asia and east Africa. Edinb. J. Bot. 49: 251 - 283.

Farjon, A. 1998. World checklist and bibliography of conifers. Edition 1, Royal Botanic Gardens, Kew, UK.

Farjon, A. 2001. World checklist and bibliography of conifers. Ed. 2. Royal Botanic Gardens, Kew, UK.

Farjon, A. 2005. A monograph of Cupressaceae and Sciadopitys. Kew Press, London.

Fassett, N. C. 1944. *Juniperus virginiana, J. horizontalis* and *J. scopulorum* - II. Hybrid swarms of *J. virginiana* and *J. scopulorum*. Bull. Torrey Bot. Club 71: 475-483.

Fassett, N. C. 1945a. *Juniperus virginiana, J. horizontalis* and *J. scopulorum* - III. Possible hybridization of *J. horizontalis* and *J. virginiana* . Bull. Torrey Bot. Club 72: 42-46.

Fassett, N. C. 1945b. *Juniperus virginiana, J. horizontalis* and *J. scopulorum* - IV. Hybrid swarms of *J. virginiana* and *J. horizontalis*. Bull. Torrey Bot. Club 72: 379-384.

Fassett, N. C. 1945c. *Juniperus virginiana, J. horizontalis* and *J. scopulorum* - V. Taxonomic treatment. Bull. Torrey Bot. Club 72: 480-482.

Fernald, M.L. 1931. Specific segregations and identities in some floras of eastern North America and the Old World. Rhodora 33: 25-62.

Fernald, M. L. and L. Griscom. 1935. Three days of botanizing in southeastern Virginia. Rhodora 37: 129-134.

Fischer, N., S. Nitz and F. Drawert, F. 1987. Original flavour compounds and the essential oil composition of Marjoram (*Majorana hortensis* Moench). Fl. Frag. J. 2: 55-61.

Flake, R. H. and B. L. Turner. 1973. Volatile oil constituents, especially terpenes, and their utility and potential as taxonomic characters in populational studies. In: Nobel Symposium 25, G. Bendz and J. Santeson (eds.). Chemistry in botanical classification, pp. 123 - 128. NY.

Flake, R.H., E. von Rudloff and B.L. Turner. 1969. Quantitative study of clinal variation in *Juniperus virginiana* using terpenoid data. Proceedings of the National Academy of Science (USA) 62: 487-494.

Flake, R. H., E. von Rudloff and B. L. Turner. 1973. Confirmation of clinal pattern of chemical differentiation in *Juniperus virginiana* from terpenoid data obtained in successive years. pp. 215-228. In: Terpenoids: Structure, Biogenesis, and distribution, Recent Advances in Phytochemistry, Vol. 6, V. C. Runeckles and T. J. Mabry, eds., Academic Press, NY.

Flint, R. F. 1971. Glacial and quaternary geology. John Wiley & Sons, NY. 892 p.

Flint, H. 1998. Landscape plants for eastern North America. John Wiley & Sons, NY.

Flora of Taiwan. 1975. (editors, H-L. Li, T-S. Liu, T-C. Huang, T. Koyama, C. E. De Vol), Epoch Publ. Co., Ltd., Taipei, Taiwan, pp. 538-544.

Flora of Taiwan. 1994. (editors, T.-C. Huang, C.-F. Hsieh, H. Keng, W.-C. Hsieh and J.-L. Tsai), Editorial committee, Taipei, Taiwan, pp. 591-595.

Florin, R. 1933. Die von E.L. Ekman in Westindien gesammelten Koniferen. Arkiv For Botanik 25A(5): 1-22.

Florin, R. 1963. The distribution of conifer and taxad genera in time and space. Acta Hori Bergiani 20: 121-312.

Fournier, G., N. Pages, C. Fournier, and G. Callen. 1991. Comparisons of volatile leaf essential oils of various *Juniperus pfitzeriana*. Pharmaciea Acta Helv. 66: 74-75.

Franco, J. A. 1962. Taxonomy of the common juniper. Bol. Soc. Broteriana 36: 101-120.

Franco, J. A. 1964. Flora Europaea. Tutin TG, Heywood VH, Valentine, VH, Walters SM, Webb DA, eds. Vol. 1, p. 38. Cambridge Univ. Press, Cambridge.

Freeman, D. C., E. D. McArthur, K. T. Harper and A. C. Blauer. 1981. Influence of environment on the floral sex ratio of monecious plants. Evolution 35: 194-197.

Friedlander, M. D., D. Atsmon and E. Galun. 1977. Sexual differentiation in cucumber: abscisic acid and gibberellic acid contents of various sex genotypes. Plant Cell Physiol. 18: 681-692.

Gadek, P. A., D. L. Alpers, M. M. Heslewood and C. J. Quinn. 2000. Relationships within *Cupressaceae sensu lato*: A combined morphological and molecular approach. Am. J. Bot. 87: 1044-1057.

Gelderen, D. M. van, and J. R. P. van Hoey Smith. 1989. Conifers, 2nd ed., Timber Press, Portland, OR.

Gillis, W.T. 1974. Name changes for the seed plants in the Bahama flora. Rhodora 76: 67-138.

Garriga, M., A. P. Thurow, T. Thurow, R. Conner and D. Brandenberger. 1997. Commercial value of Juniper on the Edwards Plateau, Texas. Texas Agric. Expt. Station Juniper Symposium, 1997, TAMU Res. and Ext. Station, San Angelo, TX. see http://texnat.tamu.edu/symposia/juniper/MATT.htm

Gauquelin, T., M. I. Hassani and P. Lebreton. 1988. Le genevier thurifere, *Juniperus thurifera L.* (Cupressacees): analyse biometrique et biochimique; propositions systematiques. Ecologia Mediterranea, 14: 31-42.

Gauquelin, T., V. Bertaudiere, N. Montes, W. Badri and J. F. Asmode. 1999. Endangered stands of thuriferous juniper in the western Mediterranean basin: Ecological status, conservation and management. Biodiversity and Conservation, 8: 1479-1498.

Gaussen, H. 1968. *Les Cupressacees* Fasc. X in *Les Gymnospremes*, Actuelles et Fossiles. Lab. Forest. Univ. Toulouse, France.

Gehring, C. and T. G. Whitham. 1992. Reduced mycorrhizae on *Juniperus monosperma* with mistletoe: the influence of environmental stress and tree gender on a plant parasite and a plant-fungal mutualism. Oecologia 89: 298-303.

Goriaev, M. I. and L. A. Ignatova. 1969. Chemistry of the Junipers. Science, Kazak SSR, 1969: 1-80.

Gower, J. C. 1966. Some distance properties of latent root and vector methods used in multivariate analysis. Biometrika 53: 326-338.

Grace, J. 1983. Ornamental Conifers. David and Charles Co., London.

Greuter, W. C. et al. eds. 1994. International Code of Botanical Nomenclature Tokyo Code, Keoltz Scientific Co., Koenigstien, Germany (Reg. Veg. Vol. 131).

Groves, G.R. 1955. The Bermuda cedar. World Crops 7: 1-5.

Graham, A. 1999. Late cretaceous and cenozoic history of north American vegetation. Oxford University Press, NY.

Guenther, E. 1952. The essential oils, Vol. 6. Reprinted 1976 by Robert E. Krieger Publ. Co., Huntington, New York.

Hall, M. T. 1952a. Variation and hybridization in *Juniperus*. Ann. Missouri Bot. Gardens 39: 1-64.

Hall, M. T. 1952b. A hybrid swarm in *Juniperus*. Evolution 6: 347-366.

Hall, J. B. 1984. *Juniperus excelsa* in Africa: a biogeographical study of an Afromontane tree. J Biogeography 11: 47-61.

Hall, M. T., and C. J. Carr. 1968. Variability in *Juniperus* in the Palo Duro Canyon of Western Texas. Southwestern Nat. 13: 75-98.

Hall, M. T., J. F. McCormick, and G. G. Fogg. 1962. Hybridization between *J. ashei* and *J. pinchotii* in southwest Texas. Butler Univ. Bot. Studies 14: 9-28.

Heslop-Harrision, J. 1972. Sexuality of angiosperms. In: Physiology of development: from seeds to sexuality. F. C. Steward (ed.), Academic Press, NY.

Herwitz, S. R. 1992. Quaternary vegetation change and dune formation on Bermuda: a discussion. Global Ecology and Biogeography Letters 2: 65 - 70.

Hemsley, W.B. 1883. The Bermuda cedar. Gardner's Chronicle 19 (n.s.) May 26, 656-567.

Holthuijzen, A. M. A. and C. S. Adkisson. 1984. Passage rates, energetics, and utilization efficiency of the cedar waxwing. Wilson Bull. 96: 680-684.

Holthuijzen, A. M. A. and T. L. Sharik. 1985. The avian seed dispersal system of eastern red cedar (*Juniperus virginiana*). Can. J. Bot. 63: 1508-1515.

Holthuijzen, A. M. A., T. L. Sharik and J. D. Fraser. 1987. Dispersal of eastern red cedar (*Juniperus virginiana*) into pastures: an overview. Can. J. Bot. 65: 1092-1095.

Horncastle, V. J. and E. C. Hellgren. 2004. Differential consumption of eastern red cedar (*Juniperus virginiana*) by avian and mammalian guilds: Implications for tree invasion. Amer. Midl. Nat. 152: 255-267.

Irving, R. S. and Adams, R. P. 1973. Genetic and biosynthetic relationships of monoterpenes. pp. 187-214. In Terpenoids: Structure, biogenesis, and distribution. Vol. 6. Recent Advances in Phytochemisty Genetics Series. V. C. Runeckles and T. J. Mabry, eds., Academic Press, NY.

Issakainen, J. 1999. Dear Mr. Code .. Taxon 47: 341-348.

Jackson, A. B. 1932. The coffin Juniper. New Flora and Silva 17: 31-34.

Johnsen, T. N. 1962. One-seed juniper invasion of northern Arizona grasslands. Ecol. Monog. 32: 187-206.

Jordano, P. 1991. Gender variation and expression of monoecy in *Juniperus phoenicea* L. (Cupressaceae). Bot. Gaz. 152: 476-485.

Karryev, M. O. 1967. Comparative characteristics of the essential oils from Central Asian species of *Juniperus*. Izv. Acad. Nauk Turkm., SSR, Ser. Biol. Nauk 1967(1): 40-43.

Kerfoot, O. 1966. Distribution of Coniferae: The Cupressaceae in Africa. Nature 5065: 961.

Kerfoot, O. and J. J. Lavranos 1984. Studies in the flora of Arabia X: *J. phoenicea* L. and *J. excelsa* M. Bieb. Notes Royal Botanic Garden Edinb 41: 483-489.

Kerfoot, O 1975. Origin and speciation of the *Cupressaceae* in Sub-Sahara Africa. Boissiera 24: 145-150.

Kim, S.-C. and Rieseberg, L. H. 1999. Genetic architecture of species differences in annual sunflowers: Implications for adaptive trait introgression. Genetics 153: 965-977.

Kistler, J. R. 1976. Detection of hybridization in natural populations. Unpubl. M.S. thesis, Colorado State Univ., Fort Collins, 215 pp.

Kitamura, S. and G. Murata. 1979. Colored illustrations of woody plants of Japan. Vol. II. Hoikusha Publ. Co., Osaka, Japan. pp. 405- 411.

Kitchens, G. D., J. Dorsky and K. Kaiser. 1971. Cedarwood oil and derivatives. Givaudanian 1: 3-9.

Koedam, A. and A. Looman. A. 1980. Effect of pH during distillation on the composition of the volatile oil from *Juniperus sabina*. Planta Med Suppl. 22-28.

Koedam, A., Scheffer, J. J. C. and A. B. Svendsen. 1981. Comparison of isolation procedures for essential oils. IV. Leyland cypress. Perfume and Flavor 5: 56-65.

Komarov, V. L. 1934. Flora of the U. S. S. R., Vol. I, Archegoniatae and embrophyta. Izdatel'stvo Akademii Nauk SSR, Leningrad.

Krussmann, G. 1991. Manual of Cultivated Conifers. Timber Press. Portland, OR.

Kurth, E. F. and H. B. Lackey. 1948. The constituents of Sierra juniper wood (*Juniperus occidentalis* Hooker). J. Am. Chem. Soc. 70: 2206-2209.

Kusumi, J. Y. Tsumura, H. Yoshimaru and H. Tachida. 2000. Phylogenetic relationships in Taxodiaceae and Cupressaceae sensu stricto based on matK gene, chlL gene, trnL-trnF IGS region, and trnL intron sequences. Am. J. Bot. 87: 1480-1488.

Lawrence, B. M. 1985. A review of the world production of essential oils (1984). Perfumer and Flavorist 10: 1-16.

LeBreton, P. and Thivend, S. 1981. Sur une sous-espece du genevrier de phenicie *Juniperus phoenicea* L, definie a partir de criteres biochimiques. Naturalia monspeliensia, ser. Bot. 47: 1-12.

LeBreton, P. 1983. Nouvelles donnees sur la distribution au Portugal et en Espagne de sous-especes du genevrier de phenicie (*Juniperus phoenicea* L.). Agronomia Lusit. 42: 55-62.

Le Duc, A., R. P. Adams and Ming Zhong. 1999. Using random amplification of polymorphic DNA for a taxonomic reevaluation of Pfitzer junipers. HortScience 34(6): 1123-1125.

Levin, D. A. 1968. The structure of a polyspecies hybrid swarm in *Liatris*. Evolution, 22: 352-372.

Lewis, J. 1995. Proposal to conserve the name, *Juniperus x media* Melle (Cupressaceae). Taxon 44: 229-231.

Mao, K-s., G. Hao, J-q. Liu, R. P. Adams and R. I. Milne. 2010. Diversification and biogeography of *Juniperus* (Cupressaceae): variable diversification rates and multiple intercontinental dispersals. New Phytologist 188: 254-272.

Marion, C. and G. Houle. 1996. No differential consequences of reproduction according to sex in *Juniperus communis* var. *depressa* (Cupressaceae). Amer. J. Bot. 83: 480-488.

Mastretta-Yanes, A. 2009. Los poblaciones solitarias: fragmentacion, diferenciacion y rareza de un junipero ribereno y sus implicaciones para la conservation. Thesis. UNAM, Mexico, DF.

McAtee, W. L. 1947. Distribution of seed by birds.

McDonald, J. A. 1990. The alpine-subalpine flora of northeastern Mexico. Sida 14: 21-28.

McDonald, J. A. 1993. Phytogeography and history of the alpine-subalpine flora of northeastern Mexico. pp. 681-703 in: Biological diversity of Mexico: origins and distribution. T. P. Ramamorthy, R. Bye, A. Lot and J. Fa, eds., Oxford Univ. Press, NY.

Narasimhachari, N. and E. von Rudloff. 1961. The chemical composition of the wood and bark extractives of *Juniperus horizontalis* Moench. Can. J. Chem. 939: 2571-2581.

Oda, J., N. Ando, Y. Nakajima and Y. Inouye. 1977. Studies on insecticidal constituents of *Juniperus recurva* Buch. Agric. Biol. Chem. 41: 201-204.

Ownbey, M. 1950. Natural hybridization and amphiploidy in the genus *Tragopogon*. Am. J. Bot. 27: 487-499.

Li, H. L. and Keng, H. 1954. Icones gymnospermum Formosanarum. Taiwania 5: 25-83.

Ohwi, J. 1965. Flora of Japan. (edited by F. G. Meyer and E. H. Walker), Smithsonian Institute, Washington, D.C.

Linnaeus, C. 1753. Species plantarum. Stockholm.

Little, E. L. Jr. 1971. Atlas of United States trees. Vol. 1. Conifers and important hardwoods. USDA Misc. Publ. No. 1146, Washington, D. C.

Lozhkin, A. V., P. M. Anderson, W. R. Eisner, L. G. Ravako, D. M. Hopkins, L. B. Brubaker, P. A. Colinvaux and M. C. Miller. 1993. Late quaternary lacustrine pollen records from Southwestern Beringia. Quaternary Research 39: 314-324.

Martinez, M. (1946) Los *Juniperus* Mexicanos. Anal. Inst. Biol. Univ. Nac. Mexico 17: 3-128.

Martinez, M. 1963. Las pinaceas mexicanas. Tercera edicion. Universidad Nacional Autonoma de Mexico. Mexico City.

Melville, R 1960. Flora of Tropical East Africa. (Turrill, W. B. and Milne-Redhead, E., eds) p.1. Crown Agents for Overseas Govt., London.

McAtee, W. L. 1947. Distribution of seeds by birds. Amer. Midl. Nat. 38: 214-223.

Mehra, P. N. 1988. Indian Conifers, Gnetophytes and Phylogeny of Gymnosperms. New Delhi.

Meyer, E. R. 1973. Late quaternary paleoecology of the Cuatro Cienegas basin, Coahuila, Mexico. Ecology 54: 982-995.

Occhietti, S. 1989. Géologie quaternaire de la sous-région de la vallée du Saint-Laurent et des Appalaches. In: Le Quaternaire du Canada et du Groenland, chap. 4, sous la direction de R. J. Fulton. Commission géologique du Canada. Vol. 1, Ottawa. pp. 374-407.

Owens, M. K. 1996. The role of leaf and canopy-level gas exchange in the replacement of *Quercus virginiana* (Fagaceae) by *Juniperus ashei* (Cupressaceae) in semiarid savannas. Am. J. Bot. 83: 617-623.

Ownbey, M. 1950. Natural hybridization and amphiploidy in the genus *Tragopogon*. Amer. J. Bot. 37: 487-499.

Palma-Otal, M., W. S. Moore, R. P. Adams and G. R. Joswiak. 1983. Morphological, chemical, and biogeographical analyses of a hybrid zone involving *Juniperus virginiana* and *J. horizontalis* in Wisconsin. Can. J. Bot. 61: 2733-2746.

Pettersson, E. and J. Runeberg. 1961. The chemistry of the order Cupressales 34. Heartwood constituents of *Juniperus procera* Hochst. and *Juniperus californica* Carr. Acta Chem. Scand 15: 713-720.

Phillips, F. J. 1910. The dissemination of Junipers by birds. For. Quart. 8: 11-16.

Pilo, C, Runeberg J. 1960. The chemistry of the order Cupressales 25. Heartwood constituents of *Juniperus chinensis* L. Acta Chem. Scand 14: 353-358.

Pisaric, M. F. J., G. M. MacDonald, A. A. Velichko and L. C. Cwynar. 2001. The late glacial and postglacial vegetation history of the northwestern limits of Beringia, based on pollen, stomate and tree stump evidence. Quaternary Science Reviews 20: 235-245.

Pilger, R. 1913. IX. Juniperi species antillanae. Symbolae Antillanae 7: 478-481.

Poddar, S. and R. J. Lederer. 1982. Juniper berries as an exclusive winter forage for Townsend's solitaires. Amer. Midl. Nat. 108:34-40.

Pyle, M. M. and R. P. Adams. 1989. Interim preservation of plant specimens for DNA utilization. Phytochem. Bull. 21: 5-7.

Rafique, M., M. Hanif and F. M. Chaudary. 1993. Evaluation and commercial exploitation of essential oil of Juniper berries of Pakistan. Pak. J. Sci. Ind. Res. 36: 107 - 109.

Rezzi, S., Cavaleiro, C., Bighelli, A., Salgueiro, L., da Cunha, A. P. and Casanova, J. 2001. Intraspecific chemical variability of the leaf essential oil of *Juniperus phoenicea* var. *turbinata* from Corsica. Biochem. Syst. Ecol. 29: 1175-1184.

Rissing, S. W. Dietary similarity and foraging range of two seed-harvester ants during resource fluctuations. Oecologia 75: 362-366.

Romo, A., O. Hidalgo, A. Boratynski, K. Sobierajska, A. K. Jasinska, J. Valles and T. Garnatje. 2012. Genome size and ploidy levels in highly fragmented habitats: the case of western Mediterranean *Juniperus* (Cupressaceae) with special emphasis on *J. thurifera* L. Tree Genetics and Genomes DOI 10.1007/s11295-012-0581-9.

Rosen, D. E. 1978. Vicariant patterns of historical explanation in biogeography. Syst. Zool. 27: 159-188.

Rumeu, B., D. P. Padilla and M. Nogales. The key role of a Ring Ouzel Turdus torquatus wintering population in seed dispersal of the endangered endemic *Juniperus cedrus* in an insular environment. Acta Ornithologia 44: 199-204.

Runeberg, J. 1960a. The chemistry of the order Cupressales 30. Heartwood constituents of *Juniperus cedrus* Webb. & Benth. Acta Chem. Scand 14: 1991-1994.

Runeberg, J. 1960b. The chemistry of the order Cupressales 27. Heartwood constituents of *Juniperus utahensis* Lemm. Acta Chem. Scand 14: 797-804.

Runeberg, J. 1960c. The chemistry of the order Cupressales 31. Heartwood constituents of *Juniperus phoenicea* L. Acta Chem. Scand 14: 1995-1998.

Runeberg, J. 1960d. The chemistry of the order Cupressales 29. Heartwood constituents of *Juniperus thurifera* L. Acta Chem. Scand 14: 1985-1990.

Runeberg, J. 1960e. The chemistry of the order Cupressales 28. Heartwood constituents of *Juniperus virginiana* L. Acta Chem. Scand 14: 1288-1294.

Runeberg, J. 1961. The chemistry of the order Cupressales 35. Heartwood constituents of *Juniperus foetidissima* Willd. Acta Chem. Scand 15: 721-726.

Rushforth, K. 1987. Conifers. Christopher Helm Co., London.

Sadri, H. A. and M. Assadi. 1994. Preliminary studies on monoterpene composition of *Juniperus polycarpos*. Iran J. Bot. 6: 219-225.

San Feliciano, A. M. Medarde, J. L. Lopez, J. M. Miguel del Corral, P. Puebla and A. F. Barrero. 1988. Terpenoids from leaves of *Juniperus thurifera*. Phytochemistry 27: 241-248.

Salomonson, M. G. 1978. Adaptations for animal dispersal of one-seed juniper seeds. Oecologia 32: 333-339.

Santos, T., J. L. Telleria and E. Virgos. 1999. Dispersal of Spanish juniper *Juniperus thurifera* by birds and mammals in a fragmented landscape. Ecography, 22: 193-204.

Savoskul, O. S. 1999. Holocene glacier advances in the headwaters of Sredniaya Avacha, Kamchatka, Russia. Quaternary Research 52: 14-26.

Schmaus, G. and K-H. Kubeczka. 1985. The influence of isolation conditions on the composition of essential oils containing linalool and linalyl acetate. In: Svendsen A. B., Scheffer J. J. C. (eds). Essential oils and aromatic plants. Nijfoff/Junk Publ. Dordrecht.

Schmidt, P. 1983. Invalid name for pfitzeriana and plumosa group of *Juniperus chinensis*. Folia Dendrologica 10: 291-296.

Schultz, R. H. 1971. A taxonomic analysis of a tri-parental hybrid swarm in *Juniperus* L. Unpubl. Ph.D. dissert., Univ. of Nebraska, Lincoln, 90 pp.

Schupp, E. W., J. M. Gomez, J. E. Jimenez and M. Fuentes. 1997. Dispersal of *Juniperus occidentalis* (western juniper) seeds by frugivorous mammals on Juniper Mountain, southeaster Oregon. Great Basin Naturalist 57: 74-78.

Silba, J. 1986. Encyclopaedia coniferae. Phytologia Memoirs 8: 1-217.

Silba, J. 1990. A supplement to the international census of the coniferae, II. Phytologia 68: 7-78.

Smith, H. N. and C. A. Rechenthin. 1964. Grassland restoration - The Texas brush problem. USDA-SCS bulletin 4-19114 (5-64).

Somerville, C. and S. Somerville. 1999. Plant Functional Genomics. Science 285: 380-383.

Steele, C. L., J. Crock, J. Bohlmann, and R. Croteau, R. 1998. Sesquiterpene synthases from Gand fir (*Abies grandis*). J. Biol. Chem. 273: 2078-2089.

Stefanovic, S., M. Jager, J. Deutsch, J. Broutin and M. Masselot. 1998. Phylogenetic relationships of conifers inferred from partial 28S rDNA gene sequences. Am. J. Bot. 85: 688-697.

Stuessy, T. F. 1990. Plant Taxonomy: The systematic evaluation of comparative data. Columbia Univ. Press, N. Y.

Teresa, J. de P., A. F. Barrero, A. S. Feliciano and M. C. Caballero. 1980. Componentes de las arcestidas de *Juniperus thurifera* Linnaeus aceite esencial. Rivisita Italiana E.P.P.O.S. 62: 116-120.

Terrab, A., P. Schonswetter, S. Talavera, E. Vela and T. Stuessy. 2008. Range-wide phytogeography of *Juniperus thurifera* L., a presumptive keystone species of western Mediterranean vegetation during cold stages of the Pleistocene. Molec. Phylogen. Evol. 48: 94-102.

Terry, R. G., R. S. Nowak, and R. J. Tausch. 2000. Genetic variation in chloroplast and nuclear ribosomal DNA in Utah Juniper (*Juniperus osteosperma*, Cupressaceae): Evidence for interspecific gene flow. Am. J. Bot. 87: 250-258.

Terry, R. G. 2010. Re-evaluation morphological and chloroplast DNA variation in *Juniperus osteosperma* (Torr.) Little and *Juniperus occidentalis* Hook and their putative hybrids. Biochem. Syst. Ecol. 38: 349-360.

Tinge, S. V. and J. P. del Tufo. 1993. Genetic analysis with amplified polymorphic DNA markers. Plant Physiol. 101: 349-352.

Tunalier, Z., N. Kirimer and K. H. C. Baser. 2004. A potential new source of cedarwood oil: *Juniperus foetidissima* Willd. J. Essent. Oil Res. 16: 233-235.

Ucar, G. and M. Balaban. 2002. The composition of volatile extractives from the wood of *Juniperus excelsa, Juniperus foetidissima* and *Juniperus oxycedrus*. Holz als Roh-und Werkstoff 60: 356-362.

van Haverbeke, D. F. 1968. A taxonomic analysis of *Juniperus* in the central and northern Great Plains. Proc. Sixth Central States Forest Tree Improvement Conf., 6: 48-52.

van Haverbeke, D. E. 1968. A population analysis of *Juniperus* in the Missouri River Basin. New Series No. 38, Univ. of Nebraska Studies, Dec., Lincoln, NB.

van Melle, P. J. 1947. Review of *Juniperus chinensis* et al. New York Botanical Garden 108.

van Riel, P., K. Jordaens, N. van Houtte, A. M. Frias Martins, R. Verhagen and T. Backeljau. 2005. Molecular systematics of the endemic *Leptaxini* (Gastropoda: Pulmonata) on the Azores Islands. Molec. Phylogen. Evol. 37: 132-143.

Vasek, F. C. 1966. The distribution and taxonomy of three western junipers. Brittonia 18: 350-372.

Vasek, F. C. and R. W. Scora. 1967. Analysis of the oils of western north American junipers by gas-liquid chromatography. Amer. J. Bot. 54: 781-789.

Vasiliauskas, S. A. and L. W. Aarssen. 1992. Sex ratio and neighbor effects in monospecific stands of *Juniperus virginiana*. Ecology 73: 622-632.

von Rudloff, E. 1967. Chemosystematic studies in the genus *Picea* (*Pinaceae*). Introduction. Canadian J. Bot. 45: 891-901.

von Rudloff, E. 1975a. Chemosystematic studies of the volatile oils of *Juniperus horizontalis, J. scopulorum* and *J. virginiana*. Phytochemistry 14: 1319-1329.

von Rudloff, E. 1975b. Volatile leaf oil analysis in chemosystematic studies of North American conifers. Biochem. Syst. Ecol. 2: 131-167.

von Rudloff, E., R. Irving and B. L. Turner. 1967. Reevaluation of allopatric introgression between *Juniperus ashei* and *J. virginiana* using gas chromatography. Am. J. Bot. 5: 600 (Abst.).

Walker, G. T. 1968. Cedarwood oil. Perfume and Essential Oil Res. 59: 346-350.

Welch, H. J. 1966. Dwarf Conifers. Charles T. Brandford, Co., MA.

Wells, P. V. 1966. Late Pleistocene vegetation and degree of pluvial climatic change in the Chihuahuan Desert. Science 153: 970-975.

Wenninger, J. A., R. L. Yates and M. Dolinsky. 1967. Sesquiterpene hydrocarbons of commercial copaiba balsam and American cedarwood oils. J. Am .Oil Chem. Soc. 50: 1304-1313.

Weyerstahl, P., Marschall-Weyerstahl, H., Manteuffel, E., Kaul, V. K. 1988. Constituents of *Juniperus recurva* var. *squamata* oil. Plant Medica 54: 259-261.

Willard, J. R. and H. H. Crowell. 1965. Biological activities of the harvester ant, *Pogonomyrmex owyheei*, in central Oregon. J. Econ. Entomology 58: 484-489.

Williams, G. K. , A. R. Kubelik, K. L. Livak, J. A. Rafalski and S. V. Tingey. 1990. DNA polymorphisms amplified by arbitrary primers are useful as genetic markers. Nucleic Acids Res. 18: 6531-6535.

Williams, W. T., M. B. Dale, and G. H. Lance. 1971. Two outstanding ordination problems. Australian. J. Bot. 19: 251-258.

Wilson, J. T. 1972. Continents Adrift: Reading from Scientific American. W. H. Freeman, San Francisco.

Wood, M.K., R. Scanlon and D. S. Cram. 2012. Occurrence of sprout and seedlings of pinyon pines, alligator junipers, and gray oaks following harvest of fuelwood and prescribed burning. Southwestern Naturalist 57: 51-57.

Yaltırık, F., G. Eliçin and S. Terzioğlu. 2007. *Juniperus oxycedrus* L. subsp. *oxycedrus* var. *spilinanus* Yalt., Eliçin & Terzioğlu: A new variety from Turkey. Turk. J. Bot. 31: 37-40.

Yatagai, M. and T. Takahashi 1988. Evaluation of essential oil-producing conifers. Baiomasu Henkan Keikaku Kenkyu Hokoku 9: 3-30.

Yu, Z. 1997. Late quaternary paleoecology of *Thuja* and *Juniperus* (Cupressaceae) at Crawford Lake, Ontario, Canada; pollen, stomata and macrofossils. Review of Paleobotany and Palynology 96: 241-254.

Yu, Y. F. and L. K. Fu. 1997. Notes on Gymnosperms II. New taxa and combinations in *Juniperus* (Cupressaceae) and *Ephedra* (Ephedraceae) from China. Novon 7: 443-444.

Zanoni, T.A. 1978. The American junipers of the section *Sabina* (*Juniperus*, Cupressaceae)-a century later. Phytologia 38: 433-454.

Zanoni, T. A. and R. P. Adams. 1975. The genus *Juniperus* (Cupressaceae) in Mexico and Guatemala: numerical and morphological analysis. Bull. Bot. Soc. Mexico 35: 69-92.

Zanoni, T. A. and R. P. Adams. 1976. The genus *Juniperus* (Cupressaceae) in Mexico and Guatemala: numerical and chemosystematic analysis. Biochem. Syst. Ecol. 4: 147-158.

Zanoni, T. A. and R.P. Adams. 1979. The genus *Juniperus* (Cupressaceae) in Mexico and Guatemala. Synonymy, key and distributions of the taxa. Boletin de la Sociedad Botanica de Mexico 38: 83-121.

Zhu, B. C. R., G. Henderson, F. Cheng, H. Fei and R. A. Lane. 2001. Evaluation of vetiver oil and seven insect-active essential oils against the Formosan subterranean termite. J. Chem. Ecol. 27: 1617-1625.

Zhu, B. C-R. , G. Henderson, R. P. Adams, Li-Xin Mao, Ying Yu and R. A. Laine. 2003. Repellency of vetiver oils from different biogenetic and geographic origins against Formosan subterranean termites (Isoptera: Phinotermitidae). Sociobiology 42: 623-638.

Appendix II: Leaf essential oils by species in sections

Table A.1. Essential oils of *Juniperus* sect *Juniperus*, plus the monotypic section *Caryocedrus* (*Juniperus drupacea*). Comparisons of the per cent total oil for leaf essential oils for *J. brevifolia* (BR), *J. cedrus* (CE), *J. maderensis* (MD), *J. communis* var. *communis* (CC), *J. communis* var. *saxatilis* (CS), *J. communis* var. *kelleyi* (CK), *J. jackii* (JK), *J. oxycedrus* (OX), *J. macrocarpa* (MC), *J. navicularis* (NA), *J. formosana* (FM), *J. mairei* (MA), *J. rigida* (RG), *J. rigida* var. *conferta* (CF), *J. taxifolia* (TX), *J. taxifolia* var. *lutchuensis* (TL), *J. deltoides* (DT) and *J. drupacea* (sect. Caryocedrus, DP, from Adams, 1997. J. Essent. Oil Res. 9: 541-544). Components that tend to separate the species are highlighted in boldface.

KI	Compound	BR	MD	CE	CC	CS	CK	JK	OX	MC	NA	FM	MA	RG	CF	TX	TL	DT	DP
854	(E)-2-hexenal	0.1	0.4	0.2	0.7	1.2	0.3	0.2	0.1	-	0.2	1.2	0.2	2.4	0.3	-	-	0.6	0.1
926	tricyclene	t	0.1	0.3	0.3	t	t	t	0.1	0.1	0.1	t	0.1	0.1	0.1	0.8	0.2	0.1	t
931	α-thujene	0.1	0.1	0.1	0.1	4.1	-	t	0.1	2.8	2.1	t	-	t	-	t	-	t	t
939	**α-pinene**	**4.8**	**24.9**	**70.7**	**56.8**	**14.1**	**56.5**	**16.1**	**41.3**	**22.6**	**22.9**	**44.0**	**47.7**	**39.7**	**53.2**	**47.5**	**46.6**	**24.8**	**4.2**
953	α-fenchene	t	0.3	0.4	0.4	0.1	0.5	0.3	0.1	0.1	-	t	-	0.6	0.2	-	-	0.3	0.9
953	camphene	0.1	0.1	0.6	0.6	0.2	0.5	0.3	0.2	0.1	0.2	0.4	0.6	1.0	0.8	1.6	0.8	0.4	-
957	thuja-2,4(10)-diene	-	-	-	-	-	-	-	0.1	0.1	-	-	0.1	t	-	t	t	0.2	-
967	verbenene	-	-	-	-	0.1	-	0.3	-	-	-	-	1.5	0.1	-	-	-	-	-
976	**sabinene**	**1.5**	**0.1**	**1.0**	**0.7**	**32.8**	**0.3**	**0.1**	**0.6**	**26.5**	**8.2**	**0.1**	**0.2**	**t**	**0.4**	**0.2**	**0.5**	**0.2**	**0.4**
978	1-octen-3-ol	0.8	0.3	1.0	-	-	-	0.1	-	-	-	-	-	-	1.0	-	-	t	-
980	**β-pinene**	**0.2**	**0.7**	**4.1**	**4.4**	**1.9**	**5.4**	**1.9**	**1.7**	**0.8**	**3.5**	**4.2**	**2.9**	**1.9**	**8.0**	**8.8**	**8.3**	**2.7**	**0.4**
991	**myrcene**	**4.0**	**1.2**	**6.3**	**5.2**	**5.0**	**4.5**	**3.2**	**4.7**	**2.9**	**8.6**	**6.4**	**7.2**	**11.2**	**10.4**	**11.2**	**10.0**	**3.0**	**2.8**
1001	δ-2-carene	-	-	t	0.2	0.4	0.1	0.2	0.3	0.4	1.2	2.6	0.8	0.8	t	t	-	0.4	t
1005	**α-phellandrene**	**-**	**-**	**0.5**	**2.1**	**0.5**	**0.1**	**2.2**	**8.2**	**0.4**	**8.0**	**0.7**	**1.2**	**1.0**	**0.3**	**t**	**t**	**1.1**	**t**
1011	**δ-3-carene**	**-**	**5.6**	**-**	**4.7**	**0.5**	**11.5**	**17.9**	**t**	**0.5**	**-**	**-**	**-**	**-**	**3.2**	**3.8**	**t**	**5.2**	**17.5**
1018	α-terpinene	0.2	-	t	-	1.9	-	-	0.2	1.8	0.9	-	t	t	-	t	0.1	0.1	-
1026	**p-cymene**	**0.1**	**0.1**	**0.2**	**0.3**	**0.3**	**t**	**1.1**	**6.2**	**3.4**	**2.6**	**0.3**	**0.9**	**0.6**	**0.2**	**t**	**t**	**1.4**	**0.4**
1031	**limonene**	**54.5**	**10.9**	**4.5**	**6.9**	**6.7**	**2.1**	**6.6**	**4.5**	**2.5**	**14.3**	**3.5**	**4.0**	**4.2**	**1.9**	**3.2**	**17.5**	**20.0**	**47.1**
1031	β-phellandrene	-	7.3	4.6	6.9	0.6	3.1	13.4	5.0	2.5	3.5	5.3	1.4	2.1	4.0	3.1	0.1	5.9	t
1040	(Z)-β-ocimene	t	-	-	0.2	-	-	-	-	-	-	-	-	t	-	t	t	-	-
1050	**(E)-β-ocimene**	**2.7**	**0.1**	**-**	**-**	**0.1**	**-**	**0.3**	**-**	**-**	**0.4**	**0.2**	**-**	**t**	**-**	**t**	**t**	**0.2**	**-**
1057	amyl isobutyrate	-	-	-	0.2	-	-	-	-	-	0.1	-	0.2	0.6	0.2	-	0.3	-	-
1062	γ-terpinene	0.3	0.1	0.1	t	3.4	0.1	0.1	0.4	3.0	1.6	t	0.1	0.2	0.1	t	0.1	0.2	t

KI	Compound	BR	MD	CE	CC	CS	CK	JK	OX	MC	NA	FM	FM	RG	CF	TX	TL	DT	DP
1064	3-me-3-butenol butanoate	-	-	-	-	-	-	-	-	-	-	0.6	-	-	-	-	-	-	-
1068	cis-sabinene hydrate	t	-	-	-	-	-	-	-	1.7	0.1	-	-	-	-	-	-	-	t
1087	fenchone	-	-	-	-	1.8	-	-	-	-	-	-	-	1.6	-	-	0.1	-	t
1088	terpinolene	3.0	0.7	0.6	1.1	3.0	1.8	3.2	2.9	1.2	2.9	0.6	1.0	0.8	0.5	0.8	0.8	1.6	0.8
1095	α-pinene oxide	-	-	-	-	-	-	-	-	-	-	-	1.4	-	-	-	-	-	-
1097	ipsenol	-	-	-	-	-	-	-	-	-	-	-	0.7	-	-	-	-	-	t
1097	trans-sabinene hydrate	-	-	-	-	1.3	-	-	-	1.7	-	-	-	-	0.7	-	-	-	t
1098	linalool	0.5	-	-	-	t	0.2	-	t	-	0.2	-	-	0.6	0.8	0.8	0.2	0.4	0.1
1102	n-nonanal	t	-	-	-	-	-	-	-	-	-	-	-	-	-	-	-	0.4	-
1110	1-octen-3-yl acetate	-	0.5	0.2	-	-	-	-	-	-	-	-	-	-	-	-	-	-	-
1112	**endo-fenchol**	-	-	-	-	-	-	-	-	-	-	-	0.5	0.8	0.5	-	-	-	-
11124	trans-thujone	-	-	-	-	0.6	-	-	-	-	-	-	-	-	t	t	-	t	t
1121	cis-p-menth-2-en-1-ol	-	-	-	-	-	t	0.2	0.1	0.7	0.1	0.1	-	-	-	-	t	t	0.1
1123	trans-p-mentha-2,8-dien-1-ol	-	-	-	-	-	-	-	-	-	-	-	0.2	-	-	-	-	0.3	-
1125	α-campholenal	-	-	-	-	-	-	0.2	0.5	0.7	-	-	0.3	0.2	0.2	0.2	0.3	0.8	t
1132	cis-limonene oxide	-	-	-	-	-	-	0.1	-	-	-	-	-	-	-	-	-	-	0.4
1134	trans-limonene oxide	-	-	-	-	-	-	-	-	-	-	-	-	-	-	-	-	0.4	0.3
1139	**trans-pinocarveol**	-	-	-	-	-	0.1	-	0.5	1.2	-	-	0.3	0.2	t	t	0.3	0.6	-
1140	trans-p-menth-2-en-1-ol	-	-	-	-	0.3	-	-	0.5	-	-	-	t	-	-	-	-	-	-
1140	cis-verbenol	-	-	-	-	-	-	-	t	0.2	-	-	0.1	t	-	-	-	0.2	t
1141	camphor	-	t	-	-	-	0.2	-	-	-	-	-	-	0.6	0.4	0.4	-	-	t
1143	trans-verbenol	-	-	-	-	t	-	0.2	0.5	1.3	0.1	0.1	0.5	t	t	0.4	t	1.1	t
1147	**camphene hydrate -**	-	-	-	-	-	-	-	-	-	0.3	-	0.2	0.2	t	0.1	-	-	-
1156	sabina ketone	-	-	-	-	-	-	-	-	0.5	-	-	-	-	-	-	-	-	-
1160	trans-pinocamphone -	-	-	-	-	-	-	-	-	0.3	t	-	-	t	t	t	t	-	-
1162	pinocarvone -	-	-	-	-	-	-	-	-	0.1	-	-	t	t	t	t	t	-	-
1165	borneol	0.1	-	0.1	0.2	0.1	0.1	-	-	-	-	-	0.4	0.5	0.4	0.5	0.1	0.3	-
1166	δ-terpineol	-	-	-	-	-	-	-	-	0.3	-	-	-	0.4	-	-	-	-	-
1170	p-mentha-1,5-dien-8-ol	-	-	-	-	-	-	-	-	t	-	-	-	-	-	-	-	0.3	0.6
1177	**terpinen-4-ol**	0.5	0.1	t	0.2	7.3	0.3	0.7	1.5	7.3	2.8	t	0.5	0.6	0.2	0.2	0.3	t	0.6

KI	Compound	BR	MD	CE	CC	CS	CK	JK	OX	MC	NA	FM	FM	RG	CF	TX	TL	DT	DP
1179	naphthalene	0.2	0.1	0.1	t	0.3	-	-	0.4	-	0.1	0.6	-	t	-	1.8	0.2	0.2	-
1183	p-cymen-8-ol	t	t	0.3	t	t	-	0.3	0.4	0.7	t	0.1	t	t	t	-	-	0.3	0.2
1189	**α-terpineol**	**0.2**	**0.1**	**0.3**	**0.2**	**0.4**	**0.5**	**0.3**	**5.0**	**1.4**	**1.1**	**0.1**	**0.6**	**1.1**	**0.5**	**0.4**	**0.7**	**0.3**	**0.8**
1193	myrtenal	-	-	-	-	-	0.4	-	-	0.4	-	-	0.2	-	-	-	-	0.2	-
1194	myrtenol	-	-	-	0.3	-	-	0.3	t	0.4	-	-	0.1	t	t	t	t	0.3	0.2
1204	verbenone	-	-	-	-	-	t	0.3	t	0.4	-	-	0.1	t	t	t	0.1	0.8	0.2
1217	trans-carveol	-	-	-	-	-	t	0.4	0.2	0.3	-	-	0.1	t	t	t	0.1	-	-
1220	**endo-fenchyl acetate**	-	-	-	-	-	0.2	-	-	-	-	-	0.2	0.6	0.6	-	-	-	-
1228	citronellol	-	-	-	t	0.2	-	-	-	-	0.3	0.3	0.4	0.2	0.2	-	0.4	-	-
1229	cis-carveol	-	-	-	-	-	0.1	0.2	-	0.2	-	-	-	-	-	-	-	0.2	0.2
1235	thymol, methyl ether	-	-	-	-	0.1	-	-	-	t	-	-	-	-	-	-	-	-	-
1235	trans-chrysanthenyl acetate	-	-	-	-	-	-	-	-	-	-	-	-	-	-	-	-	-	-
1239	carvone	-	-	-	-	0.1	0.1	-	t	t	-	-	0.3	-	-	-	0.1	0.7	-
1244	methyl carvacrol	0.2	-	-	-	-	-	-	-	-	-	-	-	-	-	-	-	-	-
1249	piperitone	-	-	-	-	-	0.2	t	t	0.2	-	-	0.6	t	-	-	-	-	0.1
1255	geraniol	-	-	-	-	0.3	0.3	-	-	-	-	-	0.2	-	0.1	-	-	-	-
1257	methyl citronellate	-	-	-	-	-	-	-	-	-	-	-	-	-	-	-	-	-	-
1262	3-methyl-3-butenol hexanoate	-	-	-	-	-	-	-	-	-	0.1	-	1.2	0.8	-	-	-	-	-
1264	(2E)-decenal	-	-	-	-	0.7	0.5	-	-	-	-	-	-	-	-	-	-	0.2	-
1285	bornyl acetate	0.4	0.4	0.7	0.9	0.2	0.7	-	-	t	0.7	-	1.6	1.3	2.0	9.4	1.1	0.7	-
1291	2-undecanone	-	-	-	-	-	-	-	-	-	-	-	-	4.8	t	-	-	-	-
1292	prenyl hexanoate	-	-	-	-	-	-	-	-	-	-	-	1.4	-	-	-	-	-	-
1298	carvacrol	0.1	0.3	-	-	0.3	0.3	1.0	-	-	-	-	-	-	-	-	-	-	-
1302	α-terpinyl formate	-	t	-	-	0.6	-	0.9	t	-	-	-	-	-	-	-	-	-	-
1350	α-terpinyl acetate	-	1.4	1.0	-	0.5	0.6	-	-	-	-	-	0.2	-	0.6	0.2	0.6	-	0.6
1351	α-cubebene	0.1	-	-	-	0.1	0.1	t	t	-	-	-	-	-	-	-	-	-	-
1354	citronellyl acetate	-	-	-	0.1	0.1	0.1	-	-	-	0.3	0.3	-	0.1	0.1	t	0.3	-	-
1376	α-copaene	-	-	-	-	-	-	-	-	t	t	0.3	0.2	-	-	t	-	0.2	0.6
1379	geranyl acetate	-	-	-	-	0.3	-	-	-	-	-	-	0.2	-	-	-	-	t	-

KI	Compound	BR	MD	CE	CC	CS	CK	JK	OX	MC	NA	FM	FM	RG	CF	TX	TL	DT	DP
1384	β-bourbenene	-	-	-	-	-	-	-	0.2	t	-	-	-	0.2	-	-	-	0.2	0.1
1390	β-cubebene	0.1	0.1	-	-	-	-	-	-	-	-	-	-	-	-	-	0.1	-	t
1391	β-elemene	-	-	-	0.2	t	0.1	0.3	-	-	-	0.2	-	-	-	t	t	-	-
1400	β-longipinene	-	0.5	-	-	-	-	-	-	-	-	-	-	-	-	-	-	-	-
1400	sibirene	-	-	-	-	-	-	-	-	-	-	0.7	-	-	-	-	-	-	-
1418	(E)-caryophyllene	0.2	4.3	0.4	0.7	t	0.1	0.4	0.2	0.1	0.7	0.9	1.0	1.6	0.5	1.0	0.4	0.9	0.3
1431	cis-thujopsene	-	1.0	-	-	-	-	-	-	-	-	-	1.0	-	-	-	-	t	0.1
1454	α-humulene	0.3	1.0	t	0.5	t	0.1	0.5	t	t	0.3	0.4	0.6	1.3	0.4	0.6	0.4	0.6	0.4
1458	(E)-β-farnesene	-	-	-	-	t	-	-	t	-	t	-	0.2	0.4	t	t	t	-	t
1471	sesquiterpene	-	-	-	-	-	-	-	-	-	-	-	-	1.3	-	-	-	-	-
1478	γ-muurolene	-	-	-	-	-	-	-	-	-	0.3	-	-	-	-	-	-	-	-
1480	germacrene D	0.1	0.7	-	0.7	0.4	0.6	4.1	1.0	0.5	0.2	1.5	2.3	2.3	t	t	0.4	0.7	0.3
1486	**ar-curcumene**	-	-	-	-	-	-	-	-	-	-	0.6	t	-	-	-	-	**0.2**	**2.8**
1493	epi-cubebol	-	0.7	-	-	-	-	0.3	-	-	-	-	-	-	-	-	-	-	0.4
1495	β-alaskene	-	-	-	-	0.2	-	-	-	-	-	-	-	-	-	-	-	-	-
1495	(E)-methyl isoeugenol	-	-	-	-	-	-	-	-	-	-	0.2	0.2	0.7	-	-	-	-	-
1499	**α -muurolene**	t	**0.1**	-	**0.2**	**0.2**	**0.2**	**0.6**	-	**t**	**0.3**	**0.6**	**0.2**	**t**	t	**t**	**t**	**0.2**	**t**
1508	α-farnesene	-	-	-	-	-	-	-	-	-	-	-	-	0.7	0.7	-	-	-	-
1513	α-alaskene	-	-	-	-	-	-	-	-	-	-	-	-	-	-	-	-	-	0.3
1513	**γ-cadinene**	**0.3**	**0.4**	-	**0.2**	**0.4**	**0.3**	**1.2**	**0.5**	**t**	**0.6**	**1.3**	**2.4**	**0.6**	-	**t**	**0.7**	**0.7**	**0.9**
1513	cubebol	-	0.6	0.2	-	-	-	-	-	-	-	1.3	0.5	-	-	-	-	-	-
1524	**δ-cadinene**	**0.2**	**0.6**	-	**0.5**	**0.8**	**0.7**	**2.2**	**t**	**0.3**	**1.9**	**2.9**	**0.9**	**0.1**	**0.1**	**t**	**0.4**	**0.7**	**0.3**
1541	α-copaen-11-ol	-	-	-	-	-	-	-	-	-	-	-	-	-	-	t	-	0.5	-
1546	**α-calacorene**	-	-	-	-	-	-	-	-	-	-	-	-	-	-	-	-	**0.6**	-
1556	germacrene B	0.1	-	-	0.3	0.3	1.2	0.5	-	-	-	-	-	0.2	0.1	-	-	-	-
1562	geranyl butyrate	-	-	-	-	-	-	-	-	-	-	-	0.5	0.3	-	-	-	-	-
1566	β-calacorene	-	-	-	-	-	-	-	-	-	-	-	-	-	-	-	-	0.3	-
1564	**(E)-nerolidol**	-	**0.1**	-	-	-	-	**0.9**	**3.3**	-	**4.2**	**3.3**	**0.3**	**1.0**	**0.2**	**t**	**t**	**t**	-
1574	**germacrene D-4-ol**	-	-	-	**0.8**	**1.8**	**1.2**	**0.9**	-	**0.1**	**0.1**	**3.3**	**0.9**	-	**0.2**	-	**0.4**	-	-
1577	spathulenol	-	-	-	-	-	0.1	0.2	-	0.4	-	0.3	0.3	-	-	0.3	-	-	-
1581	caryophyllene oxide	-	2.3	-	-	-	-	0.2	0.2	0.4	-	0.3	0.3	0.6	t	0.4	0.1	2.2	0.7

KI	Compound	BR	MD	CE	CC	CS	CK	JK	OX	MC	NA	FM	FM	RG	CF	TX	TL	DT	DP
1601	cedrol	-	0.1	-	-	-	-	-	-	-	-	-	-	0.3	-	-	-	0.3	t
1606	humulene epoxide II	0.6	0.1	-	-	-	-	t	0.2	0.4	0.3	0.5	-	-	-	-	1.0	0.1	-
1609	C15OH, 41,55,91, 131,220	-	-	-	-	-	-	-	-	-	-	-	-	-	-	-	-	-	0.7
1627	1-epi-cubenol	0.1	-	-	-	t	t	1.5	-	-	0.8	t	-	-	-	-	0.3	-	-
1640	epi-α-cadinol	-	-	-	t	0.5	0.2	0.7	t	t	1.6	1.5	1.6	0.6	-	-	0.3	0.2	0.6
1640	epi-α-muurolol	-	-	-	0.4	0.5	0.3	0.8	-	-	-	15.	-	-	-	-	0.1	0.2	-
1645	α-muurolol	-	-	-	-	0.1	0.1	0.4	-	-	0.3	0.6	0.2	-	-	-	-	-	-
1651	C15OH, 41,159,99, 177,220	-	-	-	-	-	-	-	-	-	-	-	-	-	-	-	-	-	0.7
1653	**α-cadinol**	-	**t**	-	**0.5**	**1.3**	**0.8**	-	**t**	**0.3**	**2.2**	**4.1**	**1.1**	**0.3**	-	-	**0.4**	**t**	-
1677	cadalene	-	-	-	-	-	-	-	-	-	-	-	-	-	-	-	-	0.2	-
1685	eudesma-4(15),7-dien-1-β-ol	-	-	-	-	-	0.2	-	-	1.2	-	-	-	-	-	-	-	-	-
1686	epi-α-bisabolol	-	-	-	-	-	-	-	-	-	-	-	-	0.7	2.0	0.4	1.2	-	-
1686	germacra-4(15),5, 10,(14)-triene-1-al	-	0.3	-	-	-	-	0.3	-	-	0.6	-	-	-	-	-	-	t	-
1688	shyobunol	-	-	-	-	0.3	0.3	t	0.3	-	-	-	-	-	-	-	-	-	-
1718	(2Z,6Z)-farnesol	-	-	-	-	-	-	-	0.3	-	-	-	-	t	-	-	-	t	-
1725	**(2E,6E)-farnesol**	-	-	-	-	-	-	-	**0.9**	**0.2**	-	**0.7**	-	**1.4**	**0.2**	-	**0.9**	**t**	-
1746	(2E,6Z)-farnesol	-	-	-	-	-	-	-	0.4	-	0.4	-	-	-	-	-	-	t	-
1773	(E)-α-atlantone	-	-	-	0.3	-	-	-	-	-	-	-	-	-	-	-	-	t	-
1941	pimaradiene	0.2	0.8	-	-	-	-	-	-	-	-	-	-	-	-	-	-	t	-
1961	**sandaracopimara-8(14),15-diene**	**18.6**	**2.8**	-	-	-	-	-	**t**	**t**	-	-	-	**t**	-	-	-	-	-
1989	**manoyl oxide**	**0.8**	-	-	-	**0.2**	-	**0.2**	**5.3**	**2.6**	**0.1**	**t**	**t**	-	**0.2**	**t**	**t**	**4.1**	**2.7**
2056	manool	-	-	-	-	-	-	0.6	-	-	1.1	-	-	-	-	1.7	t	-	-
2080	abietadiene	0.2	0.1	-	-	t	-	-	0.7	0.7	-	t	t	-	t	t	0.3	0.9	-
2103	**iso-abienol**	**0.5**	**7.2**	-	-	-	-	**0.2**	-	-	-	-	-	-	**5.1**	-	-	-	-
2132	nezukol	-	0.4	-	-	-	-	-	-	-	-	-	-	-	-	-	-	-	-
2234	trans-totarol, methyl ether	0.3	-	-	-	-	-	-	-	-	-	-	-	-	-	-	-	-	-
2312	trans-totarol	1.2	-	-	-	-	-	-	-	-	0.2	-	-	t	-	-	-	-	-
2331	trans-ferruginol	-	-	-	-	-	-	-	-	-	0.5	-	-	-	-	-	-	-	-

KI = Kovat's Index on DB-5(= SE54) column. *Tentatively identified. Compositional values less than 0.1% are denoted as traces (t). Unidentified components less than 0.5% are not reported.

Table A.2. Comparisons of the per cent total oil for leaf essential oils for the one seeded, turbinate junipers of the eastern hemisphere (from Adams, 2000c). *J. convallium* (CV), *J. coxii* (CX), *J. fargesii* (FG), *J. indica* (IN), *J. komarovii* (KM), *J. morrisonicola* (MO), *J. pingii* (PG), *J. carinata* (CR), *J. przewalskii* (PZ), *J. pseudosabina* (PS), *J. recurva* (RC), *J. rushforthiana* (RF), *J. saltuaria* (SA), *J. squamata* (SQ), *J. tibetica* (TB) and *J. uncinata* (UN).

KI	Compound	CV	KM	TB	SA	PZ	RF	IN	PS	PG	CR	RC	UN	CX	FG	SQ	MO
926	tricyclene	0.1	t	t	-	t	t	t	0.1	t	0.1	t	0.1	0.1	-	0.1	t
931	α-thujene	0.2	2.4	1.7	1.3	0.3	1.7	1.4	0.2	0.9	2.2	0.5	-	t	0.3	1.0	t
939	α-pinene	47.6	3.4	9.5	2.2	9.4	9.4	2.8	52.1	19.5	20.9	6.9	2.8	18.4	17.1	26.0	0.8
953	α-fenchene	t	t	t	-	t	t	-	t	0.4	0.1	1.0	1.8	0.4	0.2	t	t
953	camphene	0.4	t	t	-	0.1	t	t	0.5	t	0.2		-	0.1	-	0.1	t
976	sabinene	1.5	32.5	23.0	38.2	7.2	31.8	26.1	5.8	19.2	23.7	13.4	0.4	0.5	13.7	15.5	0.3
980	β-pinene	1.1	0.2	0.3	0.2	0.2	0.3	t	4.5	1.4	0.6	0.2	0.2	0.4	2.2	0.6	t
991	myrcene	8.4	2.6	2.4	2.1	1.2	3.7	3.3	3.8	8.7	3.4	2.4	1.2	1.9	2.1	1.3	0.4
1001	δ-2-carene	t	-	3.8	-	4.1	t	t	-	0.1	0.2	-	-	t	-	0.6	0.6
1005	α-phellandrene	0.1	t	0.1	0.1	0.1	0.3	0.1	t	-	0.1	-	-	t	t	0.1	t
1011	δ-3-carene	t	t	0.1	-	0.2	t	t	0.3	8.3	0.1	23.7	32.1	7.4	-	-	t
1018	α-terpinene	t	2.2	1.8	1.2	0.4	1.6	1.7	0.2	0.7	2.3	0.8	0.1	t	0.2	0.7	t
1026	p-cymene	0.8	0.3	0.3	0.9	0.6	0.2	0.2	t	t	1.1	0.2	0.5	t	0.3	0.6	0.5
1031	limonene	5.1	9.4	8.5	2.0	11.6	1.7	0.4	0.9	8.2	2.1	18.4	3.0	15.7	3.0	3.3	27.3
1031	β-phellandrene	1.7	-	0.2	0.1	-	1.6	1.6	1.8	0.4	0.2	t	3.1	-	0.5	13.6	-
1032	1,8-cineole	t	-	-	-	-	0.2	1.3	t	-	-	t	-	-	-	-	-
1050	(E)-β-ocimene	-	-	-	-	-	t	-	-	-	-	-	-	0.3	-	-	0.1
1062	γ-terpinene	0.2	3.4	2.8	1.9	0.5	2.6	2.7	0.3	1.3	4.2	1.3	t	0.1	0.5	1.2	0.2
1068	cis-sabinene hydrate	0.1	2.3	1.3	1.6	0.2	1.2	1.6	0.1	0.3	1.3	0.6	-	-	0.5	0.5	t
1088	terpinolene	0.7	1.1	1.2	0.7	0.6	1.0	0.9	0.3	1.0	1.5	1.8	4.8	0.9	0.4	0.7	0.2
1091	2-nonanone	-	-	-	0.8	-	-	-	3.4	-	-	-	-	-	0.2	-	-
1097	trans-sabinene hydrate	t	2.3	1.3	2.0	0.2	1.0	1.0	t	0.2	1.3	0.3	-	-	0.9	0.3	t
1098	linalool	0.5	1.1	t	0.7	0.3	t	-	2.1	t	t	-	0.3	t	1.0	-	0.4
1102	cis-thujone	-	-	t	0.3	0.7	t	2.3	-	-	-	-	-	-	3.0	-	-
1103	isopentyl-isovalerate	-	-	-	-	-	0.1	-	-	0.4	0.3	0.1	-	0.1	-	3.2	0.1

KI	Compound	CV	KM	TB	SA	PZ	IR	IN	PS	PG	PC	RC	UN	CX	FG	SQ	MO
1114	trans-thujone	-	1.4	t	1.3	3.0	0.1	16.0	-	-	-	-	-	-	11.9	-	t
1116	3-me- butanoate, 3-me-3-butenyl-	-	-	0.3	-	-	-	-	-	-	-	-	-	-	-	2.9	-
1121	cis-p-menth-2-en-1-ol	0.1	0.5	0.6	0.2	0.5	0.3	0.6	0.6	0.6	0.5	0.2	0.1	0.2	-	0.2	-
1134	cis-limonene oxide	-	-	-	-	-	-	-	-	0.1	0.5	0.2	t	-	-	-	-
1139	trans-pinocarveol	-	-	-	-	-	-	-	-	-	-	-	-	-	0.8	-	-
1140	trans-sabinol	-	-	-	0.2	0.7	-	1.4	-	-	t	-	-	-	-	-	-
1140	trans-p-menth-2-en-1-ol	0.2	0.2	0.4	-	-	0.2	0.7	0.1	0.1	0.4	t	-	-	-	0.1	-
1143	cis-sabinol*	-	-	-	-	-	0.1	-	-	-	t	-	-	-	-	0.2	-
1143	trans-verbenol	0.2	-	-	-	0.2	-	-	-	-	-	-	-	-	0.5	-	-
1148	camphene hydrate	-	-	-	-	-	-	-	0.2	-	-	-	-	-	0.3	-	-
1153	citronellal	-	-	-	-	-	-	-	-	-	-	-	-	-	0.4	-	-
1159	p-mentha-1,5-dien-8-ol	-	-	-	-	0.2	-	-	-	-	-	0.5	t	t	-	-	-
1165	borneol	t	t	t	-	-	-	-	0.1	t	0.1	-	t	-	t	-	t
1167	δ-terpineol	-	-	-	-	-	-	-	-	-	-	0.3	-	-	-	-	-
1171	umbellulone	-	-	-	-	-	0.1	0.2	-	-	-	-	-	-	-	-	t
1177	terpinen-4-ol	0.2	8.1	6.2	3.9	1.4	4.5	7.2	0.8	2.9	8.8	3.7	0.2	0.2	1.3	3.3	0.1
1180	m-cymen-8-ol	-	-	-	-	-	-	-	-	-	-	0.2	-	-	-	-	-
1183	p-cymen-8-ol	t	t	t	0.1	0.2	-	-	-	-	-	0.1	-	-	-	-	-
1189	α-terpineol	t	0.2	0.2	0.1	0.1	0.1	0.2	0.1	0.1	0.3	0.3	0.5	0.3	0.1	t	t
1204	verbenone	0.1	-	-	-	-	-	-	t	-	-	-	-	-	0.2	-	-
1228	citronellol	-	-	-	-	0.6	-	t	0.2	-	-	-	-	-	-	-	t
1252	piperitone	t	-	4.2	-	6.8	-	0.1	t	-	-	-	-	-	t	0.2	t
1257	methyl citronellate	-	-	-	-	-	0.1	0.1	-	-	-	0.1	0.4	-	1.0	-	-
1277	pregeijerene B	-	4.5	5.6	0.9	3.7	t	-	-	2.9	4.9	2.2	-	-	-	5.5	1.3
1285	bornyl acetate	0.9	t	0.1	0.1	0.2	0.2	0.2	0.8	0.2	0.7	0.1	0.2	0.5	0.4	0.2	t
1287	trans- linalool oxide acetate (pyranoid)	t	-	-	-	0.1	t	0.1	-	0.4	-	-	0.2	-	-	-	t
1290	trans-sabinyl acetate	-	t	t	t	0.7	0.1	15.7	0.2	-	-	-	-	t	t	0.4	t
1291	2-undecanone	0.2	t	-	0.2	-	-	-	t	-	-	-	-	-	-	-	-
1322	methyl geranate	-	-	-	-	-	-	-	-	-	-	t	-	-	0.4	-	-
1345	α-cubebene	0.4	-	-	-	-	0.4	t	-	-	-	-	0.4	0.5	-	-	-
1374	α-copaene	t	-	t	-	-	0.2	t	-	-	-	-	0.2	0.1	-	-	-

KI	Compound	CV	KM	TB	SA	PZ	IR	IN	PS	PG	PC	RC	UN	CX	FG	SQ	MO
1387	β-cubebene	0.8	-	-	-	-	0.8	t	-	t	-	-	0.5	0.9	-	-	-
1409	α-cedrene	-	t	t	-	-	-	-	0.1	-	-	-	-	-	-	-	-
1417	(E)-caryophyllene	0.1	t	0.1	t	t	0.3	t	-	0.1	0.9	t	0.3	0.4	-	0.6	t
1429	cis-thujopsene	-	t	-	-	0.8	-	-	0.3	0.2	t	-	-	-	4.0	-	-
1448	cis-muurola-3,5-diene	0.9	-	-	-	0.1	1.6	0.3	-	0.1	-	-	3.6	3.6	-	-	-
1454	α-humulene	0.2	-	-	-	-	0.2	t	-	-	-	-	0.3	0.2	-	-	-
1461	cis-muurola-4(14),5-diene	-	t	t	-	0.2	-	-	-	-	-	t	-	0.1	-	-	-
1475	trans-cadina-1(6),4-diene	0.6	-	-	-	t	1.2	0.2	-	t	-	-	3.5	3.2	-	-	-
1475	β-chamigrene	-	-	-	-	-	-	-	-	-	-	-	-	-	0.8	-	-
1478	γ-muurolene	-	t	t	-	0.2	0.1	t	0.1	-	-	t	t	0.1	-	-	t
1480	germacrene D	0.4	t	t	-	0.3	t	-	-	0.2	-	t	t	0.1	-	-	-
1483	ar-curcumene	-	-	-	-	-	-	-	-	-	0.3	-	-	-	-	-	-
1493	trans-murrola-4(14),5-diene	2.4	t	t	-	0.1	3.9	0.9	-	0.1	-	-	9.3	8.6	-	-	-
1493	cis-cadina-1,4-diene	-	t	-	-	-	-	-	t	-	-	-	-	2.2	-	-	-
1493	4-epi-cubebol	0.9	-	-	-	0.1	0.8	-	-	-	-	t	-	-	0.3	-	-
1500	α-muurolene	0.2	0.1	0.2	-	0.3	0.2	0.1	0.4	t	-	0.2	0.7	0.4	0.3	-	0.3
1512	β-curcumene	-	-	-	-	-	-	-	-	-	0.2	-	-	-	-	-	-
1513	γ-cadinene	-	0.3	0.3	0.2	0.7	3.8	0.7	0.4	0.1	-	0.4	0.7	6.0	0.3	t	-
1513	cubebol	3.3	-	-	-	-	-	-	-	-	-	-	6.1	-	1.0	-	-
1522	δ-cadinene	1.4	0.7	0.8	0.3	1.3	1.6	0.8	1.4	0.1	t	0.8	6.4	3.6	1.4	0.1	-
1528	zonarene	0.4	-	-	-	-	0.8	t	-	-	-	-	0.8	2.3	-	-	-
1533	trans-cadina-1,4-diene	0.3	-	-	-	-	0.3	t	-	-	-	-	0.6	0.4	-	-	-
1548	elemol	-	2.5	3.9	7.6	2.0	1.3	0.6	t	3.0	2.5	3.9	t	0.3	2.7	2.9	3.8
1556	germacrene B	-	t	t	-	-	-	-	-	t	0.4	t	-	-	0.4	0.4	t
1574	germacrene D-4-ol	0.3	1.2	1.2	0.6	1.2	0.7	0.4	2.0	0.1	t	1.0	1.8	0.3	1.2	0.1	-
1587	trans-muurol-5-en-4-α-ol	-	-	-	-	0.1	-	-	-	-	-	-	1.3	-	-	-	-
1592	longiborneol(juniperol)	-	-	-	-	0.2	-	-	-	-	-	-	-	0.5	-	-	t
1596	cedrol	t	7.7	1.0	3.2	0.2	-	t	10.7	5.4	t	-	-	0.4	0.4	0.4	0.1
1607	β-oplopenone	t	0.1	t	-	t	0.1	-	t	-	-	0.1	t	t	-	-	-
1627	1-epi-cubenol	1.9	-	-	-	0.2	2.4	0.3	-	-	-	-	3.6	4.8	0.7	-	-
1630	acorenol	-	t	-	-	-	-	t	0.3	-	-	-	-	-	-	-	-
1630	γ-eudesmol	-	0.2	0.5	0.5	0.3	0.3	t	-	0.5	0.6	0.1	-	-	0.5	0.7	0.6

KI	Compound	CV	KM	TB	SA	PZ	IR	IN	PS	PG	PC	RC	UN	CX	FG	SQ	MO
1638	epi-α-cadinol	t	0.4	0.2	t	1.2	0.2	0.2	0.4	-	-	0.3	0.8	1.0	0.4	0.4	-
1638	epi-α-muurolol	-	0.5	0.7	-	1.2	0.6	t	0.6	-	-	0.5	0.8	0.1	0.7	-	-
1641	cubenol	0.4	-	-	-	-	-	0.2	-	-	-	-	-	-	-	-	-
1644	α-muurolol	t	t	t	-	0.6	0.1	t	0.1	-	-	0.1	0.2	0.3	t	-	-
1649	β-eudesmol	-	0.7	1.0	1.8	0.8	0.3	0.1	-	0.7	1.1	1.5	-	t	0.5	1.5	1.2
1652	α-eudesmol	-	0.5	1.0	4.3	t	0.5	0.2	-	0.9	1.1	1.0	0.6	0.1	0.3	2.2	1.1
1653	α-cadinol	0.4	1.3	1.3	-	3.2	0.7	0.2	1.3	-	-	0.8	0.6	0.5	1.6	-	-
1666	bulnesol	-	0.5	0.9	1.8	0.4	0.4	0.1	-	0.5	0.5	0.8	t	t	0.4	0.5	0.7
1685	eudesma-4(15),7-dien-1-β-ol	-	-	-	-	0.2	-	-	-	-	-	-	-	-	-	-	-
1733	oplopenone	-	-	-	-	0.1	-	-	-	-	-	-	-	-	0.4	-	-
1789	8-α-acetoxyelemol	-	2.0	3.6	0.3	3.9	0.8	-	-	1.4	2.0	1.9	-	-	2.4	4.7	4.4
1894	rimuene	-	-	-	0.2	0.1	-	-	-	t	t	-	-	0.1	-	-	0.2
1898	isopimara-9(11),5-diene	-	-	-	0.2	0.4	-	-	-	t	t	-	-	0.2	-	-	0.3
1909	pimara-9(11),5-diene*	-	-	t	1.0	0.3	0.2	0.2	-	0.4	-	-	-	0.6	-	-	-
1933	rosa-5,15-diene (ent)	t	-	t	t	0.5	0.2	t	-	0.2	t	-	-	3.0	-	-	3.7
1961	sandaracopimara-8(14),15-diene	0.2	t	0.3	2.5	0.8	0.8	0.3	t	0.4	t	-	-	-	-	-	-
2000	dolabradiene	1.6	1.1	0.1	0.5	-	-	-	-	0.2	0.6	-	-	1.9	-	t	2.3
2011	phyllocladene	-	-	0.3	-	-	-	-	-	-	0.5	-	-	0.1	0.3	0.3	0.5
2011	abieta-8,12-diene	-	-	0.1	-	t	-	-	-	-	-	-	-	0.3	-	0.3	0.4
2054	abietatriene	4.7	0.3	1.0	0.3	6.3	t	t	t	t	0.5	0.5	0.1	0.4	t	t	0.4
2056	manool	4.7	-	0.9	-	-	-	0.8	t	1.8	2.0	t	1.2	t	-	-	-
2080	abietadiene	0.4	0.3	4.5	5.3	5.1	t	0.3	0.2	0.3	0.6	1.3	0.2	t	t	t	t
2126	nezukol	-	-	0.5	-	4.0	0.5	-	t	3.5	-	-	-	3.3	-	-	42.8
2147	abieta-8(14),13(15)-diene	-	-	0.1	-	0.4	4.0	-	-	-	-	-	-	-	-	-	-
2269	sandaracopimarinol	0.3	-	-	-	-	-	-	-	-	-	-	-	-	-	-	0.4
2278	cis-totarol	-	-	-	-	0.2	-	-	-	-	-	-	-	-	-	-	t
2282	sempervirol	-	-	-	-	-	-	-	-	-	-	-	-	-	t	t	t
2288	4-epi-abietal	-	-	-	-	-	0.4	t	-	-	-	t	0.2	-	-	t	-
2302	trans-totarol	0.7	t	t	0.7	1.6	0.2	t	-	-	t	0.2	0.2	0.1	-	-	t
2325	trans-ferruginol	t	t	t	-	0.3	-	-	-	-	-	-	-	-	-	-	t

KI = Kovat's Index on DB-5(= SE54) column. *Tentatively identified. Compositional values less than 0.1% are denoted as traces (t).

Table A.3. Leaf essential oils for the multi-seeded junipers of the eastern hemisphere (from Adams, 1999). Comparisons of the per cent total oil for leaf essential oils for *J. davurica* (DV), *J. sabina* (SA), *J. erectopatens* (ER), *J. microsperma* (MS), *J. semiglobosa* (SM), *J. chinensis* (CH), *J. excelsa* (EX), *J. polycarpos* var. *polycarpos* (PC), *J. seravschanica* - Kazakhstan (SK), *J. seravschanica* - Pakistan (SP), *J. polycarpos* var. *turcomanica* (PT), *J. foetidissima* (FT), *J. thurifera* (TH), *J. phoenicea* (PH), *J. turbinata* (TB) and *J. procera* (PR). Components that tend to separate the species are highlighted in boldface.

KI	Compound	DV	SA	ER	MS	SM	CH	EX	PC	SK	SP	PT	FT	TH	PH	TB	PR
921	tricyclene	-	t	-	-	t	t	0.1	0.1	0.3	0.1	01	-	t	0.3	0.2	t
924	**α-thujene**	**1.3**	**1.7**	**0.8**	**0.9**	**2.4**	**0.7**	**-**	**t**	**0.6**	**0.4**	**t**	**1.3**	**t**	**-**	**-**	**t**
932	**α-pinene**	**1.9**	**17.2**	**1.5**	**0.9**	**3.7**	**16.7**	**22.5**	**68.4**	**44.4**	**15.5**	**68.8**	**2.6**	**2.2**	**41.8**	**45.3**	**12.5**
945	α-fenchene	-	-	-	t	-	-	0.2	t	-	0.2	t	t	t	t	0.1	0.1
946	camphene	-	0.2	-	t	t	0.2	0.5	0.2	0.5	0.2	0.1	t	t	0.6	0.3	0.1
969	**sabinene**	**35.4**	**36.8**	**33.3**	**33.9**	**44.1**	**17.8**	**t**	**0.2**	**0.9**	**0.5**	**0.1**	**15.9**	**1.0**	**0.1**	**t**	**t**
973	1-octen-3-ol	-	-	-	-	-	-	-	-	-	-	-	-	0.5	-	-	0.3
974	β-pinene	0.1	1.9	0.1	0.2	0.3	0.4	0.6	0.5	2.2	1.2	0.6	-	0.2	1.3	1.1	1.2
988	**myrcene**	**2.8**	**4.8**	**1.0**	**0.5**	**4.8**	**3.2**	**1.9**	**1.2**	**19.2**	**20.7**	**1.5**	**2.0**	**3.4**	**4.5**	**4.1**	**1.2**
992	**hexanoic acid, 4-me, methyl ester**	**-**	**-**	**0.8**	**-**	**0.3**	**-**	**-**	**-**	**-**	**-**	**-**	**-**	**-**	**-**	**-**	**-**
1001	δ-2-carene	0.3	0.2	0.2	0.5	-	-	-	-	-	0.1	-	-	2.2	0.1	0.2	-
1002	α-phellandrene	t	t	-	t	0.1	-	0.1	-	0.1	-	-	0.2	t	0.6	1.6	-
1008	δ-3-carene	-	0.2	-	-	t	-	2.3	t	-	3.5	t	t	1.1	-	2.0	6.1
1014	α-terpinene	0.8	0.9	0.4	1.0	2.1	0.4	0.1	t	0.1	0.1	t	4.4	0.1	t	t	t
1020	p-cymene	0.3	0.1	t	0.3	2.8	0.3	0.4	0.1	0.1	0.7	0.1	0.4	t	0.8	0.9	t
1024	**limonene**	**1.6**	**3.2**	**37.1**	**0.8**	**2.4**	**15.1**	**22.6**	**1.2**	**4.4**	**9.0**	**1.5**	**0.8**	**51.5**	**4.7**	**1.5**	**0.2**
1025	**β-phellandrene**	**t**	**0.9**	**-**	**0.7**	**0.4**	**t**	**t**	**-**	**0.5**	**1.0**	**-**	**0.6**	**-**	**3.5**	**13.4**	**0.8**
1044	(E)-β-ocimene	0.4	0.1	0.2	0.3	0.3	0.2	t	-	t	0.2	-	-	0.3	0.1	t	t
1054	γ-terpinene	1.4	1.5	0.5	1.8	3.3	0.7	0.6	0.2	1.4	1.3	0.3	6.5	0.2	0.7	0.4	t
1065	cis-sabinene hydrate	1.9	0.6	0.3	0.7	1.3	0.6	-	-	0.1	0.2	-	1.9	0.1	-	-	-
1074	trans-linalool oxide (furanoid)	-	-	-	-	-	-	-	-	-	-	-	-	0.4	-	-	-
1086	terpinolene	0.7	1.0	0.6	0.6	1.5	0.6	0.9	0.4	1.5	1.7	0.5	1.9	1.0	1.3	0.7	1.1
1088	2-nonanone	0.9	0.6	-	t	t	-	-	-	-	-	-	-	-	-	-	-
1095	**trans-sabinene hydrate**	**-**	**-**	**0.1**	**0.5**	**0.9**	**0.4**	**-**	**-**	**-**	**t**	**t**	**1.9**	**-**	**-**	**-**	**-**

KI	Compound	SD	SA	ER	TB	SM	CH	EX	PC	SK	SP	TU	FT	TH	PH	MS	PR
1095	linalool	0.9	0.8	-	0.4	1.1	1.0	-	0.1	0.5	0.7	t	t	7.1	0.7	0.5	0.5
1101	**cis-thujone**	-	-	-	-	-	-	-	-	-	-	-	**18.6**	-	-	-	-
1102	isopentyl-isovalerate	0.3	-	-	-	-	-	-	t	-	-	t	-	-	-	-	-
1106	cis-rose oxide	0.2	-	-	-	-	-	-	-	-	-	-	-	-	-	-	-
1112	trans-thujone	-	-	-	-	t	-	-	t	0.1	-	t	3.5	-	-	-	-
1114	endo-fenchol	-	-	-	-	0.1	-	0.2	-	-	-	-	-	-	-	-	-
1116	3-me butanoate, 3-me-3-butenyl-	-	0.4	-	-	-	-	-	-	-	-	-	-	-	-	-	-
1118	cis-p-menth-2-en-1-ol	0.4	0.3	0.1	0.3	0.4	0.2	0.1	-	t	-	-	1.2	0.4	0.1	0.2	t
1122	α-campholenal	-	-	-	-	-	-	0.1	0.2	-	0.1	t	-	-	0.6	0.2	t
1135	trans-pinocarveol	-	-	-	-	-	-	0.2	0.1	-	t	0.1	-	-	0.6	0.3	t
1137	cis-verbenol	0.2	-	-	-	-	-	-	-	-	-	-	1.1	-	-	-	-
1137	trans-p-menth-2-en-1-ol	-	0.1	-	0.2	0.2	-	-	t	-	-	-	-	0.3	-	-	0.2
1140	trans-verbenol	-	-	-	-	-	-	0.5	0.5	t	0.2	0.6	-	-	-	0.3	t
1141	camphor	-	-	-	-	-	-	-	t	t	-	t	-	-	0.4	-	-
1146	neo-3-thujanol	-	-	-	-	-	-	-	-	-	-	t	0.2	-	-	-	-
1148	citronellal	0.6	-	-	-	0.4	-	-	-	-	-	-	-	-	-	-	-
1154	sabina ketone	-	-	-	0.3	t	-	-	-	-	-	-	0.2	-	-	-	-
1165	borneol	-	-	-	-	-	-	-	t	t	t	t	t	-	-	-	0.2
1164	δ-terpineol	-	-	-	-	-	-	-	t	-	t	t	-	-	0.6	-	-
1166	p-mentha-1,5-dien-8-ol	-	-	-	-	-	-	-	-	-	-	-	-	-	-	0.2	0.1
1166	coahuilensol	-	-	-	0.1	0.1	-	-	-	-	-	-	-	0.2	-	-	-
1174	**terpinen-4-ol**	**5.6**	**4.1**	**0.9**	**2.0**	**6.8**	**2.1**	**0.2**	**t**	**0.4**	**0.3**	**-**	**17.6**	**0.6**	**0.2**	**t**	**0.1**
1178	naphthalene	-	-	-	-	-	-	-	0.1	-	-	0.4	-	t	-	t	-
1179	p-cymen-8-ol	-	-	-	0.1	-	-	-	t	-	t	-	0.1	t	0.2	t	t
1186	α-terpineol	0.2	0.2	-	t	0.2	0.1	t	t	0.1	-	t	0.7	0.7	0.8	1.1	0.5
1193	4Z-decenal	0.2	-	-	-	0.2	-	-	-	-	-	-	-	-	-	-	-
1195	cis-piperitol	-	t	-	-	-	-	-	-	-	-	-	0.3	0.1	-	0.1	-
1195	methyl chavicol	-	-	-	-	-	0.3	-	-	-	-	-	-	-	-	-	-
1207	trans-piperitol	t	t	-	0.1	t	-	0.3	-	-	-	-	0.3	-	-	t	-
1218	endo-fenchyl acetate	-	-	-	-	-	-	-	-	-	-	-	-	-	-	-	-
1223	citronellol	4.8	0.3	-	-	0.9	-	-	-	-	-	-	-	0.4	0.6	0.2	-

KI	Compound	SD	SA	ER	TB	SM	CH	EX	PC	SK	SP	TU	FT	TH	PH	MS	PR
1233	(3Z)-hexenyl 3-methyl butanoate	-	-	-	-	-	-	-	-	-	-	-	-	-	-	-	-
1249	piperitone	t	0.1	-	0.6	-	-	-	0.1	-	t	0.2	-	3.2	-	0.2	-
1252	linalyl acetate	t	0.2	-	t	-	-	-	-	-	-	-	-	7.6	2.0	-	-
1255	4Z-decen-1-ol	1.4	-	-	-	-	-	-	-	-	-	-	-	-	-	1.0	-
1257	methyl citronellate	5.4	t	-	-	2.1	t	-	-	t	-	-	-	-	-	-	-
1271	isopulegol acetate	-	-	-	-	-	-	-	-	-	-	-	-	-	3.3	-	-
1274	**pregeijerene B**	-	-	**13.2**	**16.3**	-	-	-	-	-	-	-	-	-	-	-	**0.3**
1278	iso-isopulegol acetate	-	-	-	-	-	-	-	-	-	-	-	-	-	0.2	-	-
1284	bornyl acetate	t	0.5	0.2	0.2	0.4	0.1	0.4	0.2	1.0	0.6	0.2	0.1	0.3	-	-	0.4
1285	safrole	-	-	-	-	-	5.1	-	-	-	-	-	-	-	-	-	-
1287	trans-linalool oxide acetate (pyranoid)	-	-	-	-	t	-	0.2	0.1	-	-	-	-	0.3	0.5	-	-
1288	trans-sabinyl acetate	-	-	-	-	t	-	-	-	-	0.1	-	0.9	-	-	-	-
1293	2-undecanone	4.4	0.1	-	-	0.2	-	-	-	-	-	-	-	-	-	-	-
1312	decadienal isomer?	-	-	-	-	-	-	3.3	-	-	-	-	-	1.2	-	-	-
1315	(2E,4E)-decadienal	-	-	-	-	0.1	-	-	-	t	t	-	-	-	-	0.7	-
1319	aromatic, 149, 91, 164	-	-	-	-	-	-	-	-	-	-	-	0.1	-	0.3	-	-
1322	methyl geranate	-	0.6	-	-	0.6	-	-	-	-	-	-	-	-	-	-	-
1346	**α-terpinyl acetate**	-	**0.2**	-	-	-	-	-	-	-	-	-	-	**3.2**	**4.6**	**4.1**	-
1359	neryl acetate	-	-	-	-	-	-	-	-	-	-	-	-	0.2	-	-	-
1374	α-copaene	-	-	-	-	-	-	0.2	-	-	t	t	-	-	-	-	-
1379	geranyl acetate	-	-	-	-	-	-	-	-	-	-	-	-	0.3	t	-	-
1387	β-bourbonene	-	-	-	0.1	-	-	0.1	-	-	-	-	-	-	-	-	-
1403	methyl eugenol	-	-	-	-	-	3.4	-	-	-	-	-	-	-	-	-	-
1410	α-cedrene	-	0.2	-	-	-	t	-	t	-	-	-	-	-	-	-	-
1411	1,7-di-epi-β-cedrene	-	0.1	-	-	0.3	0.1	1.6	1.3	0.2	1.4	-	-	-	-	-	-
1417	(E)-caryophyllene	-	0.1	t	0.8	-	-	-	0.3	0.1	0.2	0.4	0.3	-	0.5	1.0	0.5
1418	β-cedrene	-	-	-	-	0.2	0.2	0.9	-	0.1	0.2	-	-	-	-	-	-
1429	cis-thujopsene	-	0.3	-	-	0.3	0.3	0.3	0.2	0.2	0.4	-	-	-	-	-	-
1448	cis-muurola-3,5-diene	t	-	-	-	-	-	0.2	-	t	-	-	-	-	-	-	-
1452	α-humulene	t	-	-	-	-	-	0.2	-	-	-	0.1	0.1	-	0.3	0.5	0.7

KI	Compound	SD	SA	ER	TB	SM	CH	EX	PC	SK	SP	TU	FT	TH	PH	MS	PR
1454	E-β-farnesene	-	-	-	-	-	-	0.2	0.1	-	0.1	-	-	-	-	-	-
1465	cis-muurola-4(14),5-diene	0.2	-	-	-	-	-	-	0.1	-	0.1	0.2	-	-	-	-	-
1475	trans-cadina-1(6),4-diene	-	-	-	-	-	-	0.4	-	-	-	-	-	-	-	-	-
1478	γ-muurolene	0.1	-	-	-	0.1	-	-	-	t	-	0.2	-	-	-	-	-
1480	**germacrene D**	0.2	-	0.2	2.4	-	0.1	0.9	0.2	0.1	0.2	0.8	-	-	0.8	2.2	0.3
1490	cis-β-guaiene	-	-	-	-	-	-	-	-	-	-	-	-	-	0.9	-	-
1493	trans-murrola-4(14),5-diene	0.1	-	-	-	-	-	0.4	-	t	-	0.1	-	-	-	-	-
1493	epi-cubebol	0.2	-	-	-	-	t	-	-	t	-	-	-	-	0.6	-	-
1495	γ-amorphene	-	-	-	-	-	-	-	-	-	t	-	-	-	-	0.5	-
1500	α-muurolene	0.3	0.2	-	t	-	0.1	0.2	-	0.2	0.2	0.3	-	0.1	0.2	0.2	-
1513	α-alaskene	-	-	-	-	-	t	0.3	t	-	0.4	-	-	-	-	-	-
1513	γ-cadinene	1.1	0.3	-	t	t	0.2	-	0.3	0.4	0.4	1.1	t	0.1	0.8	0.3	-
1513	cubebol	-	-	-	-	-	-	0.8	-	-	-	-	t	0.1	0.8	-	-
1522	δ-cadinene	1.5	0.8	t	t	0.2	0.5	0.7	0.3	1.0	0.7	1.4	t	0.4	1.3	0.7	-
1529	E-γ-bisabolene	-	-	-	-	-	-	0.2	0.1	-	-	-	-	-	-	-	-
1537	α-cadinene	0.2	-	-	-	-	t	-	-	0.1	0.1	0.2	-	t	-	-	-
1539	α-copaen-11-ol	-	-	-	0.3	-	-	-	-	-	-	-	-	-	-	-	-
1548	**elemol**	-	-	3.7	14.6	t	-	-	t	0.5	0.5	0.2	t	1.2	-	0.4	4.3
1555	elemicin	-	-	-	-	-	0.5	-	-	-	-	-	-	-	-	-	-
1559	germacrene B	t	-	-	0.2	t	-	-	0.6	0.4	0.6	1.6	-	0.2	t	0.8	-
1570	Z-hexenyl benzoate	-	-	-	-	-	0.6	-	-	-	-	-	-	-	-	-	-
1574	germacrene D-4-ol	5.7	0.8	0.1	-	0.2	0.6	-	0.4	0.5	2.0	3.0	-	0.6	-	0.2	0.1
1582	caryophyllene oxide	-	-	-	-	-	-	-	-	-	-	-	-	-	0.3	0.5	0.5
1589	**allo-cedrol**	-	-	-	-	1.1	1.1	1.9	0.8	1.0	1.7	-	t	-	-	-	-
1600	**cedrol**	t	15.2	-	-	14.8	20.1	28.1	17.0	14.6	26.4	t	3.2	t	-	-	0.5
1608	humulene epoxide II	0.5	-	-	-	0.1	0.9	t	-	-	-	0.2	0.5	0.2	t	-	-
1608	β-oplopenone	-	-	-	-	-	-	t	-	-	-	0.2	-	-	-	-	-
1627	1-epi-cubenol	-	-	-	-	-	-	1.6	-	-	-	-	0.2	0.2	2.3	-	-
1630	α-acorenol	-	0.2	-	-	-	-	-	-	-	-	-	-	-	-	t	-
1630	γ-eudesmol	-	-	-	0.8	-	-	-	-	-	-	-	-	0.3	-	t	1.4
1638	epi-α-cadinol	0.7	0.3	-	0.2	t	0.3	t	t	0.5	0.4	1.4	0.3	0.3	t	t	-
1638	epi-α-muurolol	0.7	0.3	-	0.2	t	0.4	t	t	0.1	0.1	t	0.3	0.3	-	t	-

KI	Compound	SD	SA	ER	TB	SM	CH	EX	PC	SK	SP	TU	FT	TH	PH	MS	PR
1640	cubenol	-	-	-	-	-	-	-	-	-	-	-	-	-	0.7	-	-
1642	α-muurolol	0.2	t	-	-	-	0.1	t	-	t	t	0.2	t	t	0.2	-	-
1649	β-eudesmol	-	-	0.4	1.5	-	-	-	t	t	0.1	t	-	0.6	-	0.3	2.3
1652	α-eudesmol	-	-	0.9	1.4	-	-	-	t	0.1	t	t	-	0.7	-	0.4	3.8
1653	α-cadinol	0.8	-	-	-	0.2	1.3	t	0.2	0.9	0.8	1.6	1.0	0.8	0.7	0.3	-
1670	bulnesol	-	-	0.4	0.5	-	-	t	t	-	0.1	t	-	0.3	-	-	1.3
1666	(2E,4E)-decadienol	-	-	-	-	-	-	0.6	-	-	-	-	-	-	-	-	-
1687	eudesma-4(15),7-dien-1-β-ol	-	-	-	-	-	-	-	-	-	-	-	-	-	-	0.7	0.2
1688	shyobunol	-	-	-	-	-	-	-	0.3	0.4	-	0.9	-	0.5	0.7	1.0	-
1688	cadinol isomer	-	-	-	-	-	-	-	-	-	1.2	-	-	-	-	-	-
1735	oplopenone	-	-	-	-	-	0.6	-	-	-	-	-	-	-	-	-	-
1792	**8-α-acetoxyelemol**	-	-	1.5	7.1	-	-	-	-	-	-	-	-	0.3	-	-	**3.5**
1930	rosa-5,15-diene	-	-	-	-	-	-	-	-	-	-	-	-	-	-	-	0.4
1968	sandaracopimara-8(14),15-diene	-	-	-	-	-	-	-	-	-	-	-	-	-	-	-	-
1978	manoyl oxide	-	-	-	-	-	t	t	0.1	-	0.3	0.1	0.2	-	-	1.2	0.3
2009	epi-13-manoyl oxide	-	-	-	-	-	-	t	-	-	-	t	0.4	0.3	-	-	0.5
2055	abietatriene	-	-	t	-	-	-	-	-	-	-	-	-	t	-	-	1.3
2087	abietadiene	-	-	t	-	-	-	-	-	-	-	-	-	-	-	-	15.4
2103	**iso-abienol**	-	-	-	-	-	-	-	-	-	-	-	-	-	-	-	**2.6**
2147	**abieta-8(14),13(15)-diene**	-	-	-	-	-	-	-	-	-	-	-	-	-	-	-	**0.3**
2181	**sandaracopimarinal**	-	-	-	-	-	-	-	-	-	-	-	-	-	-	-	**0.8**
2282	**sempervirol**	-	-	-	-	-	-	-	-	-	-	-	-	-	-	-	**0.6**
2298	4-epi-abietal	0.9	-	-	0.3	t	0.2	0.1	0.5	-	1.0	1.1	-	-	t	t	1.8
2299	diterpene, 245, 107,269,288	-	-	-	-	-	-	-	-	-	-	-	-	-	-	-	1.0
2314	**trans-totarol**	-	-	-	-	t	t	-	-	-	-	-	0.4	-	1.9	t	**21.4**
2331	**trans-ferruginol**	-	-	-	-	-	-	-	-	-	-	-	-	-	0.3	t	**3.4**

KI = Kovat's Index on DB-5(=SE54) column. *Tentatively identified. Compositional values less than 0.1% are denoted as traces (t). Unidentified components less than 0.5% are not reported.

Table A.4. Leaf essential oils for the smooth leaf margin junipers of the western hemisphere (Adams, 2000a). Comparisons of the per cent total oil for leaf essential oils for *J. barbadensis* (BA), *J. b.* var. *lucayana* (LU), *J. bermudiana* (BR), *J. blancoi* (BL), *J. blancoi* var. *mucronata* (BM),), *J. gracilior* (GR), *J. gracilior* var. *ekmanii* (EK), *J. gracilior* var. *saxicola* (SX), *J. gracilior* var. *urbaniana* (UR), *J. horizontalis* (HZ), *J. maritima* (MR), *J. scopulorum* (SC), *J. virginiana* (VG) and *J. virginiana* var. *siliciola* (SI). Components that tend to separate the species are highlighted in boldface.

KI	Compound	GR	EK	SX	UR	BR	BA	LU	SI	VG	MR	SC	BL	BM	HZ
854	**2E-hexenal**	**0.1**	t	-	t	-	**0.1**	**0.5**	-	t	-	-	-	-	-
857	**3Z-hexenol**	**0.7**	t	-	t	-	**0.2**	-	-	-	-	-	-	-	-
899	**heptanal**	-	-	-	-	-	-	-	-	-	**0.2**	**0.2**	**0.5**	**1.0**	**0.3**
901	**unknown, FW 125, 43, 55, 67**	**0.7**	t	-	-	-	-	-	-	-	-	-	-	-	-
912	**unknown, FW 125, 43, 55, 67**	**0.6**	t	-	-	-	-	-	-	-	-	-	-	-	-
926	tricyclene	0.7	3.1	t	1.5	0.3	t	0.2	t	t	t	t	-	t	-
931	α-thujene	0.3	0.7	0.2	0.8	0.1	0.5	0.6	0.1	0.2	0.6	1.1	1.5	2.6	1.0
939	**α-pinene**	**1.2**	**6.4**	**4.6**	**1.6**	**14.3**	**4.7**	**36.7**	**0.9**	**1.3**	**0.6**	**4.5**	**1.9**	**3.9**	**3.9**
953	camphene	0.9	2.4	t	0.6	0.5	0.1	0.4	0.1	0.1	-	t	-	t	t
957	thuja-2,4(10)-diene	-	-	-	-	0.1	0.2	0.2	-	-	-	-	-	-	-
976	sabinene	6.3	15.0	6.2	12.2	2.7	26.2	8.5	0.6	6.4	20.0	35.6	34.5	36.8	38.6
978	1-octen-3-ol	t	t	-	t	1.6	0.9	0.1	0.9	-	-	t	t	-	t
980	β-pinene	t	0.3	-	0.2	t	t	1.2	t	0.1	t	0.2	t	0.3	0.5
991	myrcene	1.6	4.5	0.6	3.6	3.0	2.7	3.7	0.5	0.7	0.9	1.5	2.5	4.9	3.5
996	**hexanoic acid, 4-methyl, methyl ester***	-	-	-	-	-	-	-	-	**0.1**	-	-	**0.2**	**1.4**	-
1001	δ-2-carene	0.1	0.3	-	0.6	-	-	0.1	t	0.1	-	-	0.7	0.4	0.1
1005	α-phellandrene	0.1	0.1	t	t	t	0.1	-	-	-	-	t	0.1	-	t
1011	δ-3-carene	-	-	-	t	t	-	-	t	0.1	-	0.1	-	t	0.5
1018	α-terpinene	1.4	0.8	0.6	1.4	0.4	0.8	0.7	t	0.2	0.7	1.1	1.6	1.8	1.0
1026	p-cymene	1.2	0.4	0.2	2.2	0.5	1.2	0.1	t	0.1	t	0.4	0.1	0.3	0.1
1031	limonene	6.2	15.8	1.0	12.2	23.5	42.6	20.8	13.1	15.3	11.7	7.9	1.7	4.9	6.8
1031	β-phellandrene	-	-	t	t	t	-	-	-	-	-	t	0.1	t	0.1
1032	1,8-cineole	-	-	-	-	-	-	-	-	0.3	-	-	-	-	t
1050	(E)-β-ocimene	-	0.1	0.1	-	-	0.4	0.2	-	t	-	0.1	0.1	0.7	0.2
1062	γ-terpinene	2.7	1.8	1.2	2.8	0.7	1.4	1.1	0.1	0.3	1.3	1.9	2.6	2.9	1.6
1068	cis-sabinene hydrate	1.0	0.7	0.5	1.0	0.1	1.2	0.4	-	0.3	0.5	1.9	1.5	2.6	1.1

KI	Compound	GR	EK	SX	UR	BR	BA	LU	SI	VG	MR	SC	BL	BM	HZ
1074	trans-linalool oxide(furanoid)	0.3	-	-	-	-	-	-	-	0.1	-	-	-	-	-
1088	terpinolene	0.7	0.6	0.5	0.9	0.8	0.8	1.3	0.3	0.4	0.7	1.1	1.1	1.6	0.8
1091	**2-nonanone**	-	-	-	-	-	-	-	t	-	-	0.4	2.6	1.8	-
1091	6,7-epoxymyrcene	0.2	t	-	t	-	-	-	-	-	-	-	-	-	-
1097	trans-sabinene hydrate	0.7	0.3	0.5	0.4	t	1.0	0.3	-	-	0.2	1.3	1.3	1.7	0.9
1098	**linalool**	2.4	0.3	-	-	1.3	-	-	0.6	5.7	0.1	2.0	1.0	3.0	0.5
1102	**n-nonanal**	-	-	-	-	-	-	t	0.1	0.1	t	-	0.3	0.2	t
1111	1-octen-3-yl acetate	-	-	-	-	-	0.2	-	-	-	-	-	-	-	-
1116	**C₁₀OH, 43,79,67,53,152**	1.9	0.7	-	0.1	-	-	-	-	-	-	-	-	-	-
1121	cis-p-menth-2-en-1-ol	0.8	0.3	0.4	0.5	0.3	0.6	0.2	-	0.1	0.1	0.5	0.6	0.5	0.5
1125	**α-campholenal**	-	-	0.1	-	0.5	-	0.1	-	-	-	-	-	-	-
1139	trans-pinocarveol	-	-	-	-	0.8	-	-	-	-	-	-	-	-	-
1140	trans-p-menth-2-en-1-ol	0.7	0.1	0.4	0.3	-	0.5	0.2	-	-	0.1	0.3	0.3	0.2	0.2
1143	camphor	1.0	3.5	2.4	1.9	10.5	-	-	0.4	4.0	-	0.2	-	-	-
1144	**trans-verbenol**	1.5	-	-	-	-	0.1	0.2	t	-	-	-	-	-	-
1148	**camphene hydrate**	-	1.0	-	0.6	-	-	-	t	0.2	-	-	-	-	-
1156	sabina ketone	0.4	t	0.1	t	0.2	0.1	-	-	-	-	-	-	-	-
1163	pinocarvone	-	-	-	-	3.1	-	-	-	-	-	-	-	-	-
1165	borneol	2.2	2.4	-	2.0	-	-	0.1	t	0.8	0.1	-	-	-	-
1170	coahuilensol	-	-	-	-	-	-	t	t	0.7	-	-	-	-	-
1177	terpinen-4-ol	11.5	3.2	4.6	6.0	1.8	9.2	2.7	0.2	1.7	1.8	8.4	7.6	6.5	5.7
1179	naphthalene	-	-	-	0.2	0.3	0.1	-	-	-	0.5	-	-	-	-
1183	p-cymen-8-ol	0.4	-	0.1	-	0.2	0.2	-	-	-	-	-	-	-	-
1184	**dill ether**	0.7	0.1	-	t	-	t	-	-	-	-	-	-	-	-
1189	α-terpineol	t	0.2	0.3	1.2	0.2	0.3	0.3	t	0.1	0.1	0.3	0.3	0.2	0.2
1191	myrtenal	t	-	t	-	0.2	0.1	-	-	-	-	-	-	-	-
1191	myrtenol	0.4	-	-	-	0.2	-	t	-	-	-	-	t	t	t
1193	cis-piperitol	-	-	-	-	-	0.1	0.1	-	-	-	t	t	-	-
1195	**methyl chavicol(=estragole)**	0.9	-	-	-	-	-	-	0.7	0.1	0.4	t	t	-	-
1199	γ-terpineol	-	-	-	-	-	-	-	-	-	-	-	-	-	-
1204	verbenone	-	-	0.3	-	-	t	t	-	-	-	-	-	-	-
1205	trans-piperitol	0.4	-	-	-	0.2	0.2	-	-	t	-	0.1	0.1	-	0.1

KI	Compound	GR	EK	SX	UR	BR	BA	LU	SI	VG	MR	SC	BL	BM	HZ
1217	trans-carveol	-	-	t	-	0.6	0.1	t	-	-	-	-	-	-	-
1228	citronellol	0.9	t	0.4	t	-	0.2	0.1	0.2	3.1	0.5	0.6	t	t	0.4
1229	cis-carveol	-	-	-	-	0.4	-	-	-	-	-	-	-	-	-
1252	**piperitone**	**0.4**	**0.2**	-	**1.1**	-	-	-	**t**	**0.2**	-	-	**t**	**0.2**	-
1257	**4Z-decen-1-ol**	-	-	**0.3**	-	-	**0.1**	-	-	-	**t**	**0.2**	**t**	**t**	**t**
1258	trans-myrtanol	-	-	-	-	-	-	-	-	0.2	-	-	-	-	-
1261	methyl citronellate	-	-	-	-	-	-	0.2	-	0.1	0.2	t	t	-	-
1265	C$_{10}$OH, 152,123,91,77,109	-	-	-	-	-	-	-	-	0.5	-	-	-	-	-
1267	C$_{10}$OH, 43,91,79,119,152	0.7	t	-	t	-	-	-	-	-	-	-	-	-	-
1277	**pregeigerene B**	-	-	**0.9**	-	-	-	-	**3.9**	**5.3**	**3.1**	**7.6**	**3.0**	**2.2**	-
1285	**bornyl acetate**	**31.3**	**30.8**	**0.3**	**38.4**	**7.8**	**0.6**	**1.7**	**t**	**t**	**t**	**1.0**	**0.3**	**0.4**	**0.2**
1285	**safrole**	-	-	-	-	**1.2**	-	-	**27.2**	**18.6**	**3.8**	-	-	-	-
1297	trans-pinocarvyl acetate	-	-	-	-	0.2	0.1	t	-	t	t	-	-	-	-
1323	methyl geranate	-	-	-	-	-	-	0.2	-	-	t	t	-	-	-
1354	citronellyl acetate	-	-	-	-	-	-	-	-	0.1	-	-	-	-	-
1356	eugenol	-	-	-	-	-	-	-	-	0.1	-	-	-	-	-
1401	**methyl eugenol**	**0.6**	**t**	-	**0.9**	**t**	-	-	**15.7**	**3.3**	**8.4**	-	-	**t**	-
1418	(E)-caryophyllene	0.1	-	0.2	t	0.2	0.3	t	0.2	0.1	0.5	0.2	t	0.7	-
1429	**cis-thujopsene**	-	-	**0.2**	-	**2.7**	**0.3**	-	-	-	-	-	-	-	-
1450	trans-muurola-3,5-diene	-	-	-	0.3	-	-	-	-	-	-	-	-	-	0.2
1461	cis-muurola-4(14),5-diene	-	-	-	-	-	-	-	-	-	-	-	-	-	-
1473	trans-cadina-1(6),4-diene	t	t	-	0.2	-	-	-	-	-	0.5	-	-	-	0.2
1477	γ-muurolene	-	-	-	-	-	-	-	-	-	-	-	-	t	-
1480	germacrene D	-	t	3.9	0.3	-	0.1	1.6	0.4	-	-	0.2	0.1	0.2	0.1
1491	**trans-murrola-4(14),5-diene**	**0.1**	**0.1**	-	**0.8**	-	-	-	-	-	**1.0**	-	-	**0.2**	**0.1**
1493	epi-cubebol	t	-	-	0.1	-	-	-	-	-	0.5	t	-	-	0.3
1495	(E)-methyl isoeugenol	-	-	-	t	-	-	-	0.2	-	-	-	-	-	-
1499	α-muurolene	t	t	t	t	t	-	0.2	t	0.1	0.4	t	-	0.3	0.4
1505	β-bisabolene	-	-	-	-	-	-	-	-	-	0.3	-	-	-	-
1513	γ-cadinene	t	t	0.1	0.6	-	-	0.2	t	0.3	0.8	0.2	-	0.5	1.1
1513	cubebol	t	-	-	-	-	-	-	0.2	-	0.8	-	-	-	-
1524	δ-cadinene	t	0.3	0.6	0.6	t	t	0.8	0.3	0.5	1.0	0.3	-	1.0	2.1

KI	Compound	GR	EK	SX	UR	BR	BA	LU	SI	VG	MR	SC	BL	BM	HZ
1538	α-cadinene	-	-	-	-	-	-	-	-	-	-	-	-	-	0.2
1549	**elemol**	-	-	8.2	0.7	-	-	t	11.5	7.2	1.0	3.1	1.9	1.7	5.0
1554	**elemicin**	0.7	t	-	t	-	-	-	2.7	1.9	20.2	-	-	-	-
1564	(E)-nerolidol	-	-	0.5	-	-	-	-	-	-	-	-	-	-	-
1574	germacrene D-4-ol	-	0.1	0.5	t	-	t	0.6	0.4	1.1	0.9	1.0	t	3.6	5.5
1581	**caryophyllene oxide**	t	-	t	-	t	0.2	0.3	0.2	0.1	-	-	-	-	-
1596	cedrol	0.1	t	-	t	0.2	0.4	-	-	-	0.1	-	-	-	-
1606	β-oplopenone	t	t	0.7	t	0.2	0.1	0.3	0.2	0.2	0.1	0.2	-	0.7	3.2
1610	C15OH, 105,91,41,131,220?	-	-	-	-	-	-	0.5	-	-	-	-	-	-	-
1627	1-epi-cubenol	0.2	-	-	0.6	-	-	-	-	-	0.8	-	-	-	0.2
1630	α-acorenol	-	-	-	-	-	-	-	0.5	0.2	-	-	-	-	-
1630	γ-eudesmol	t	-	3.0	t	-	-	-	1.4	0.9	t	0.3	0.3	t	0.2
1640	epi-α-cadinol	t	t	0.7	t	-	t	0.4	0.4	0.4	0.6	0.2	-	1.1	1.2
1640	epi-α-muurolol	t	t	1.0	t	-	t	0.5	0.4	0.5	0.6	0.2	-	t	1.6
1645	α-muurolol (=torreyol)	-	-	t	t	-	-	0.3	-	0.2	0.2	-	-	t	0.4
1649	β-eudesmol	t	t	3.6	0.2	-	-	-	2.5	1.2	0.3	0.3	0.4	0.2	0.6
1652	α-eudesmol	t	t	3.3	0.2	-	-	-	3.9	1.4	0.7	0.3	0.5	t	0.5
1653	α-cadinol	t	t	3.3	t	0.2	0.1	2.3	t	1.5	0.7	0.6	-	1.7	4.5
1666	bulnesol	-	-	1.6	t	-	-	-	1.5	1.6	0.2	0.6	0.4	0.3	0.9
1685	**eudesma-4(15),7-dien-1-β-ol**	-	-	3.2	-	0.5	0.3	1.7	-	-	-	-	-	t	-
1789	**8-α-acetoxyelemol**	-	-	3.3	-	-	-	-	2.7	-	6.1	8.7	6.8	2.3	-
1881	oplopanonyl acetate	-	-	-	-	-	-	-	-	-	-	-	-	-	0.6
1930	rosa-5,15-diene	-	-	-	-	-	-	-	-	-	-	-	1.1	-	-
2054	abietatriene	t	-	t	t	0.5	t	0.2	t	t	t	-	t	t	-
2055	**manool**	t	-	3.7	-	-	-	-	-	t	-	-	16.2	1.4	-
2080	abietadiene	t	-	2.1	-	1.6	t	t	-	-	-	-	-	t	-
2147	**abieta-8(14),13(15)-diene**	-	-	0.4	-	-	-	-	-	-	-	-	-	-	-
2180	**sandaracopimarinal**	-	-	0.5	-	-	-	-	-	-	-	-	-	-	-
2220	**diterpene, 243,286,271,41,91**	-	-	1.0	-	-	-	-	-	-	-	-	-	-	-
2227	**diterpene, 91,43,105,133,187,286**	-	-	0.6	-	-	-	-	-	-	-	-	-	-	-
2278	sempervirol	0.3	-	1.2	t	2.2	0.2	2.5	0.4	t	-	-	0.8	t	-
2288	**4-epi-abietal**	-	-	8.9	-	-	0.1	-	0.1	0.2	0.3	-	0.6	0.6	1.5

KI	Compound	GR	EK	SX	UR	BR	BA	LU	SI	VG	MR	SC	BL	BM	HZ
2302	trans-totarol	0.6	-	1.7	t	2.1	0.2	1.5	0.3	0.1	-	-	0.4	t	-
2325	trans-ferruginol	t	-	-	-	t	-	0.2	-	-	-	-	t	-	-
2361	**diterpene, 135,286,91,148,187**	-	-	**0.6**	-	-	-	-	-	-	-	-	-	-	-
2477	**diterpene, 43,255,270,187,131**	-	-	**2.7**	-	-	-	-	-	-	-	-	-	-	-

KI = Kovat's Index on DB-5(=SE54) column. *Tentatively identified. Compositional values less than 0.1% are denoted as traces (t). Unidentified components less than 0.5% are not reported

Table A.5. Comparisons of the per cent total oil for leaf essential oils for the serrate leaf junipers of the western hemisphere. *J. pinchotii*(PN), *J. saltillensis*(SA), *J. zanonii* (ZA), *J. californica*(CA), *J. osteosperma*(OS), *J. standleyi*(ST), *J. deppeana* var. *gamboana* (GM), *J. deppeana*(DP), *J. jaliscana*(JA), *J. durangensis*(DR), *J. coahuilensis* (CH), *J. arizonica* (AZ), *J. monticola*(MT), *J. monosperma*(MN), *J. angosturana*(AN), *J. flaccida* (FF), *J. martinezii* (MR), *J. poblana* (PB), *J. comitana*(CM), *J. ashei*(AS), *J. occidentalis*(OC), *J. grandis* (GR). For *J. ovata* oil, see Adams and Baker (2007).

KI	Compound	PN	SA	ZA	CA	OS	ST	GM	DP	JA	DR	CH	AZ	MT	MN	AN	FL	MR	PB	CM	AS	OC	GR
926	tricyclene	0.4	0.6	-	0.2	0.5	0.4	0.3	3.1	0.1	0.2	t	0.4	0.8	0.1	0.1	0.2	0.5	t	0.1	2.9	2.5	0.2
931	α-thujene	0.5	0.2	1.8	0.2	0.2	1.4	t	0.5	t	t	1.3	0.8	-	t	t	0.1	0.6	t	-	-	1.4	0.5
939	**α-pinene**	**2.0**	**2.7**	-	**40.0**	**3.0**	**4.1**	**31.0**	**13.2**	**43.2**	**47.5**	**2.3**	**13.9**	**9.9**	**49.2**	**23.1**	**52.4**	**13.5**	**58.8**	**65.7**	**0.6**	**9.7**	**1.3**
953	α-fenchene	t	-	-	t	-	t	t	0.7	t	0.8	-	-	-	t	0.6	-	-	t	-	-	0.1	0.1
953	camphene	0.4	0.6	t	0.3	0.5	0.5	0.6	1.5	0.5	0.5	-	0.5	1.0	0.5	0.2	0.7	0.6	0.5	1.0	2.3	1.1	0.2
957	thuja-2,4(10)-diene	-	-	-	t	0.1	-	t	0.1	t	0.1	-	t	-	t	t	0.1	0.1	0.1	0.1	t	-	-
967	verbenene	-	1.6	0.2	-	-	0.3	1.8	0.2	0.1	1.8	-	0.3	1.2	0.4	1.7	1.4	1.8	0.5	-	-	t	0.2
976	**sabinene**	**22.0**	**7.1**	**30.9**	**7.6**	**5.3**	**31.9**	**0.1**	**8.1**	t	t	**20.5**	**0.3**	t	**0.2**	t	**0.2**	**8.5**	**0.1**	**0.1**	-	**23.3**	**10.8**
980	β-pinene	0.1	1.1	0.6	0.9	0.1	0.2	3.9	0.9	2.6	2.0	0.2	0.9	0.7	1.5	1.1	5.9	1.1	4.4	6.4	0.1	0.4	0.1
985	6-methyl-5-hepten-2-one	-	-	-	-	t	-	-	-	-	-	-	-	-	-	-	-	-	-	-	t	-	-
991	myrcene	2.0	2.0	1.6	1.4	0.8	3.4	4.6	2.9	4.6	2.8	2.4	2.1	2.4	2.1	2.8	5.6	4.0	5.0	6.8	0.8	1.7	1.0
1000	hexanoic acid, 4-methyl, methyl ester*	-	t	-	-	-	-	-	-	-	-	-	-	0.2	-	-	0.2	-	-	-	-	-	-
1001	δ-2-carene	-	-	-	0.2	t	0.4	t	0.2	-	-	-	0.1	0.1	t	0.1	-	t	1.7	-	t	0.1	0.1
1005	α-phellandrene	t	t	0.3	t	0.1	0.1	0.3	1.3	t	-	0.2	-	0.1	0.6	0.1	0.1	0.9	0.1	0.1	0.5	0.8	-
1011	**δ-3-carene**	**0.5**	**0.4**	-	**0.1**	**0.1**	**1.3**	**1.6**	**7.6**	t	**14.9**	-	t	-	**1.1**	**11.5**	t	t	**1.3**	**0.1**	t	**1.1**	**2.1**
1018	α-terpinene	1.3	0.4	0.3	0.6	0.3	1.9	t	0.7	0.1	t	3.1	t	0.1	0.1	0.1	t	1.1	t	t	2.6	2.2	0.3
1026	p-cymene	0.1	0.1	0.2	0.3	1.5	0.1	t	1.8	0.2	0.2	0.2	t	0.8	0.7	0.2	0.1	1.2	0.7	t	2.6	13.5	5.1
1027	sylvestrene	-	-	-	-	-	-	-	-	-	0.1	-	-	-	-	0.1	-	-	-	-	-	-	-
1031	**limonene**	**3.6**	**3.0**	**1.0**	**2.9**	**3.6**	**3.0**	**0.2**	**0.3**	**24.0**	**1.6**	**2.2**	**2.8**	**10.0**	**1.5**	**0.5**	**3.5**	**1.6**	**2.0**	**1.4**	**4.6**	**0.1**	**1.8**
1031	**β-phellandrene**	**0.1**	**0.3**	**1.1**	**0.4**	t	**0.8**	**4.9**	**10.7**	**0.1**	**1.7**	**0.5**	**2.6**	**0.4**	**6.2**	**0.6**	**3.6**	**5.0**	**4.2**	**1.9**	-	**5.0**	**1.9**
1032	1,8-cineole	0.1	0.3	-	-	-	-	t	t	-	0.1	0.5	0.1	-	-	0.4	0.1	-	-	-	-	-	-
1050	(E)-β-ocimene	-	0.1	-	t	0.1	0.2	0.8	0.3	0.1	0.3	0.2	0.4	0.1	t	0.2	1.7	0.4	0.7	0.1	-	0.1	-
1062	γ-terpinene	2.2	0.7	5.6	1.1	0.6	3.1	t	1.1	0.4	0.3	4.8	1.8	0.4	0.7	0.5	0.2	2.0	t	0.1	0.3	3.4	2.7
1068	cis-sabinene hydrate	1.2	0.4	0.9	0.4	0.8	1.4	-	0.7	-	-	1.8	0.8	-	t	-	0.2	0.5	-	-	-	0.9	-
1072	unknown,41,55,101,159	-	0.2	-	-	-	-	-	-	-	-	-	-	1.1	-	-	-	-	-	-	-	-	-
1074	trans-linalool oxide(fur.)	-	-	-	-	0.2	-	-	-	-	-	t	-	-	t	t	0.1	-	-	-	t	-	t

KI	Compound	PN	SA	ZA	CA	OS	ST	GM	DP	JA	DR	CH	AZ	MT	MN	AN	FL	MR	PB	CM	AS	OC	GR
1088	fenchone	-	-	-	-	-	-	-	-	-	-	-	-	-	0.6	-	-	-	-	-	t	-	-
1088	terpinolene	0.8	0.6	1.7	0.5	0.4	1.3	0.6	1.6	0.9	2.1	1.6	1.0	0.5	0.6	2.3	0.5	0.9	t	0.7	0.3	1.1	0.2
1088	p-cymenene	-	-	-	-	-	-	-	-	-	-	-	0.5	-	-	-	-	-	-	-	-	-	0.2
1094	96, 109, 67, 137, 152	-	-	-	-	-	-	-	-	-	-	-	-	-	-	-	-	-	-	-	-	-	-
1095	α-pinene oxide	-	3.2	-	-	-	-	0.9	-	-	0.9	-	0.4	0.5	0.6	0.6	1.5	1.9	0.2	-	-	-	-
1097	trans-sabinene hydrate	1.0	0.1	1.0	0.5	1.0	1.0	-	0.4	-	-	2.6	-	-	-	-	-	0.5	-	-	0.7	0.7	2.5
1098	linalool	-	0.1	-	t	0.4	2.0	5.8	1.7	0.2	1.9	-	-	1.1	0.1	0.3	4.6	3.0	1.4	1.4	1.4	0.3	-
1102	n-nonanal	-	-	0.1	-	-	-	-	-	-	-	-	-	-	0.1	0.1	-	-	-	-	-	-	-
1103	isopentyl isovalerate	-	-	0.1	0.1	-	-	-	-	-	-	-	-	-	-	-	-	-	-	-	0.2	-	-
1112	endo-fenchol	-	-	-	-	-	-	-	-	-	-	-	-	t	t	-	0.1	-	0.2	t	-	-	-
1114	trans-thujone	t	-	-	-	0.2	t	-	-	-	t	t	-	-	t	-	-	-	-	-	-	t	0.1
1116	3-methyl butanoate, 3-methyl-3-butenyl-	-	-	0.2	-	-	-	-	-	-	-	-	-	-	0.6	-	0.2	-	-	-	0.1	-	-
1121	cis-p-menth-2-en-1-ol	0.4	0.2	0.8	0.2	0.7	0.7	0.1	1.0	t	t	1.0	0.5	0.1	0.4	0.1	0.2	0.4	0.7	t	0.3	0.6	1.4
1124	chrysanthone	-	-	-	-	-	-	0.1	0.1	-	-	-	-	-	t	0.1	-	-	-	-	-	-	-
1125	α-campholenal	t	t	-	t	0.3	-	t	0.6	0.1	0.3	-	-	-	0.3	-	0.3	0.4	0.7	0.2	0.1	-	-
1139	cis-pinene hydrate	-	-	-	-	-	-	-	-	-	-	-	0.4	-	-	-	-	-	-	-	-	-	-
1139	trans-pinocarveol	-	-	-	t	t	-	t	-	0.1	0.2	0.6	0.3	t	0.8	-	0.3	0.7	0.8	0.2	-	-	-
1140	trans-p-menth-2-en-1-ol	0.2	-	0.5	0.2	t	0.5	0.9	-	-	-	0.6	t	-	-	-	-	-	0.2	-	0.1	0.5	1.0
1141	cis-verbenol	-	-	-	-	-	-	-	-	-	-	-	-	t	-	0.1	0.1	-	0.2	-	-	-	-
1143	**camphor**	**31.4**	**42.1**	**0.1**	**16.2**	**33.4**	**0.3**	**1.1**	**19.0**	**0.1**	**0.2**	**0.4**	**9.9**	**0.9**	**1.9**	**0.9**	**0.9**	**11.4**	**t**	**0.1**	**64.9**	**1.7**	**6.1**
1144	trans-verbenol	-	-	-	-	-	-	0.2	-	-	-	-	-	-	-	-	-	0.5	1.3	0.3	-	-	-
1146	isopulegol	-	-	-	-	-	-	-	-	-	1.5	0.5	-	0.5	-	-	0.1	-	t	-	-	-	-
1148	camphene hydrate	0.8	1.2	t	0.3	1.7	0.1	0.4	0.8	0.1	-	-	0.4	1.0	0.2	-	0.3	-	1.1	0.2	1.7	0.2	0.3
1153	citronellal	0.7	0.9	4.3	1.0	0.8	t	0.1	-	t	0.1	4.8	t	t	-	-	0.1	-	-	-	-	-	-
1156	sabina ketone	-	-	-	-	-	t	-	0.2	-	-	-	-	-	t	-	-	-	-	-	-	0.3	1.1
1156	iso-isopulegol	-	-	0.2	0.1	-	-	0.1	-	-	0.5	0.2	-	t	-	-	0.1	-	-	-	-	-	-
1156	isoborneol	0.1	0.1	-	-	-	-	-	-	-	-	-	-	-	-	-	-	-	-	-	-	-	-
1159	p-mentha-1,5-dien-8-ol	-	-	-	-	-	-	-	0.5	-	0.2	-	-	-	-	0.2	-	-	-	-	-	-	-
1160	trans-pinocamphone	-	-	-	0.1	-	-	-	-	-	-	-	-	0.2	0.2	-	-	-	-	-	-	-	-
1162	pinocarvone	-	-	-	-	t	-	-	t	-	t	-	-	t	-	-	-	-	t	t	-	-	-
1165	borneol	1.6	0.8	-	0.2	5.8	0.3	0.2	0.1	0.2	0.1	-	0.5	1.8	0.4	t	0.8	0.9	0.8	t	2.0	1.3	4.8

KI	Compound	PN	SA	ZA	CA	OS	ST	GM	DP	JA	DR	CH	AZ	MT	MN	AN	FL	MR	PB	CM	AS	OC	GR
1167	δ-terpineol	-	-	-	-	-	-	0.1	0.5	0.1	0.4	-	-	-	-	-	-	-	-	-	-	-	-
1173	cis-pinocamphone	-	0.1	-	0.1	-	-	t	t	-	-	-	-	0.1	-	t	0.2	0.3	0.2	t	-	t	-
1177	**coahuilensol**	**0.2**	**-**	**-**	**-**	**-**	**-**	**-**	**-**	**-**	**-**	**t**	**2.5**	**-**	**-**	**0.3**	**-**	**-**	**-**	**-**	**-**	**-**	**-**
1177	**terpinen-4-ol**	**7.5**	**1.9**	**13.5**	**3.4**	**7.4**	**9.7**	**0.4**	**2.7**	**0.5**	**0.4**	**13.9**	**4.6**	**0.5**	**0.5**	**0.3**	**0.4**	**8.2**	**1.1**	**0.2**	**0.2**	**4.6**	**13.2**
1180	m-cymen-8-ol	-	-	-	t	-	-	-	0.6	-	0.2	-	-	-	-	-	-	-	-	-	-	-	t
1183	p-cymen-8-ol	-	t	-	t	1.2	-	-	1.1	t	0.2	-	t	t	t	t	t	0.4	0.1	t	0.3	0.5	2.3
1186	m-terpineol*	t	t	-	-	-	-	-	0.2	-	0.2	-	-	-	-	0.7	-	-	-	-	-	-	-
1189	α-terpineol	0.3	0.2	0.5	0.2	0.4	0.4	0.3	1.9	0.5	0.4	0.6	0.2	0.2	0.2	0.4	0.4	0.7	0.3	0.5	0.1	0.2	0.6
1191	myrtenol	-	-	-	-	0.2	-	-	-	-	-	-	-	-	t	-	-	0.2	0.2	t	-	-	-
1193	4Z-decenal	-	0.1	-	-	-	-	-	-	-	-	-	-	-	-	0.3	0.1	-	-	-	-	-	-
1193	myrtenal	t	-	-	t	0.2	-	t	t	-	t	-	t	-	-	-	-	0.2	-	t	t	-	-
1193	cis-piperitol	t	-	0.2	t	-	0.1	-	0.2	-	-	0.2	t	-	-	t	-	-	-	-	-	-	0.6
1195	methyl chavicol	-	-	-	-	-	-	0.2	-	-	-	-	-	-	-	t	t	0.3	0.2	-	-	-	-
1200	trans-dihydrocarvone	-	0.3	-	t	-	-	-	-	-	-	-	-	-	-	t	0.1	-	-	-	-	-	-
1204	verbenone	-	0.2	-	0.1	1.4	-	t	0.7	t	0.2	-	t	0.4	0.4	0.6	0.2	0.4	1.2	0.1	0.1	-	-
1205	trans-piperitol	-	0.1	0.4	-	-	0.3	-	-	-	-	0.3	t	-	t	-	-	-	-	-	-	-	1.0
1217	trans-carveol	-	-	-	t	1.3	-	t	0.2	-	0.2	-	-	-	t	-	0.1	0.1	0.3	t	0.5	-	-
1220	endo-fenchyl acetate	-	-	-	-	-	-	t	-	0.1	-	-	0.1	0.1	-	-	-	-	0.1	0.2	t	-	-
1222	**coahuilensol, methyl ether 0.1**	-	-	-	-	-	-	-	-	-	-	**13.8**	**-**	**-**	**0.9**	**-**	**-**	**-**	**-**	**-**	**-**	**-**	**-**
1228	citronellol	3.0	1.5	1.4	9.0	-	0.2	0.3	-	-	0.4	6.4	t	t	-	-	-	-	-	-	-	-	-
1229	cis-carveol	-	-	-	0.2	0.2	-	-	-	-	-	-	-	-	-	-	0.2	-	-	-	-	-	-
1235	trans-chrysanthenyl acetate	-	-	-	-	-	-	-	-	-	-	-	-	-	-	-	-	-	-	-	-	-	-
1235	acetate	-	0.5	-	-	-	0.2	0.5	0.2	-	-	-	-	1.1	-	1.3	0.1	-	-	-	-	-	-
1235	thymol, methyl ether	-	-	-	-	-	-	-	-	0.1	t	-	-	-	-	-	-	-	-	-	-	-	-
1236	aromatic, 119,134,FW152	-	-	-	-	-	-	-	-	-	-	-	0.6	-	-	-	-	-	-	-	-	-	-
1239	cumin aldehyde	-	-	-	-	-	-	-	-	-	-	-	-	-	-	-	-	-	-	-	-	-	-
1242	carvone	-	-	-	t	0.2	-	-	0.2	-	-	-	-	-	-	0.1	0.1	0.1	0.1	-	0.8	t	0.5
1251	alcohol, 119,109,91,152	-	-	-	-	0.2	-	-	0.2	-	0.4	-	-	-	t	0.3	0.3	-	-	-	-	-	t
1252	carvenone	-	0.5	-	-	0.1	-	-	-	-	-	-	-	-	-	-	-	-	-	-	-	-	t
1252	piperitone	-	-	-	0.2	0.1	-	t	0.7	-	-	-	t	0.1	0.3	0.1	0.6	0.4	0.5	-	-	t	1.1
1254	cis-myrtenol	-	-	-	-	-	-	-	-	-	-	-	-	-	-	-	0.4	0.4	-	-	-	-	t

KI	Compound	PN	SA	ZA	CA	OS	ST	GM	DP	JA	DR	CH	AZ	MT	MN	AN	FL	MR	PB	CM	AS	OC	GR
1257	linalyl acetate	-	-	-	-	-	-	-	0.4	-	-	-	-	0.1	-	-	-	0.4	-	-	-	-	-
1258	4Z-decen-1-ol	-	t	-	-	-	-	-	-	-	0.2	-	t	-	-	-	0.2	-	-	-	-	-	t
1261	trans-myrtanol	-	-	-	-	-	-	-	-	-	-	-	-	-	-	-	0.1	-	t	-	-	-	-
1264	methyl citronellate	-	-	-	t	-	-	-	-	-	1.0	-	-	0.1	-	-	t	-	-	-	-	-	-
1272	alcohol, FW 152, 123, 79	-	-	-	-	-	-	0.4	0.4	-	-	-	0.6	-	0.8	-	-	-	-	-	-	-	-
1277	p-menth-1-en-7-al	-	0.4	1.0	-	-	-	-	0.2	-	-	-	-	-	-	-	-	-	-	-	-	-	-
1281	pregeijerene B	-	-	-	-	-	-	-	-	-	-	0.3	0.2	0.7	2.8	-	-	-	-	-	-	-	-
1283	isopulegyl acetate	-	-	-	0.3	-	-	0.7	-	-	-	-	-	0.8	-	-	-	-	-	-	-	-	-
1285	**bornyl acetate**	7.1	1.0	0.1	0.9	15.0	13.0	11.4	1.9	1.7	0.3	t	2.2	39.1	1.3	1.3	0.7	2.3	0.9	3.2	9.2	8.1	9.4
1287	p-cymen-7-ol	-	-	-	-	0.6	-	-	-	-	-	-	-	-	-	-	-	-	-	-	-	-	-
1293	trans-verbenyl acetate	-	-	-	-	-	-	-	-	-	-	-	-	-	-	-	-	0.9	-	-	-	-	0.2
1298	carvacrol	-	-	-	-	-	1.0	t	-	-	-	-	0.3	0.1	-	0.1	t	0.1	-	-	-	-	-
1302	alcohol, 135, 77, FW 150	-	-	-	t	-	-	-	-	-	-	-	0.1	-	-	-	-	-	-	-	-	-	-
1314	2E, 4E-decadienal	-	-	-	-	t	-	-	-	-	-	0.3	-	0.2	-	-	t	-	-	-	-	-	-
1321	aromatic, 149, FW 164	-	-	-	t	0.5	0.5	-	-	-	0.2	-	-	-	t	-	0.7	0.5	-	-	-	0.5	0.9
1322	methyl geranate	-	-	-	-	-	-	-	3.7	-	-	-	-	-	-	-	-	-	-	-	-	-	t
1350	α-terpinyl acetate	-	-	-	-	-	-	-	-	-	-	-	0.2	-	-	-	-	0.5	-	-	-	t	t
1351	α-cubebene	0.1	-	-	-	-	-	-	-	-	-	-	-	-	-	0.3	-	-	-	-	-	-	-
1374	151, 121, FW166	-	-	-	-	-	-	-	-	-	-	-	0.6	-	-	-	-	-	-	-	-	t	t
1376	α-copaene	t	-	-	-	-	0.2	t	t	-	t	-	0.1	-	-	-	-	t	-	-	-	0.5	0.2
1384	β-bourbonene	-	-	-	-	-	t	-	-	-	t	-	-	-	-	-	-	-	-	-	-	t	0.2
1390	β-cubebene	0.1	-	-	-	-	0.1	0.2	0.2	-	t	-	0.2	-	-	0.4	-	t	-	-	-	t	-
1401	methyl eugenol	-	-	-	0.4	-	-	0.2	-	-	t	t	0.2	-	-	-	0.1	-	-	0.2	-	-	-
1402	longifolene	-	-	-	-	-	0.3	-	t	-	t	-	-	-	-	-	-	-	-	-	-	-	-
1409	α-cedrene	-	-	-	t	-	-	-	-	-	-	t	-	-	-	-	-	-	-	-	-	-	-
1418	(E)-caryophyllene	t	t	-	t	-	0.1	0.4	t	0.2	1.4	t	0.2	t	t	-	0.3	0.2	0.2	0.2	-	t	0.3
1429	cis-thujopsene	-	-	-	0.7	-	-	-	-	-	t	-	t	-	-	-	-	-	0.2	0.2	-	-	0.3
1444	cis-murrola-3,5-diene	-	-	-	-	-	0.3	0.3	-	0.4	t	0.2	0.7	-	-	-	-	-	-	-	-	-	-
1450	trans-muurola-3,5-diene	0.2	-	-	-	-	t	t	t	0.3	t	0.2	0.2	-	0.9	-	-	-	-	-	-	-	-
1454	α-humulene	t	t	-	t	-	t	t	t	0.3	1.0	-	0.2	t	-	-	-	-	-	-	-	t	-
1461	cis-muurola-4(14),5-diene	-	-	-	-	-	0.9	0.9	-	1.1	0.3	-	-	-	-	-	-	-	-	-	-	-	-
1470	**pinchotene acetate**	t	-	-	-	-	-	-	-	-	-	t	2.6	-	-	0.2	-	-	-	-	-	-	-

KI	Compound	PN	SA	ZA	CA	OS	ST	GM	DP	JA	DR	CH	AZ	MT	MN	AN	FL	MR	PB	CM	AS	OC	GR
1471	unknown, 43,133,105,149	-	-	-	-	0.5	-	-	-	-	-	-	-	-	-	-	-	-	-	-	-	-	-
1473	trans-cadina-1(6),4-diene	0.1	-	-	-	-	-	0.1	t	-	0.2	0.1	0.7	-	-	0.8	-	-	-	-	-	0.3	0.3
1477	γ-muurolene	-	-	-	-	-	-	0.1	-	0.1	t	-	-	-	-	-	-	-	-	-	-	-	-
1480	germacrene D	t	-	-	-	-	0.4	0.4	-	1.5	1.3	-	-	-	-	-	-	0.2	0.7	-	-	0.3	-
1491	trans-muurola-4(14),5-diene	0.5	-	-	-	-	t	0.2	t	-	0.3	0.4	1.8	-	-	2.2	-	-	-	-	-	0.1	t
1493	epi-cubebol	-	-	-	-	-	t	0.1	-	-	0.2	0.2	0.7	-	-	0.9	-	-	-	-	-	-	0.7
1499	epi-zonarene	-	-	-	-	-	-	t	0.2	0.2	-	-	-	-	-	-	-	-	-	-	-	-	-
1499	α-muurolene	t	-	-	-	-	t	0.5	0.2	0.2	0.2	-	0.1	-	-	t	t	t	-	-	-	0.4	0.8
1512	β-curcumene	-	-	-	-	-	t	-	-	-	0.2	-	-	-	-	-	-	-	-	-	-	-	-
1513	γ-cadinene	0.4	-	-	-	t	-	1.7	t	0.9	1.1	-	1.8	-	-	2.0	-	1.0	-	-	-	1.7	1.5
1513	cubebol	-	t	-	-	0.5	0.5	-	-	-	-	0.4	-	-	-	-	-	-	-	-	-	-	-
1515	endo-1-bourbonanol	-	-	-	-	-	-	-	-	-	0.1	-	-	-	-	2.0	-	-	-	-	-	t	2.2
1521	trans-calamenene	-	-	-	-	-	-	t	t	-	-	-	-	-	-	-	-	t	-	-	-	-	1.3
1524	δ-cadinene	0.1	-	-	-	t	0.3	2.5	t	1.1	1.1	0.2	1.6	-	-	1.4	t	0.9	-	0.2	-	1.4	-
1526	zonarene	t	-	-	-	-	-	t	-	t	t	0.1	0.4	-	-	0.6	-	-	-	-	-	-	-
1532	trans-cadina-1(2),4-diene	t	-	-	-	-	-	0.1	-	-	t	t	0.1	-	-	0.2	-	-	-	-	-	t	0.4
1538	α-cadinene	-	-	-	-	-	-	-	-	-	t	-	0.2	-	-	-	-	-	-	-	-	-	-
1540	α-copaen-11-ol	-	-	-	-	-	-	-	0.1	-	-	-	-	-	-	-	-	-	-	-	-	t	t
1542	α-calacorene	-	-	-	-	-	-	-	-	-	-	-	-	-	-	-	0.2	-	-	-	-	-	-
1542	sesquiterpene, 43,95,207	-	-	-	-	-	-	0.5	-	-	-	-	-	-	-	-	-	-	-	-	-	-	-
1549	**elemol**	**2.6**	**1.5**	**2.5**	**2.1**	**0.6**	**3.5**	**t**	**t**	**-**	**t**	**7.6**	**1.5**	**5.4**	**3.7**	**8.5**	**0.1**	**0.7**	**-**	**0.8**	**-**	**-**	**0.1**
1554	elemicin	0.1	-	-	1.2	-	0.1	0.1	-	-	-	-	-	-	t	10.2	t	t	-	-	-	-	-
1556	germacrene B	-	-	-	-	-	-	-	-	-	0.2	-	-	-	-	-	-	-	-	0.2	-	-	-
1564	(E)-nerolidol	-	-	-	-	-	-	-	-	-	-	-	-	-	-	-	-	-	1.1	-	-	-	-
1574	germacrene D-4-ol	-	-	-	-	t	-	0.5	t	0.2	-	-	-	-	-	-	-	0.2	-	-	-	0.3	-
1581	caryophyllene oxide	-	-	-	-	t	-	0.2	t	0.1	1.2	-	0.4	-	-	-	0.3	0.2	0.2	-	-	-	0.4
1588	muurolol isomer,161,204	-	-	-	t	-	-	-	-	0.1	-	-	0.3	-	-	-	-	-	0.2	-	-	-	-
1596	cedrol	-	-	-	-	-	-	-	-	-	-	-	-	t	-	-	-	-	-	t	t	-	-
1606	β-oplopenone	-	-	-	-	-	-	0.5	t	t	0.1	-	-	-	-	0.1	-	-	-	-	-	0.2	1.4
1606	humulene epoxide II	-	-	-	-	-	-	-	t	0.1	-	-	-	-	-	-	-	-	-	-	-	-	-
1610	unknown, 43, 119, 195	-	0.2	-	-	0.1	-	-	-	-	-	-	-	0.9	-	-	-	-	-	-	-	-	-
1627	1-epi-cubenol	0.4	0.1	-	-	t	0.1	0.4	t	0.1	0.4	0.6	1.6	-	-	3.1	-	-	-	t	-	0.7	0.9

KI	Compound	PN	SA	ZA	CA	OS	ST	GM	DP	JA	DR	CH	AZ	MT	MN	AN	FL	MR	PB	CM	AS	OC	GR
1630	eremoligenol	-	-	-	-	-	-	-	-	-	-	-	-	0.1	0.5	-	-	-	-	-	-	-	-
1630	γ-eudesmol	0.4	0.3	0.4	0.4	t	0.6	-	t	-	t	1.7	0.2	0.7	2.4	1.6	-	0.2	-	0.2	-	-	-
1640	epi-α-cadinol	-	-	-	-	-	-	3.1	t	4.3	0.6	0.5	0.2	-	-	0.5	-	t	-	0.3	t	0.6	0.8
1640	epi-α-muurolol	-	-	-	-	t	-	3.1	-	-	0.6	0.4	t	-	-	0.5	-	0.2	-	0.2	-	0.7	0.9
1642	cubenol	-	-	-	-	t	-	-	-	-	-	-	0.2	-	-	-	-	1.0	-	-	-	-	-
1645	α-muurolol (= torreyol)	-	-	-	-	-	-	1.0	-	0.7	t	-	-	-	-	0.3	-	t	-	t	-	0.2	t
1649	β-eudesmol	0.6	0.4	0.4	0.8	t	0.8	-	t	-	t	2.8	0.4	1.1	6.5	1.7	-	0.3	-	0.5	-	-	t
1652	α-eudesmol	0.7	0.6	0.5	0.7	t	1.7	-	t	-	-	7.1	0.4	1.9	4.8	3.0	-	0.3	-	0.5	-	-	-
1653	α-cadinol	-	-	-	-	t	-	6.1	-	3.4	1.5	-	-	-	-	-	-	-	-	1.0	-	0.9	1.9
1666	bulnesol	0.6	0.2	0.1	0.1	t	0.9	-	-	-	-	1.6	-	1.1	0.5	0.9	-	-	-	0.2	-	-	-
1685	eudesma-4(15), 7-dien-1-β-ol	-	-	-	-	-	-	-	-	0.2	-	-	t	-	-	-	-	-	-	-	-	-	0.8
1733	oplopenone	-	-	-	-	t	-	-	-	-	-	-	t	-	-	-	-	-	-	-	-	-	-
1741	8-α-11-elemodiol	-	0.2	0.2	-	-	-	-	-	-	-	0.3	-	-	-	-	-	-	-	-	-	-	-
1789	8-α-acetoxyelemol	-	0.5	0.7	0.1	-	-	-	-	-	-	1.3	0.3	2.0	1.7	t	-	-	-	-	-	-	-
1958	iso-pimara-8(14),15-diene	-	-	0.6	-	-	-	-	-	t	-	-	-	-	-	-	-	-	-	-	-	-	-
1960	sandaracopimara-8(14),15-diene	-	-	-	-	-	-	-	-	-	-	-	-	-	-	-	0.2	-	-	t	t	-	-
1989	**manoyl oxide**	**0.7**	**3.6**	**11.7**	**0.2**	**-**	**1.4**	**0.4**	**0.8**	**0.4**	**0.6**	**t**	**-**	**1.6**	**t**	**0.3**	**6.4**	**0.9**	**-**	**0.2**	**2.6**	**2.7**	**-**
1989	isokaurene	-	-	-	-	-	-	-	-	-	-	-	-	1.5	-	-	-	-	-	-	-	-	-
2043	kaurene	-	-	-	-	-	-	-	-	-	-	-	-	-	-	-	-	2.5	-	-	-	-	-
2054	abietatriene	t	0.4	3.4	t	-	t	0.2	-	0.7	0.3	t	-	1.6	t	0.2	0.4	0.3	-	t	t	-	t
2054	manool	-	-	-	-	-	-	0.2	-	-	-	-	-	0.8	-	-	-	0.3	-	-	-	-	-
2080	abietadiene	0.3	-	0.3	-	-	t	-	t	t	t	t	-	-	-	-	-	-	-	-	0.1	-	-
2103	iso-abienol	-	0.5	-	-	-	-	-	-	-	-	t	-	-	-	0.9	-	-	-	t	-	-	-
2278	sempervirol	-	0.4	0.1	-	-	-	-	-	-	-	t	-	-	-	-	-	-	-	-	-	-	-
2288	4-epi-abietal	0.3	-	-	-	-	0.2	-	-	t	0.4	t	-	-	-	-	-	-	-	t	-	-	-
2302	abieta-7,13-dien-3-one	0.1	-	-	-	-	-	-	-	-	t	-	-	-	-	-	-	-	-	t	-	-	-
2302	trans-totarol	-	0.2	-	-	-	1.1	-	-	-	-	-	-	1.3	-	0.3	-	-	-	t	-	-	-
2325	trans-ferruginol	-	0.1	-	-	-	t	-	-	-	-	t	-	-	-	0.3	-	-	-	t	-	-	-

KI = Kovat's Index on DB-5(=SE54) column. *Tentatively identified. Compositional values less than 0.1% are denoted as traces (t).
Unidentified components less than 0.5% are not reported

Appendix III: Cross Indexed Synonymy of *Juniperus*
Note: the genus *Sabina* is abbreviated *S.* to save space

Arceuthos drupacea (Labill.)) Antoine & Kotschy **see** *J. drupacea* Labill.
Chamaecyparis boursieri Decne. **see** *J. occidentalis* Hook.
Chamaecyparis thurifera (Kunth) Endl. **see** *J. poblana* (Mart.) R. P. Adams
Cupressus sabinoides H.B.K. **see** *J. monticola* f. *compacta* (Mart.) R. P. Adams
Cupressus sabinoides Kunth in Humboldt et al. **see** *J. ashei* Buch.
Cupressus sabinoides Kunth **see** *J. monticola* Mart. f. *monticola*
Cupressus thurifera Kunth **see** *J. poblana* (Mart.) R. P. Adams
Cupressus thurifera Schltdl. **see** *J. poblana* (Mart.) R. P. Adams
J. abyssinica hort. ex K. Koch **see** *J. procera* Hochst. ex Endl.
J. aegaea Griseb **see** *J. excelsa* M.-Bieb.
J. africana (Maire) Villar see *J. thurifera* var. *africana* Maire
J. albanica Penzes **see** *J. communis* L. var. communis
J. alpina (Smith) S. F. Gray **see** *J. communis* L. var. *saxatilis* Pall.
J. andina Nutt **see** *J. occidentalis* Hook.
***J. angosturana* R.P. Adams**
J. arenaria (E. H. Wilson) Florin **see** *J. davurica* var. *arenaria* (E.H. Wilson) R. P. Adams
***J. arizonica* (R. P. Adams) R. P. Adams**
***J. ashei* Buchholz**
J. ashei Buchholz var. *saltillensis* (M.T. Hall) Silba **see** *J. saltillensis* M. T. Hall
J. ashei var. *ovata* R. P. Adams **see** *J. ovata* (R. P. Adams) R. P. Adams
J. australis (Endl.) Pilg. in Urban see *J. barbadensis* L. var. *lucayana* (Britton) R. P. Adams
J. bacciformis Carriere **see** *J. phoenicea* L. var. phoenicea
J. badia H. Gay **see** *J. oxycedrus* L.
J. baimashanensis Y.F. Yu & L.K. Fu **see** *J. squamata* D. Don in Lambert
J. barbadensis C. Mohr *non* Linnaeus **see** *J. virginiana* L. var. *silicicola* (Small) E. Murray
J. barbadensis L. subsp. *saxicola* (Britt. & P. Wils.) Borhidi **see** *J. gracilior* var. *saxicola*
J. barbadensis L. subsp. *urbaniana* (Pilg. & Ekm.)Borhidi **see** *J. gracilior* var. *urbaniana*
J. barbadensis* L. var. *barbadensis
J. barbadensis L. var. *jamaicensis* Silba, see *J. barbadensis* L. var. *lucayana* (Britton) R. P. Adams
***J. barbadensis* L. var. *lucayana* (Britton) R. P. Adams**
J. barbadensis L. var. *saxicola* (Britt. & P. Wils.) Silba **see** *J. gracilior* var. *saxicola*
J. barbadensis L. var. *urbaniana* (Pilg. & Ekm.)Silba **see** *J. gracilior* var. *urbaniana*
J. barbadensis L. ssp. *urbaniana* (Pilg. & Ekm.) A. Borhidi **see** *J. gracilior* var. *urbaniana*
J. barbadensis Thunb. **see** *J. chinensis* L.
***J. bermudiana* L.**
J. biasolettii Link **see** *J. macrocarpa* Sibth. & Sm.
J. bieberstieniana hort. ex K. Koch **see** *J. oxycedrus* L.
J. blancoi* Martinez var. *blancoi
***J. blancoi* Martinez var. *huehuentensis* R. P. Adams, S. González and M. González Elizondo**
***J. blancoi* var. *mucronata* (R. P. Adams) Farjon**
J. bonatiana Vis see *J. thurifera* L.
J. borealis Salisb. **see** *J. communis* L. var. *saxatilis* Pall.
***J. brevifolia* (Seub.) Antonine**
J. butanensis Wender **see** *J. recurva* Buch.-Ham. ex D.Don
J. cabiancae Vis. **see** *J. chinensis* L.
J. caesia Regel. **see** *J. communis* var. *saxatilis* Pall.
***J. californica* Carriere**
***J. californica* Carr. f. *luthcyana* J. T. Howell & Twisselm.**
J. californica Carr. subsp. *osteosperma* (Torr.) E. Murray **see** *J. osteosperma* (Torr.) Little
J. californica Carr. var. *monosperma* (Engelm.) Lemm. **see** *J. monosperma* (Engelm.) Sarg.

J. californica Carr. var. *osteosperma* (Torr). E. Murray **see *J. osteosperma* (Torr.) Little**

J. californica Carr. var. *osteosperma* (Torr.) L. Bensen **see *J. osteosperma* (Torr.) Little**

J. californica Carr. var. *siskiyouensis* L.F. Henderson **see *J. occidentalis* Hook.**

J. californica Carr. var. *utahense* Vasey **see *J. osteosperma* (Torr.) Little**

J. californica Carr. var. *utahensis* Engelm. **see *J. osteosperma* (Torr.) Little**

J. canadensis Lodd. ex. Burgsd. **see *J. communis* L var. *depressa* Pursh.**

J. canariensis Guyot **see *J. turbinata* Guss.**

J. carinata (Y.F.Yu & L.K. Fu) R. P. Adams

J. cedrosiana Kellogg **see *J. californica* Carriere**

J. cedrus Webb & Berthol.

J. centrasiatica Kom. **see *J. pseudosabina* Fisch., Mey. & Ave-Lall.**

J. cernua Roxb. **see *J. chinensis* L.**

J. cerrosianus Kellogg **see *J. californica* Carriere**

J. changii Hu & Cheng **see *J. fargesii***

J. chekiangensis Nakai **see *J. mairei* Lemee & Lev.**

J. chengii Y.F.Yu & L.K. Fu **see *J. pingii* Cheng ex. Ferre**

J. chinensis L.

J. chinensis var. *arenaria* E. H. Wilson **see *J. davurica* var. *arenaria* (E.H. Wilson) R. P. Adams**

J. chinensis cv. *gausseni* **see *J. chinensis* L.**

J. chinensis L. var. *gaussenii* (W. C. Cheng) J. Silba **see *J. chinensis* L.**

J. chinensis L. var. *pendula* Franch. **see *J. chinensis* L.**

J. chinensis L. var. procumbens Siebold ex Endl.

J. chinensis L. var. *pyramidalis* **see *J. chinensis* L.**

J. chinensis L. var. *taiwanensis* R. P. Adams & C-F. Hsieh **see *J. tsukusiensis* var. *taiwanensis***

J. chinensis L. var. *torrulosa* **see *J. chinensis* cv. Kaizuki**

J. chinensis L. var. *tsukusiensis* (Masam.) Masamune **see *J. tsukusiensis***

J. chinensis var. sargentii Henry in Elwes & Henry

J. cinerea Carriere **see *J. thurifera* L.**

J. coahuilensis (Martinez) Gaussen ex R. P. Adams

J. coahuilensis var. *arizonica* R.P. Adams **see *J. arizonica* (R. P. Adams) R. P. Adams**

J. comitana Mart.

J. communis L.

J. communis L. f. *crispa* Browicz & Zielinsk **see *J. communis* L.**

J. communis L. subsp. *alpina* (Suter) Celak. **see *J. communis* var. *saxatilis* Pall.**

J. communis L. subsp. *brevifolia* (Sanio) Penzes **see *J. communis* L.**

J. communis L. subsp. *cupressiformis* Vict. & Sennen ex Penzes **see *J. communis* L.**

J. communis L. subsp. *depressa* (Pursh) Franco **see *J. communis* var. *depressa* Pursh**

J. communis L. subsp. *depressa* (Pursh) E. Murray **see *J. communis* var. *depressa* Pursh**

J. communis L. subsp. *eucommunis* Syme in Sowerby **see *J. communis* L.**

J. communis L. subsp. *hemisphaerica* (J. & C. Presl) Nyman **see *J. communis* L. var. *hemisphaerica***

J. communis L. subsp. *nana* (Willd.) Syme in Sowerby **see *J. communis* var. *saxatilis* Pall.**

J. communis L. subsp. *oblonga* (M.-Bieb.) Galushko **see *J. communis* var. *saxatilis* Pall.**

J. communis L. subsp. *pannonica* Penzes **see *J. communis* L.**

J. communis L. subsp. *pygmaea* (K. Koch) Imkhan. **see *J. communis* var. *saxatilis* Pall.**

J. communis L. subsp. *saxatilis* (Pall.) E. Murray **see *J. communis* var. *saxatilis* Pall.**

J. communis L. var. *aborescens* Gaudin **see *J. communis* L.**

J. communis L. var. *alpina* Suter **see *J. communis* var. *saxatilis* Pall.**

J. communis L. var. *brevifolia* Sanio in Hegii **see *J. communis* L. vat, *communis***

J. communis L. var. *canadensis* (Lodd. ex Burgsd.) Loudon **see *J. communis* var. *depressa* Pursh**

J. communis L. var. *caucasica* Endl. **see *J. communis* L. var. *communis***

J. communis L. var. charlottensis R. P. Adams

J. communis L. var. depressa Pursh.

J. communis L. var. *depressa* (Steven) Boiss. **see *J. communis* L.**

J. communis L. var. *erecta* Pursh **see *J. communis* L.**

J. communis L. var. *fastigiata* Parl. in Candolle **see *J. communis* L.**

J. communis L. var. *genuina* Formanek **see *J. communis* L.**

J. communis L. var. hemisphaerica (J. & C. Presl) Parl. in Candolle

J. communis L. var. *hispanica* Endl. **see *J. communis* L.**

J. communis L. var. *jackii* Rehder **see *J. jackii* (Rehder) R. P. Adams**

J. communis L. var. kamchatkensis R. P. Adams

J. communis L. var. kelleyi R. P. Adams

J. communis L. var. *macrocarpa* (Sibth. & Sm.) Spach **see *J. macrocarpa* Sibth. & Sm.**

J. communis var. megistocarpa Fernald & St. John

J. communis L. var. *montana* Aiton **see *J. communis* var. *saxatilis* Pall.**

J. communis L. var. *montana* Neilr., *non* Aiton (1789) **see *J. communis* L.**

J. communis L. var. *nana* (Willd.) Baumg. **see *J. communis* var. *saxatilis* Pall.**

J. communis L. var. *oblonga* (M.- Bieb.) Parl. in Candolle **see *J. communis* var. *saxatilis* Pall.**

J. communis L. var. *oblonga* hort. ex Loudon **see *J. communis* L. var. *communis***

J. communis L. var. *oblonga* (M.-Bieb.) Loud **see *J. communis* L. var. *saxatilis* Pall.**

J. communis L. var. *pannonica* (Penzes) Soo **see *J. communis* L. var. *communis***

J. communis L. var. stricta Carriere **see *J. communis* L. var. *communis***

J. communis L. var. *suecica* (Mill.) Aiton **see *J. communis* L. var. *communis***

J. communis L. var. *typica* Fomin **see *J. communis* L. var. *communis***

J. communis L. var. *vulgaris* Aiton **see *J. communis* L. var. *communis***

J. communis Thunb. **see *J. rigida* Mig. in Sieb.**

J. communis L. var. nipponica (Maxim.) E. H. Wilson

J. communis L. var. saxatilis Pall.

J. compacta (Mart.) R. P. Adams **see *J. monticola f. compacta* Mart.**

J. conferta Parl. **see *J. rigida* Siebold & Zucc. var. *conferta* (Parl.) Patschke**

J. convallium Rehder & Wils.

J. convallium f. pendula (Cheng & L.K. Fu) R.P. Adams

J. convallium var. *microsperma* (Cheng & Fu) Silba **see *J. microsperma* (Cheng & L. K. Fu) R. P. Adams**

J. coreana Nakai **see *J. rigida* Siebold & Zucc. var. *conferta* (Parl.) Patschke**

J. coxii A. B. Jacks, New Fl. Silva 5:33 (1932

J. cracovia hort. ex K. Koch **see *J. communis* L.**

J. davurica Pall. subsp. *maritima* V. M. Urussov **see *J. davurica* Pall.**

J. davurica Pall.

J. davurica Pall. var. arenaria (E. H. Wilson) R. P. Adams

J. davurcia Pall. var. mongolensis (R. P. Adams) R. P. Adams

J. deltoides R. P. Adams

J. deltoides var. spilinanus (Yalt., Elicin & Terzioglu) Terzioglu

J. densa (Carriere) Gordon **see *J. squamata* D. Don in Lambert**

J. deppeana Steud. var. deppeana

J. deppeana Steud. f. elongata R. P. Adams

J. deppeana Steud. f. sperryi (Correll) R.P. Adams

J. deppeana Steud. f. zacatecensis (Mart.) R. P. Adams

J. deppeana subsp. *sperryi* (Correll) E. Murray **see *J. deppeana* f. *sperryi* (Correll) R.P. Adams**

J. deppeana **Steud.** var. *obscura* (Martinez) Gaussen **see *J. deppeana* var. *patoniana* (Mart.) Zanoni**

J. deppeana Steud. var. *pachyphlaea* (Torr.) Martinez **see *J. deppeana* Steudl. var. *deppeana***

J. deppeana Steud. var. patoniana (Mart.) Zanoni

J. deppeana Steud. var. gamboana (Mart.) R. P. Adams

J. deppeana Steud. var. robusta Martinez

J. deppeana Steud. var. *sperryi* (Correll) E. Murray **see *J. deppeana* f. *sperryi* (Correll) R. P. Adams**

J. deppeana Steud. var. *sperryi* Correll **see *J. deppeana* f. *sperryi* (Correll) R.P. Adams**

J. deppeana Steud. var. *zacatecensis* Martinez see **J. deppeana f. zacatecensis (Mart.) R. P. Adams**

J. depressa (Pursh) Raf. **see** *J. communis* var. *depressa* Pursh

J. depressa Raf. ex M'Murtrie **see** *J. communis* L. var. *depressa* **Pursh.**

J. depressa Steven **see** *J. communis* **L.**

J. difformis Gilib. **see** *J. communis* **L.**

J. dimorpha Roxb. **see** *J. chinensis* **L.**

J. distans Florin see *J. tibetica* **Kom.**

J. drobovii Sumnev. **see** *J. semiglobosa* **Regel**

J. drupacea Labill.

J. durangensis Martinez

J. durangensis Martinez var. *topiensis* **R. P. Adams & S. Gonzalez**

J. ekmanii Florin **see** *J. gracilior* var. *ekmanii* (Florin) **R. P. Adams**

J. erectopatens (Cheng & L. K. Fu) R. P. Adams

J. erythrocarpa Cory **see** *J. pinchotii* **Sudw.**

J. erythrocarpa var. *coahuilensis* Martinez **see** *J. coahuilensis* (Martinez) **Gaussen ex. R. P. Adams**

J. excelsa auct., non M.-Bieb. **see** *J. semiglobosa* **Regel**

J. excelsa M.-Bieb.

J. excelsa M.-Bieb. subsp. *excelsa* var. *depressa* O. Schwarz **see** *J. excelsa* **M.-Bieb.**

J. excelsa M.-Bieb. subsp. *polycarpos* var. *pendula* (Mulk.) Imkhan. **see** *J. polycarpos* var. *polycarpos*

J. excelsa M.-Bieb. subsp. *polycarpos* (K. Koch) Takht. **see** *J. polycarpos* **K. Koch** var. *polycarpos*

J. excelsa M.-Bieb. subsp. *polycarpos* var. pendula (Mulk.) Imkhan. **see** *J. polycarpos* var. *polycarpos*

J. excelsa M.-Bieb. subsp. *seravschanica* (Kom.) Imkhan. **see** *J. seravschanica* **Kom.**

J. excelsa M.-Bieb. subsp. *turcomanica* (B. Fedtsch.) Imkhan. **see** *J. polycarpos* var. *turcomanica*

J. excelsa M.-Bieb. var. *densa* Endl. **see** *J. squamata* **D. Don in Lambert**

J. excelsa M.-Bieb. var. *depressa* O. Schwarz **see** *J. excelsa* **M.-Bieb.**

J. excelsa M.-Bieb. var. *farreana* P.N. Mehra **see** *J. seravschanica* **Kom.**

J. excelsa M.-Bieb. var. *polycarpos* (K. Koch) Silba **see** *J. polycarpos* **K. Koch** var. *polycarpos*

J. excelsa Pursh, *non* M.-Bieb. **see** *J. scopulorum* **Sarg.**

J. excelsa Wall. **see** *J. polycarpos* **K. Koch** var. *polycarpos*

J. excelsa Wall., sensu Aitch. **see** *J. polycarpos* **K. Koch** var. *polycarpos*

J. fargesii (Rehder & E. H. Wilson) Kom.

J. fassettii A. Boivin see *J. scopulorum* **Sarg.**

J. flaccida Schltdl.

J. flaccida var. *gigantea* (Roezl) Gaussen **see** *J. flaccida* **Schlecht.**

J. flaccida var. *martinezii* (Perez de la Rosa) J. Silba **see** *J. martinezii* **Perez de la Rosa**

J. flaccida var. *poblana* Mart. **see** *J. poblana* (Mart.) **R. P. Adams**

J. flagelliformis hort. ex Loudon **see** *J. chinensis* **L.**

J. foetida multicaulis Spach in part **see** *J. horizontalis* **Moench**

J. foetida flaccida (Schlecht.) Spach **see** *J. flaccida* **Schlecht.**

J. foetida Spach **see** *J. deppeana* **Steudel** var. *deppeana*

J. foetida Spach var. *excelsa* (M.-Bieb) Spach **see** *J. excelsa* **M.-Bieb.**

J. foetida Spach var. *squarrulosa* Spach **see** *J. foetidissima* **Willd.**

J. foetidissima Willd.

J. foetidissima Willd. var. *squarrosa* Medw. **see** *J. foetidissima* **Willd.**

J. foetidissima Willd. var. *pindicola* Formanek **see** *J. foetidissima* **Willd.**

J. formosana Hayata var. **formosana**

J. formosana f. *tenella* Handel-Mazzetti **see** *J. mairei* **Lemee & Lev.**

J. formosana var. **concolor Hayata**

J. formosana var. *mairei* (Lemee & Lev.) R. P. Adams & C-F. Hsieh, **see** *J. mairei* **Lemee & Lev.**

J. fortunei hort. ex Carriere **see** *J. chinensis* **L.**

J. franchetiana Lev. ex Kom. **see** *J. squamata* **D. Don in Lambert**

J. gallica (Coincy) Rouy **see** *J. thurifera* **L.** var. *thurifera*

J. gamboana Martinez **see** *J. deppeana* var. *gamboana* (Mart.) **R. P. Adams**

J. gaussenii W.C.Cheng **see** *J. chinensis* cv. *gaussenii*

J. gigantea Roezl in part **see** *J. flaccida* Schltdl.

J. gigantea Roezl in part **see** *J. deppeana* Steudel var. *deppeana*

J. glaucescens Florin **see** *J. komarovii* Florin

J. gracilior var. ekmanii (Florin) R.P. Adams

J. gracilior var. *saxicola* (Britt. & P. Wilson) R. P. Adams

J. gracilior var. urbaniana (Pilger & Ekman) R.P. Adams

J. gracilior Pilger var. *gracilior*

J. gracilis Endl. **see** *J. flaccida* Schlecht.

J. grandifolius Link in Buch **see** *J. cedrus* Webb & Berthol.

J. grandis R. P. Adams

J. gymnocarpa (Lemmon) Cory **see** *J. monosperma* (Engelm.) Sarg.

J. hemisphaerica (J. & C. Presl.) Parl. **see** *J. communis* var. *communis*

J. heterocarpa Timb.-Lagr. ex Loret & Barrandon **see** *J. oxycedrus* L.

J. hispanica Mill. **see** *J. thurifera* L.

J. hochstetteri Ant. **see** *J. procera* Hochst. ex Endl.

J. horizontalis Moench

J. horizontalis forma *alpina* (Loud.) Rehder **see** *J. horizontalis* Moench

J. horizontalis forma *lobata* O.W. Knight **see** *J. horizontalis* Moench

J. horizontalis var. *douglasii* Hort. **see** *J. horizontalis* Moench

J. horizontalis var. *variegata* Beissn. **see** *J. horizontalis* Moench

J. hudsonica Forbes **see** *J. horizontalis* Moench

J. humilis Salisb. **see** *J. sabina* L.

J. indica Bertol.

J. indica Bertol. var. **caespitosa** Farjon

J. indica Bertol. var. *rushforthiana* R. P. Adams **see** *J. rushforthiana* (R.P. Adams) R. P. Adams

J. intermedia Schur, **see** *J. communis* L. var. *communis*

J. isophyllos K. Koch **see** *J. excelsa* M.-Bieb.

J. jackii (Rehder) R. P. Adams

J. jaliscana Martinez

J. japonica hort. ex Carriere **see** *J. chinensis* var. *procumbens*

J. jarkendensis Kom. **see** *J. semiglobosa* var. *jarkendensis* (Kom.) R. P. Adams

J. kansuensis Kom. **see** *J. fargesii*

J. knightii A. Nelson **see** *J. osteosperma* (Torr.) Little

J. komarovii Florin

J. lemeeana Lev. & Blin in Leveille **see** *J. fargesii*

J. litoralis Maxim. **see** *J. rigida* Siebold & Zucc. var. *conferta* (Parl.) Patschke

J. lobelii Guss. **see** *J. macrocarpa* Sibth. & Sm.

J. lucayana Britton **see** *J. barbadensis* L. var. *lucayana* (Britton) R. P. Adams

J. lusitanica Mill. **see** *J. sabina* L.

J. lutchuensis Koidz. **see** *J. taxifolia* var. *lutchuensis* (Koidz) Satake

J. lycia L. **see** *J. phoenicea* L. var. phoenicea

J. lycia Pall **see** *J. excelsa* M.-Bieb.

J. macrocarpa Sibth. & Sm.

J. macrocarpa Sibth. & Sm. var. *globosa* Neilr. **see** *J. macrocarpa* Sibth. & Sm.

J. macrocarpa Sibth. & Sm. var. *lobelii* (Guss.) Parl. in Candolle **see** *J. macrocarpa* Sibth. & Sm.

J. macropoda auct. **see** *J. semiglobosa* Regel

J. macropoda Boiss. **see** *J. seravschanica* Kom.

J. maderensis (Menezes) R. P. Adams

J. mairei Lemee & H. Leveille

J. major Dioscorides ex Bubani **see** *J. macrocarpa* Sibth. & Sm.

J. maritima R. P. Adams

J. marschalliana Stev. **see** *J. oxycedrus* L.

J. martinezii Perez de la Rosa

J. media V.D. Dmitriev **see** *J. semiglobosa* **Regel**

J. megalocarpa Sudw. **see** *J. osteosperma* **(Torr.) Little**

J. mekongensis Kom. **see** *J. convallium* **Rehder & Wils. in Sargent**

J. mexicana Schiede ex Schltdl. & Cham. **see** *J. deppeana* **Steudel var.** *deppeana*

J. mexicana Schiede var. *monosperma* (Engelm.) Cory **see** *J. monosperma* **(Engelm.) Sarg.**

J. mexicana Schlect. & Chamisso **see** *J. deppeana* **Steudel var.** *deppeana*

J. mexicana Shiede & Deppe **see** *J. deppeana* **Steudel var.** *deppeana*

J. mexicana Spreng. **see** *J. monticola* **Mart. f.** *monticola*

J. mexicana Sprengel in part, **see** *J. ashei* **Buch.**

J. microphylla Antoine **see** *J. communis* **L.**

J. microsperma (Cheng & L.K. Fu) R.P. Adams

J. monosperma (Engelm.) Sarg. var. *gracilis* Mart. **see** *J. angosturana* R.P. Adams

J. monosperma (Engelm.) Sarg. var. *pinchotii* (Sudw.) Van Melle **see** *J. pinchotii* **Sudworth**

J. monosperma (Engelm.) Sargent

J. monosperma (Engelm.) Sarg. var. *knightii* (A. Nelson) Lemmon **see** *J. osteosperma* **(Torr.) Little**

J. monosperma forma *gymnocarpa* (Lemmon) Rehder **see** *J. monosperma* **(Engelm.) Sarg.**

J. montana (Aiton) Lindl. & Gordon **see** *J. communis* **var.** *saxatilis* **Pall.**

J. monticola Mart. f. monticola

J. monticola Mart. f. compacta Mart.

J. monticola Mart. f. orizabensis Mart.

J. morrisonicola Hayata

J. mucronata R. P. Adams **see** *J. blancoi* **var.** *mucronata* **(R. P. Adams) Farjon**

J. nana Willd. **see** *J. communis* **var.** *saxatilis* **Pall.**

J. nana Willd. var. *alpina* (Aiton) Endl. **see** *J. communis* **var.** *saxatilis* **Pall.**

J. navicularis Gand.

J. niemannii E. L. Wolff **see** *J. communis* **L.**

J. nipponica Maxim. **see** *J. communis* **var.** *nipponica* **(Maxim.) Wils.**

J. oblonga M.-Bieb. **see** *J. communis* **L. var.** *communis*

J. occidentalis Hook.

J. occidentalis Hook. f. corbetii R. P. Adams

J. occidentalis Hook. f. *robinsoni* O. V. Matthew **see** *J. occidentalis* **Hook.**

J. occidentalis Hook. f. *gymnocarpa* (Lemm.) Rehder **see** *J. monosperma* **(Engelm.) Sarg.**

J. occidentalis Hook. subsp. *australis* Vasek **see** *J. grandis* **R. P. Adams**

J. occidentalis Hook. var. *australis* (Vasek) A. & N. Holmgren. **see** *J. grandis* **R. P. Adams**

J. occidentalis Hook. var. *conjugens* Engelm. **see** *J. ashei* **Buch.**

J. occidentalis Hook. var. *gymnocarpa* Lemm. **see** *J. monosperma* **(Engelm.) Sarg.**

J. occidentalis Hook. var. *monosperma* Engelm. **see** *J. monosperma* **(Engelm.) Sarg.**

J. occidentalis Hook. var. *pleiosperma* Engelm. **see** *J. scopulorum* **Sarg.**

J. occidentalis Hook. var. *texana* Vasey **see** *J. ashei* **Buch.**

J. occidentalis Hook. var. *utahensis* (Engelm.) Kent **see** *J. osteosperma* **(Torr.) Little**

J. occidentalis hort. ex Carriere **see** *J. communis* **L.**

J. occidentalis sensu Parl. *non* W. J. Hooker **see** *J. californica* **Carriere**

J. olivieri Carriere **see** *J. excelsa* **M.-Bieb.**

J. oophora Kunze **see** *J. turbinata* **Guss.**

J. oppositifolia Moench **see** *J. bermudiana* **L.**

J. osteosperma (Torr.) Little

J. oxycedrus L.

J. oxycedrus L. var. *badia* H. Gay see *J. oxycedrus* **L.**

J. oxycedrina St.-Leg. **see** *J. oxycedrus* **L.**

J. oxycedrus L. f. *parvifolia* Novak **see** *J. deltoides* **R. P. Adams**

J. oxycedrus L. var. *brevifolia* Seub. **see** *J. brevifolia* **(Seub.) Ant.**

J. oxycedrus L. var. *brachyphylla* Loret in Billot **see** *J. oxycedrus* **L.**

J. oxycedrus L. var. *fastigiata* Jovan. **see** *J. deltoides* **R. P. Adams**

J. oxycedrus L. var. *macrocarpa* (Sibth. & Sm.) Silba see *J. macrocarpa* Sibth. & Sm.

J. oxycedrus L. var. *microcarpa* Neilr. see *J. deltoides* R. P. Adams

J. oxycedrus L. var. *parvifolia* (Novak) Jovan see *J. deltoides* R. P. Adams

J. oxycedrus L. var. *rufescens* (Link) Carriere see *J. oxycedrus* L.

J. oxycedrus L. var. *spilinanus* see *J. deltoides* var. *spilinanus* (Yalt., Elicin & Terzioglu) Terzioglu

J. oxycedrus L. var. *transtagana* (Franco) Silba see *J. navicularis* Gand.

J. oxycedrus L. var. *wittmanniana* hort. ex Carriere, invalid see *J. oxycedrus* L.

J. oxycedrus L. subsp. *badia* (H. Gay) Debeaux see *J. oxycedrus* var. *badia* H. Gay

J. oxycedrus L. subsp. *hemisphaerica* (J. & C. Presl) E. Schmid see *J. communis* L. var. communis

J. oxycedrus L. subsp. *rufescens* (Link) Asch. & Graebn. see *J. oxycedrus* L.

J. oxycedrus L. subsp. *rufescens* auct. lusit. see *J. navicularis* Gand.

J. oxycedrus L. subsp. *macrocarpa* (Sibth. & Sm.) Ball see *J. macrocarpa* Sibth. & Sm.

J. oxycedrus L. subsp. *maderensis* Menezes see *J. madernesis* (Menezes) R. P. Adams

J. oxycedrus L. subsp. *transtagana* Franco see *J. navicularis* Gand.

J. ovata (R. P. Adams) R. P. Adams

J. pachyderma Sitgreaves see *J. deppeana* Steudel var. *deppeana*

J. pachyphlaea Torr. see *J. deppeana* Steudel var. *deppeana*

J. patoniana Martinez see *J. deppeana* var. *patoniana* (Mart.) Zanoni

J. patoniana f. *obscura* Martinez see *J. deppeana* var. *patoniana* (Mart.) Zanoni

J. phoenicea L. var. phoenicea

J. phoenicea L. var. *turbinata* (Guss.) Parl. **see *J. turbinata* Guss.**

J. phoenicea L. subsp. *eu-mediterranea* P. Lebreton & S. Thirend see *J. phoenicea* var. *turbinata* (Guss.) Parl.

J. phoenicea L. subsp. *phoenicea* f. *prostrata* Debreczy & Racz see *J. phoenicea* L.

J. phoenicea L. subsp. *turbinata* (Guss.) Nyman see see *J. turbinata* Guss.

J. phoenicea L. var. *lobelii* Guss. see *J. phoenicea* L.

J. phoenicea L. var. *malacocarpa* Endl. see *J. phoenicea* L.

J. phoenicea L. var. *prostrata* Willk. see *J. phoenicea* L.

J. phoenicea L. var. *sclerocarpa* Endl. see *J. phoenicea* L.

J. phoenicea L. var. *turkinata* (Guss.) Parl. see *J. phoenicea* L.

J. phoenicea Pall. see *J. foetidissima* Willd.

J. pinchotii Sudworth

J. pinchotii Sudw. var. *erythrocarpa* (Cory) Silba see *J. pinchotii* Sudw.

J. pingii Cheng ex. Ferre

J. pingii Cheng ex. Ferre var. miehei Farjon

J. pingii subsp. *polycarpos* (K. Koch) Takht see *J. pingii* Cheng ex. Ferre

J. pingii var. *carinata* Y.F. Yu & L.K. Fu see *J. carinata* (Y.F.Yu & L.K. Fu) R. P. Adams

J. pingii Cheng ex Ferre var. *wilsonii* (Rehder) Silba see *J. squamata* var. *wilsonii*

J. poblana (Mart.) R. P. Adams

J. polycarpos K. Koch var. *pendula* Mulk. see *J. polycarpos* K. Koch var. *polycarpos*

J. polycarpos K. Koch var. polycarpos

J. polycarpos K. Koch var. *seravschanica* (Kom.) Kitam. see *J. seravschanica* Kom.

J. polycarpos K. Koch var. turcomanica (B. Fedtsch) P. P. Adams

J. potaninii Kom. see *J. tibetica* Kom.

J. procera Hochst. ex Endl.

J. procumbens (Sieb. ex Endl.) Miquel see *J. chinensis* var. *procumbens* Sieb. ex Endl.

J. procumbens Sarg. see *J. chinensis* var. *sargentii* Henry

J. prostrata Persoon see *J. horizontalis* Moench

J. przewalskii Kom.

J. przewalskii f. *pendula* (Cheng & L.K. Fu) R.P. Adams & Chu Ge-lin see *J. convallium* f. *pendula* (Cheng & L.K. Fu) R.P. Adams

J. pseudocupressus Dieck see *J. occidentalis* Hook.

J. pseudosabina auct., *non* Fisch., Mey. & Ave-Lall. see *J. indica* Bertol.

J. pseudosabina **Fisch., Mey. & Ave-Lall.**

J. pseudosabina var. *turkestanica* (Kom.) Silba see *J. pseudosabina* **Fisch., Mey. & Ave-Lall.**

J. pygmaea K. Koch **see** *J. communis* **var.** *saxatilis* **Pall.**

J. pyramidalis Salisb. **see** *J. bermudiana* **L.**

J. pyriformis Lindley A. Murray bis ex Lindl see *J. californica* Carriere

J. ramulosa Florin **see** *J. convallium* **Rehder & Wils.**

J. rebunensis Kudo & Suzaki **see** *J. communis* **var.** *saxatilis* **Pall.**

J. recurva **Buch.-Ham. ex D.Don**

J. recurva Buch.-Ham. ex D.Don var. *uncinata* R. P. Adams **see** *J. uncinata*

J. recurva Buch.-Ham. ex. D. Don var. *coxii* (A. B. Jacks.) Melville **see** *J. coxii* **A. B. Jacks**

J. recurva Buch.-Ham. ex D. Don var. *densa* hort. ex Carriere **see** *J. squamata* **D. Don in Lambert**

J. recurva Buch.-Ham. ex D. Don var. *squamata* Mast., non Parlat **see** *J. chinensis var. procumbens*

J. recurva Buch.-Ham. ex D. Don var. *squamata* (D. Don) Parl. **see** *J. squamata* **D. Don in Lambert**

J. recurva Buch.-Ham. ex D. Don var. *typica* Patschke **see** *J. recurva* **Buch.-Ham. ex D.Don**

J. religiosa Carr. **see** *J. semiglobosa Regel*

J. repens Nuttall **see** *J. horizontalis* **Moench**

J. rhodocarpa Steven **see** *J. oxycedrus* **L.**

J. rigida **Siebold & Zucc. var.** *rigida*

J. rigida Siebold & Zucc. f. *modesta* (Nakai) Y. C. Zhu **see** *J. rigida* **Mig. in Sieb.**

J. rigida Siebold & Zucc. subsp. *conferta* (Parl.) Silba **see** *J. rigida* **var.** *conferta* **(Parl.) Patschke**

J. rigida Siebold & Zucc. subsp. *conferta* (Parl.) Kitam. **see** *J. rigida* **var.** *conferta* **(Parl.) Patschke**

J. rigida Siebold & Zucc. subsp. *litoralis* (Maxim.) Urussov **see** *J. rigida* **Siebold & Zucc. var.** *conferta* **(Parl.) Patschke**

J. rigida Siebold & Zucc. subsp. *nipponica* (Maxim.) Franco **see** *J. communis* **var.** *nipponica* **(Maxim.) E. H. Wilson**

J. rigida **Siebold & Zucc. var.** *conferta* **(Parl.) Patschke**

J. rufescens Link **see** *J. oxycedrus* **L.**

J. rushforthiana **(R. P. Adams) R. P. Adams**

J. sabina **L.**

J. sabina L. f. *glauca* S. Priszter **see** *J. sabina* **L.**

J. sabina L. var. *arenaria* (E.H. Wilson) Farjon **see** *J. davurica* **Pall. var.** *arenaria*

J. sabina L. var. *cupressifolia* Aiton **see** *J. sabina* **L.**

J. sabina L. var. *davurica* (Pall.) Farjon **see** *J. davurica* **Pall.**

J. sabina L. var. *erectopatens* (Cheng & L.K. Fu) Y.F. Yu & L.K. Fu **see** *J. erectopatens* **(Cheng & L.K. Fu) R. P. Adams**

J. sabina L. var. *excelsa* (M.-Bied.) Georgi **see** *J. excelsa* **M.-Bieb.**

J. sabina L. var. *lusitanica* (Mill.) C. K. Schnied. **see** *J. sabina* **L.**

J. sabina var. *mongolensis* **see** *J. davurica* **Pall. var.** *mongolensis*

J. sabina L. var. *monosperma* C. Y. Yang **see** *J. sabina* **L.**

J. sabina L. var. *tamariscifolia* Aiton **see** *J. sabina* **L.**

J. sabina L. var. *taurica* Pall. **see** *J. excelsa* **M.-Bieb.**

J. sabina L. var. *vulgaris* Endl. **see** *J. sabina* **L.**

J. sabina L. var. *yulinensis* (T. C. Chang & C. G. Chen) Y. F. Yu & L. K. Fu **see** *J. sabina* **L.**

J. sabina Michx. **see** *J. horizontalis* **Moench**

J. sabina Michx. var. *humilis* Hook. **see** *J. horizontalis* **Moench**

J. sabina Michx. var. *procumbens* Pursh **see** *J. horizontalis* **Moench**

J. sabina Pall. **see** *J. pseudosabina* **Fisch., Mey. & Ave-Lall.**

J. sabina Sibth. & Sm. **see** *J. foetidissima* **Willd.**

J. sabina ß *procumbens* Pursh **see** *J. horizontalis* **Moench**

J. sabina L. var. *jarkendensis* (Kom.) J. Silba **see** *J. semiglobosa* **var.** *jarkendensis*

J. sabina L. β *humilis* Carr. in part **see** *J. horizontalis* **Moench**

J. sabina L. β *procumbens* Pursh **see** *J. horizontalis* **Moench**

J. sabinoides (H.B.K.) Nees **see** *J. monticola* **Mart..**

J. sabinoides (H.B.K.) sensu Sargent *non* Nees **see *J. ashei* Buch.**

J. sabinoides (Kunth) Nees **see *J. monticola* Mart. f. *monticola***

J. sabinoides Endl. **see *J. thurifera* L.**

J. sabinoides f. *monticola* (Martinez) M. C. Johnst. **see *J. monticola* Mart. f. *monticola***

J. sabinoides f. *orizabensis* (Martinez) M. C. Johnst. **see *J. monticola* Mart. f. *orizabensis***

J. sabinoides Griseb. **see *J. foetidissima* Willd.**

J. sabinoides Humb. ex Lindl. & Gordon **see *J. monticola* Mart.**

J. sabinoides Sarg., *non* Griseb. **see *J. ashei* Buch.**

J. saltillensis M. T. Hall

J. saltuaria Rehder & Wils.

J. sargentii (Henry) Takeda **see *J. chinensis* var. *sargentii* Henry in Elwes & Henry**

J. saxicola Britton & P. Wilson **see *J. gracilior* var. *saxicola* (Britt. & P. Wilson) R. P. Adams**

J. schugnanica Kom. **see *J. semiglobosa* Regel**

J. scopulorum Sargent

J. scopulorum Sarg. f. *columnaris* (Fassett) Rehder **see *J. scopulorum* Sarg.**

J. scopulorum Sarg. var. *columnaris* Fassett **see *J. scopulorum* Sarg.**

J. scopulorum Sarg. var. *patens* Fassett **see *J. scopulorum* Sarg.**

J. semiglobosa Regel var. *drobovii* (Sumnev.) Silba **see *J. semiglobosa* Regel**

J. semiglobosa Regel var. semiglobosa

J. semiglobosa var. jarkendensis (Kom.) R. P. Adams

J. semiglobosa var. talassica (Lipinsky) Silba.

J. seoulensis Nakai **see *J. rigida* Mig. in Sieb.**

J. seravschanica Kom.

J. sibirica Burgsd. **see *J. communis* var. *saxatilis* Pall.**

J. sibirica Burgsd. var. *montana* (Aiton) Beck **see *J. communis* var. *saxatilis* Pall.**

J. silicicola (J. K. Small) Bailey **see *J. virginiana* L. var. *silicicola* (Small) E. Murray**

J. sinensis J. F. Gmel. **see *J. chinensis* L.**

J. sphaerica Lindl. **see *J. chinensis* L.**

J. sphaerocarpa Antoine **see *J. macrocarpa* Sibth. & Sm.**

J. squamata Buch.-Ham. ex D. Don in Lambert

J. squamata Buch.-Ham. ex D. Don in Lambert f. *wilsonii* Rehder **see *J. squamata* Buch.-Ham. ex D. Don in Lambert var. *wilsonii* (Rehder) R. P. Adams**

J. squamata Buch.-Ham. ex D. Don in Lambert var. wilsonii (Rehder) R. P. Adams

J. squamata var. *fargesii* Rehder & Wils. **see *J. fargesii* (Rehder & E. H. Wilson) Kom.**

J. squamata Buch.-Ham. ex D. Don in Lambert var. *hongxiensis* Y.F. Yu & L.K. Fu **see *J. squamata* D. Don in Lambert**

J. squamata Buch.-Ham. ex D. Don var. *loderi* (Hornibr.) Hornibr. **see *J. squamata* var. *wilsonii***

J. squamata var. *morrisonicola* (Hayata) H. L. Li & H. Keng **see *J. morrisonicola* Hayata**

J. squamata sensu Kanehira **see *J. morrisonicola* Hayata**

J. squamata var. *parvifolia* Y.F. Yu & L.K. Fu **see *J. squamata* Buch.-Ham. ex D. Don in Lambert**

J. standleyi Steyermark

J. suecica Mill. **see *J. communis* L.**

J. talassica Lipsky **see *J. semiglobosa* var. *talassica* (Lipinsky) Silba**

J. taurica (Pall.) Lipsky **see *J. excelsa* M.-Bieb.**

J. taxifolia Hook. & Arn.

J. taxifolia var. lutchuensis (Koidz) Satake

J. tenella Antoine **see *J. oxycedrus* L.**

J. terminalis Salisb. **see *J. phoenicea* L. var. *phoenicea***

J. tetragona Moench **see *J. phoenicea* L. var. *phoenicea***

J. tetragona Moench var. *oligosperma* Engelm. **see *J. ashei* Buch.**

J. tetragona Schltdl. var. *osteosperma* Torr. **see *J. osteosperma* (Torr.) Little**

J. tetragona Schltdl. *non* Moench **see *J. monticola* Mart. f. *monticola***

J. tetragona Schltdl. var. *osteosperma* J. Torrey **see *J. osteosperma* (Torr.) Little**

J. texensis van Melle **see** *J. pinchotii* Sudworth

J. thunbergii Hook. & Arn. **see** *J. chinensis* L.

J. thurifera L. subsp. gallica (Coincy) E. Murray **see** *J. thurifera* L. var. *thurifera*

J. thurifera L. var. *gallica* Coincy **see** *J. thurifera* L. var. *thurifera*

J. thurifera L. var. thurifera

J. thurifera Spach *non* L. **see** *J. deppeana* Steudel var. *deppeana*

J. thurifera subsp. *africana* (Maire) Gauquelin see *J. thurifera* var. *africana* Maire

J. thurifera var. africana Maire

J. tiahschanica Sumnevicz **see** *J. semiglobosa* Regel

J. tibetica Kom.

J. tsukusiensis Masam.

J. tsukusiensis var. taiwanensis (R. P. Adams & C-f. Hsieh) R. P. Adams

J. turcomanica var. *seravschanica* (Kom.) R.P.Adams **see** *J. seravschanica* Kom.

J. turbinata Guss.

J. turbinata subsp. *canariensis* (Guyot) S. Rivas-Martinez **see** *J. turbinata* Guss.

J. turcomanica B. Fedtsch. **see** *J. polycarpos* var. *turcomanica* (B. Fedtsch) P. P. Adams

J. turkestanica Kom. **see** *J. pseudosabina* Fisch., Mey. & Ave-Lall.

J. umbilicata Gordon **see** *J. macrocarpa* Sibth. & Sm.

J. uncinata (R. P. Adams) R. P. Adams

J. urbaniana Pilg. & Ekman **see** *J. gracilior* var. *urbaniana* (Pilger & Ekman) R.P. Adams

J. utahensis (Engelm.) Lemmon var. *cosnino* Lemmon **see** *J. osteosperma* (Torr.) Little

J. utahensis (Engelm.) Lemmon **see** *J. osteosperma* (Torr.) Little

J. utahensis (Engelm.) Lemmon var. *megalocarpa* (Sudw.) Sargent **see** *J. osteosperma* (Torr.) Little

J. utilis Koidz. var. *modesta* Nakai **see** *J. rigida* Mig. in Sieb.

J. utilis Koidz. **see** *J. rigida* Mig. in Sieb.

J. virginiana var. silicicola (Small) E. Murray

J. virginiana L. subsp. *crebra* (Fernald & Griscom) E. Murray **see** *J. virginiana* L. var. *virginiana*

J. virginiana L. subsp. *scopulorum* (Sarg.) E. Murray **see** *J. scopulorum* Sarg.

J. virginiana L. subsp. *silicicola* (Small) E. Murray **see** *J. virginiana* var. *silicicola* (Small) E. Murray

J. virginiana L. var. *australis* Endl. see *J. barbadensis* L. var. *lucayana* (Britton) R. P. Adams

J. virginiana L. var. *bermudiana* (L.) Vasey **see** *J. bermudiana* L.

J. virginiana L. var. *montana* Vasey **see** *J. scopulorum* Sarg.

J. virginiana L. var. *prostrata* (Pers.) Torr. **see** *J. horizontalis* Moench

J. virginiana L. var. *scopulorum* (Sarg.) Lemmon see *J. scopulorum* Sarg.

J. virginiana L. var. *silicicola* (Small) J. Silba **see** *J. virginiana* var. *silicicola* (Small) E. Murray

J. virginiana L. var. virginiana

J. virginiana var. *ambigens* Fassett **see** *J. virginiana* L. var. *virginiana*

J. virginiana var. *barbadensis* (L.) Gordon **see** *J. barbadensis* L.

J. virginiata var. *crebra* Fernald & Griscom **see** *J. virginiana* L. var. *virginiana*

J. virginiana var. *montana* Vasey **see** *J. scopulorum* Sarg.

J. virginiana var. *prostrata* (Persoon) Torrey **see** *J. horizontalis* Moench

J. virginiana var. *scopulorum* (Sarg.) Lemmon **see** *J. scopulorum* Sarg.

J. virginiana var. *silicicola* (Small) J. Silba **see** *J. virginiana* L. var. *silicicola* (Small) E. Murray

J. virginica Thunb. **see** *J. chinensis* L.

J. vulgaris Tragus ex Bubani **see** *J. communis* L.

J. wallichiana Hook. f. & Thomson ex Brandis see *J. indica* Bertol.

J. wallichiana Hook. f. & Thomson ex Brandis var. *loderi* Hornibr **see** *J. squamata* var. *wilsonii*

J. wallichiana Hook. f. & Thomson ex Brandis var. *meionocarpa* Hand.-Mazz **see** *J. indica*

J. webbii Carriere **see** *J. cedrus* Webb & Berthol.

J. willkommii Antoine **see** *J. macrocarpa* Sibth. & Sm.

J. wittmanniana Fisch. & Lindley invalid **see** *J. oxycedrus* L.

J. x *kanitzii* Csato **see** *J. sabina* L.

J. zacatensis (Martinez) Gaussen **see** *J. deppeana* f. *zacatecensis* (Mart.) R. P. Adams

J. zaidamensis Kom. see *J. tibetica* **Kom.**

J. zaidamensis Kom. f. *squarrosa* Kom. see *J. tibetica* **Kom.**

J. zanonii **R. P. Adams**

S. squamata var. *fargesii* (Rehder & E. H. Wilson) L.K. Fu & Y. F. Yu see *J. fargesii*

S. alpestris Jord. in Jordan & Fourreau see *J. sabina* **L.**

S. bacciformis (Carriere) Antoine see *J. phoenicea* **L.** var. *phoenicea*

S. barbadensis (L.) Small see *J. barbadensis* L.

S. bermudiana (L.) Antoine see *J. bermudiana* **L.**

S. cabiancae (Vis.) Antoine see *J. chinensis* **L.**

S. californica (Carriere) Antoine see *J. californica* Carriere

S. centrasiatica (Kom.) W. C. Cheng & L.K. Fu see *J. pseudosabina* **Fisch., Mey. & Ave-Lall.**

S. changii see *J. fargesii* **Rehder & Wils.**

S. chinensis (L.) Antoine see *J. chinensis* **L.**

S. chinensis var. *sargentii* (A. Henry) W. C. Cheng & L.K. Fu see *J. chinensis* var. *sargentii* **Henry**

S. chinensis var. *procumbens* (Siebold ex Endl.) Antoine see *J. chinensis* var. *procumbens* **Sieb. ex Endl.**

S. convallium (Rehder & Wils.) W.C. Cheng & L.K. Fu see *J. convallium* **Rehder & Wils.**

S. convallium var. *microsperma* Cheng & L.K. Fu see *J. microsperma* **(Cheng & L.K. Fu) R. P. Adams**

S. cupressifolia Antoine ex K. Koch see *J. sabina* **L.**

S. davurica (Pall.) Antoine see *J. davurica* **Pall.**

S. dimorpha (Roxb.) Antoine see *J. chinensis* **L.**

S. excelsa (M.-Bieb.) Ant. see *J. excelsa* **M.-Bieb.**

S. fischeri Antoine see *J. pseudosabina* **Fisch., Mey. & Ave-Lall.**

S. flaccida (Schlecht.) A. A. Heller see *J. flaccida* **Schltdl.**

S. flaccida (Schlecht.) Antoine see *J. flaccida* **Schltdl.**

S. flaccida (Schlecht.) I.M. Lewis see *J. flaccida* **Schltdl.**

S. foetidissima (Willd.) Antoine see *J. foetidissima* **Willd.**

S. foetidissima (Willd.) Antoine var. *bonatiana* (Vis.) Antoine see *J. thurifera* **L.**

S. foetidissima Ant. see *J. foetidissima* **Willd.**

S. gaussenii (Cheng) Cheng et W.T. Wang see *J. chinensis* **L.**

S. gigantea (Roezl) Antoine see *J. deppeana* **Steudel** var. *deppeana*

S. grisebachii Antoine see *J. foetidissima* **Willd.**

S. horizontalis (Moench) Antoine see *J. horizontalis* **Moench**

S. horizontalis (Moench) Rydb. see *J. horizontalis* **Moench**

S. indica (Bertol.) L.K. Fu & Y. F. Yu see *J. indica* **Bertol.**

S. isophyllos (K. Koch) Ant. see *J. excelsa* **M.-Bieb.**

S. knightii (A. Nelson) Rydberg see *J. osteosperma* **(Torr.) Little**

S. komarovii (Florin) W. C. Cheng & W. T. Wang see *J. komarovii* **Florin**

S. lycia (L.) Antoine see *J. phoenicea* **L.**

S. megalocarpa (Sudw.) Cockerell see *J. osteosperma* **(Torr.) Little**

S. mexicana (Schltdl. & Cham.) Antonine see *J. deppeana* **Steudel** var. *deppeana*

S. microsperma (Cheng & L. K. Fu) Cheng & L. K. Fu see *J. microsperma* **(Cheng & L. K. Fu) R. P. Adams**

S. monosperma (Engelm.) Rydberg see *J. monosperma* **(Engelm.) Sarg.**

S. morrisonicola (Hayata) Nakai see *J. morrisonicola* **Hayata**

S. multiova Goodwyn see *J. communis* var. *depressa* **Pursh**

S. occidentalis (Hook.) Antoine see *J. occidentalis* **Hook.**

S. occidentalis (Hook.) Heller see *J. grandis* **R. P. Adams**

S. officinalis Garcke see *J. sabina* **L.**

S. oliviera (Carriere) Antoine see *J. excelsa* **M.-Bieb.**

S. osteosperma (Torr.) Antoine see *J. osteosperma* **(Torr.) Little**

S. pachyphlaea (Torr.) Antoine see *J. deppeana* **Steudel** var. *deppeana*

S. pacifica Nakai **see** *J. chinensis* **var.** *sargentii*

S. phoenicea (L.) Antoine **see** *J. phoenicea* **L.**

S. phoenicea (L.) Antoine var. *lobelii* (Guss.) Antoine **see** *J. macrocarpa* **Sibth. & Sm.**

S. pingii var. *wilsonii* (Rehder) W.C. Cheng & L.K. Fu **see** *J. squamata* **var.** *wilsonii* **Rehder**

S. pingii (Cheng ex. Ferre) Chen & W. T. Wang **see** *J. pingii* **Cheng ex. Ferre**

S. plochyderma Antoine **see** *J. deppeana* **Steudel var.** *deppeana*

S. polycarpos (K. Koch) Antoine **see** *J. polycarpos* **K. Koch var.** *polycarpos*

S. procera (Hochst. ex Endl.) Antoine **see** *J. procera* **Hochst. ex Endl.**

S. procumbens (Siebold ex Endl.) Iwata & Kusaka **see** *J. chinensis* **var.** *procumbens* **Sieb. ex Endl.**

S. prostrata (Persoon) Antoine **see** *J. horizontalis* **Moench**

S. przewalskii (Kom.) W. C. Cheng & L. K. Fu **see** *J. przewalskii* **Kom.**

S. przewalskii f. *pendula* Cheng & L. K. Fu **see** *J. convallium* **f.** *pendula*

S. pseudosabina var. *turkestanica* (Kom.) C. Y. Yang **see** *J. pseudosabina* **Fisch., Mey. & Ave-Lall.**

S. pseudosabina (Fisch, Mey. & Ave.-Lall.) Cheng & W. T. Wang **see** *J. pseudosabina* **Fisch.,**
    Mey. & Ave-Lall.

S. pseudosabina var. *turkestanica* (Kom.) C. Y. Yang **see** *J. pseudosabina* **Fisch.**

S. pseudothurifera Antoine **see** *J. thurifera* **L.**

S. recurva var. *densa* (Carriere) Antoine **see** *J. squamata* **D. Don in Lambert**

S. recurva (Buch.-Ham. ex D. Don) Antoine **see** *J. recurva* **Buch.-Ham. ex D.Don**

S. recurva (Buch.-Ham. ex D. Don) Antoine var. *coxii* (A.B. Jacks.) W. C. Cheng & L.K. Fu, Fl.
    see *J. coxii* **A. B. Jacks**

S. religiosa Antoine **see** *J. excelsa* **M.-Bieb.**

S. sabinoides Small **see** *J. ashei* **Buch.**

S. saltuaria (Rehder & Wils.) Cheng & W. T. Wang **see** *J. saltuaria* **Rehder & Wils. in Sargent**

S. sargentii (A. Henry) Miyabe & Tatewaki **see** *J. chinensis* **var.** *sargentii*

S. scopulorum (Sarg.) Rydberg **see** *J. scopulorum* **Sarg.**

S. semiglobosa (Regel) L.K. Fu & Y. F. Yu **see** *J. semiglobosa* **Regel**

S. seravschanica (Kom.) Nevski **see** *J. seravschanica* **Kom.**

S. silicicola J. K. Small **see** *J. virginiana* **var.** *silicicola* **(Small) E. Murray**

S. sphaerica (Lindl.) Antoine **see** *J. chinensis* **L.**

S. squamata (Buch.-Ham. ex D. Don) Antoine **see** *J. squamata* **D. Don in Lambert**

S. tetragona (Schltdl.) Antoine **see** *J. monticola* **Mart. f.** *monticola*

S. thurifera (L.) Antoine **see** *J. thurifera* **L.**

S. tibetica (Kom.) W. C. Cheng & L.K. Fu **see** *J. tibetica* **Kom.**

S. turbinata (Guss.) Antoine **see** *J. turbinata* **Guss.**

S. turcomanica (B. Fedtsch.) Nevski **see** *J. polycarpos* **var.** *turcomanica* **(B. Fedtsch) R. P. Adams**

S. utahensis (Engelm.) Rydberg **see** *J. osteosperma* **(Torr.) Little**

S. villarsii Jord. in Jordan & Fourreau **see** *J. sabina* **L.**

S. virginiana (L.) Antoine **see** *J. virginiana* **L. var.** *virginiana*

S. virginiana (L.) Antoine var. *australis* (Endl.) Antoine see *J. barbadensis* **L. var.** *lucayana*
    (Britton) R. P. Adams

S. vulgaris Ant. **see** *J. sabina* **L.**

S. vulgaris Ant. in part **see** *J. horizontalis* **Moench**

S. vulgaris Ant. var. *arborescens* Antoine **see** *J. sabina* **L.**

S. vulgaris Ant. var. *erectopatens* Cheng & L.K. Fu **see** *J. erectopatens* **(Cheng & L. K. Fu)**
    R. P. Adams

S. vulgaris Ant. var. *yulinensis* T. C. Chang & C. G. Chen see *J. sabina* **L.**

S. vulgaris var. *jarkendensis* (Kom.) C. Y. Yang **see** *J. semiglobosa* **var.** *jarkendensis*

S. wallichiana var. *meionocarpa* (Hand.-Mazz.) W. C. Cheng & L.K. Fu **see** *J. indica* **Bertol.**

S. wallichiana (Hook. f. & Thomson ex E. Brandis) W.C. Cheng & L.K. Fu **see** *J. indica* **Bertol.**

S. wilsonii (Rehder) W.C. Cheng & L.K. Fu **see** *J. squamata* **var.** *wilsonii* **Rehder)**

Sabinella phoenicea (L.) Nakai **see** *J. phoenicea* **L.**

Sabinella recurva (Buch.-Ham. ex D. Don) Naka **see** *J. recurva* **Buch.-Ham. ex D.Don**

Appendix IV: Alphabetical list of *Juniperus* taxa recognized in this edition.

Please see www.juniperus.org for the most current list. - R. P. Adams

J. angosturana R. P. Adams

J. arizonica (R. P. Adams) R. P. Adams

J. ashei Buchholz

J. barbadensis L.

J. barbadensis var. *lucayana* (Britton) R. P. Adams

J. bermudiana L.

J. blancoi var. *huehuentensis* R. P. Adams, S. Gonzalez, and M. G. Elizondo

J. blancoi var. *mucronata* (R. P. Adams) Farjon

J. blancoi Martinez var. *blancoi*

J. brevifolia (Seub.) Ant.

J. californica Carriere

J. californica f. *lutheyana* J. T. Howell & Twisselm.

J. carinata (Y. K. Yu & L. K. Fu) R. P. Adams

J. cedrus Webb & Berthol.

J. chinensis var. *procumbens* Sieb.ex Endl.

J. chinensis var. *sargentii* Henry

J. chinensis L. var. *chinensis*

J. coahuilensis (Martinez) Gaussen ex R. P. Adams

J. comitana Martinez

J. communis L. var. *communis*

J. communis var. *charlottensis* R. P. Adams

J. communis var. *depressa* Pursh

J. communis var. *hemisphaerica* (J. & C. Presl) Parl.

J. communis var. *kamchatkensis* R. P. Adams

J. communis var. *kelleyi* R. P. Adams

J. communis var. *megistocarpa* Fernald & H. St. John

J. communis var. *nipponica* (Maxim.) E. H. Wilson

J. communis var. *saxatilis* Pall. (only in eastern hemisphere)

J. convallium f. *pendula* (Cheng & L. K. Fu) R. P. Adams

J. convallium Rehder & Wilson

J. coxii A.B. Jacks

J. davurica var. *arenaria* (E. H. Wilson) R. P. Adams

J. davurica var. *mongolensis* (R. P. Adams) R. P. Adams

J. davurica Pall.

J. deltoides R. P. Adams

J. deltoides var. *spilinanus* (Yalt., Elicin & Terzioglu) Terzioglu

J. deppeana Steudel var. *deppeana*

J. deppeana var. *patoniana* (Martinez) Zanoni

J. deppeana var. *robusta* Martinez

J. deppeana forma *elongata* R. P. Adams

J. deppeana forma *sperryi* (Correll) R. P. Adams

J. deppeana forma *zacatacensis* (Mart.) R. P. Adams

J. deppeana var. *gamboana* (Mart.) R. P. Adams

J. drupacea Labill.

J. durangensis var. *topiensis* R. P. Adams & S. Gonzalez

J. durangensis Martinez

J. erectopatens (Cheng & L. K. Fu) R. P. Adams

J. excelsa M.-Bieb.

J. fargesii (Rehder & Wils.) Kom.

J. flaccida Schlecht.
J. foetidissima Willd.
J. formosana Hayata
J. gracilior Pilger var. *gracilior*
J. gracilior var. *ekmanii* (Florin) R. P. Adams
J. gracilior var. *saxicola* (Britton & P. Wilson) R. P. Adams
J. gracilior var. *urbaniana* (Pilger & Ekman) R .P. Adams
J. grandis R. P. Adams
J. horizontalis Moench
J. indica var. *caespitosa* Farjon
J. indica Bertol.
J. jackii (Rehder) R. P. Adams
J. jaliscana Martinez
J. komarovii Florin
J. macrocarpa Sibth. & Sm.
J. maderensis (Menezes) R. P. Adams
J. mairei Lemee & Lev.
J. maritima R. P. Adams
J. martinezii Perez de la Rosa
J. microsperma (Cheng & L. K. Fu) R. P. Adams
J. monosperma (Engelm.) Sarg.
J. monticola Martinez forma *monticola*
J. monticola forma *compacta* Martinez
J. monticola forma *orizabensis* Martinez
J. morrisonicola Hayata
J. navicularis Gand.
J. occidentalis f. *corbetii* R. P. Adams
J. occidentalis Hook.
J. osteosperma (Torr.) Little
J. ovata (R. P. Adams) R. P. Adams
J. oxycedrus L. *(= var. badia)*
J. phoenicea L.
J. pinchotii Sudworth
J. pingii Cheng & Ferre.
J. pingii var. *miehei* Farjon
J. poblana (Martinez) R. P. Adams
J. polycarpos K. Koch var. *polycarpos*
J. polycarpos var. *turcomanica* (B. Fedtsch.) R. P. Adams
J. procera Hochst. ex. Endl.
J. przewalskii Kom.
J. pseudosabina Fisch., Mey. & Ave-Lall.
J. recurva Buch.-Ham. ex D. Don.
J. rigida var. *conferta* (Parl.) Patschka
J. rigida Mig. in Sieb. var. *rigida*
J. rushforthiana (R. P. Adams) R. P. Adams
J. sabina L.
J. saltillensis M. T. Hall
J. saltuaria Rehder & Wils.
J. scopulorum Sarg.
J. semiglobosa Regel var. *semiglobosa*
J. semiglobosa var. *jarkendensis* (Kom.) R. P. Adams
J. semiglobosa var. *talassica* (Lipsky) Silba
J. seravschanica Kom.

J. squamata var. *wilsonii* (Rehder) R. P. Adams

J. squamata D. Don in Lambert f. *squamata*

J. standleyi Steyermark

J. taxifolia Hook. & Arn.

J. taxifolia var. *lutchuensis* (Koidz.) Satake

J. thurifera L. var. *thurifera*

J. thurifera var. *africana* Maire

J. tibetica Kom.

J. tsukusiensis var. *taiwanensis* (R. P. Adams & C-F. Hsieh) R. P. Adams

J. tsukusiensis var. *tsukusiensis* Masam.

J. turbinata Guss.

J. uncinata (R. P. Adams) R. P. Adams

J. virginiana L. var. *virginiana*

J. virginiana var. *silicicola* (Small) E. Murray

J. zanonii R. P. Adams